NANOSCALED FILMS AND LAYERS

Edited by **László Nánai**

Nanoscaled Films and Layers
http://dx.doi.org/10.5772/65465
Edited by László Nánai

Contributors
Shamim Ahmad, Dinesh Kumar, Thomas Riedl, Jörg K.N. Lindner, S V Prabhakar Vattikuti, Chan Byon, M. Federica De Riccardis, Yi Zheng, Mei Wang, Meizhen Yin, Chendong Ji, Yanbao Guo, Deguo Wang, Rodica Ghita, Constantin Logofatu, Constantin-Catalin Negrila, Lucian Trupina, Costel Cotirlan-Simioniuc, Marcela Socol, Gabriel Socol, Nicoleta Preda, Anca Stanculescu, Florin Stanculescu, Jothi Sudagar, Tamilarasan Tr, Sanjith U, Rajendran R, Ravi Kumar Nv

Published by InTech
Janeza Trdine 9, 51000 Rijeka, Croatia

Publishing Process Manager Dajana Pemac
Technical Editor SPi Global
Cover InTech Design team

First published May, 2017
Printed in Croatia
Legal deposit, Croatia: National and University Library in Zagreb

Additional hard copies can be obtained from orders@intechopen.com

Nanoscaled Films and Layers, Edited by László Nánai
p. cm.
Print ISBN 978-953-51-3143-4
Online ISBN 978-953-51-3144-1

For EU product safety concerns:
IN TECH d.o.o., Prolaz Marije Krucifikse Kozulić 3, 51000 Rijeka, Croatia,
info@intechopen.com or visit our website at intechopen.com.

Meet the editor

László Nánai was born on April 19, 1948, in Csopak (Hungary). He studied physics (MSc) at Saint Petersburg State University (RU), and his PhD degree and habilitation in the field of quantum electronics were obtained at Lebedev Physical Institute, Moscow (RU), and Szeged University (H). He is a specialist in the fields of solid-state physics, laser-matter interaction fabrication and characterization of nanostructures. He has written over 170 scientific publications including about 10 books and chapters in books and conference proceedings.

Contents

Preface

With the progress of new—mostly laser-based—technologies, the preparation, characterization and applications of nanostructures (on different scales) have become a subject of wide-range investigations. The applications of quantum-based laws and rules resulted in new discoveries and devices. These book chapters are dedicated to different methods (physical, chemical, etc.) of nanolayer fabrications of different components (metals, semiconductors, polymers and bioactive materials). The wide range of physical, chemical, biological and other properties and characteristics are also subjects of investigations focused—mostly—on possible applications in different scientific and technological fields. We hope that the book will be of interest for those specialized in nanotechnologies.

László Nánai
University of Szeged
JGYPK Department of Physics
Szeged, Hungary

Section 1

Preparation and Fabrication of Nanolayers

Chapter 1

Formation of Nanolayer on Surface of EPD Coatings Based on Poly-Ether-Ether-Ketone

Maria Federica De Riccardis

Additional information is available at the end of the chapter

http://dx.doi.org/10.5772/67570

Abstract

Poly-ether-ether-ketone (PEEK) is a high performance polymer with many intrinsic properties. When it is used in the form of coating, an improvement of some of its functional properties was achieved by forming a surface nanolayer. In this chapter, it will be described how it was possible to obtain this result. Firstly, three kinds of PEEK composite coatings were deposited by electrophoretic deposition, adding alumina particles, polytetrafluoroethylene (PTFE) and lignin to PEEK. Then, the composite coatings were thermal treated in a furnace. Therefore, surface nanostructure and chemical composition of these PEEK composite coatings were modified with respect to bulk coatings, due to interaction between PEEK chain and secondary phase, emphasised by the thermal treatment conditions. Experimental evidence of the formation of surface nanolayer was provided by SEM, TEM, GIXRD, ATR-FTIR and XPS characterisations. Functional characterisations demonstrated that wear resistance—in the presence of alumina particles—hydrophobicity—in the presence of PTFE—and corrosion resistance—in the presence of Lignin—were increased with respect to pure PEEK.

Keywords: nanolayers PEEK, wear, friction coefficient, corrosion, EPD

1. Introduction

Poly-ether-ether-ketone (PEEK) is a high performance polymer with excellent mechanical, chemical and thermal properties. It is a polyaromatic semicrystalline thermoplastic polymer (typically with 30–40% crystallinity), formed by three benzene rings and a carbonyl group (**Figure 1**). In form of a coating, PEEK can be used continuously up to 260°C, is insoluble in all common solvents, is characterised by a low moisture absorption, and shows a low outgassing.

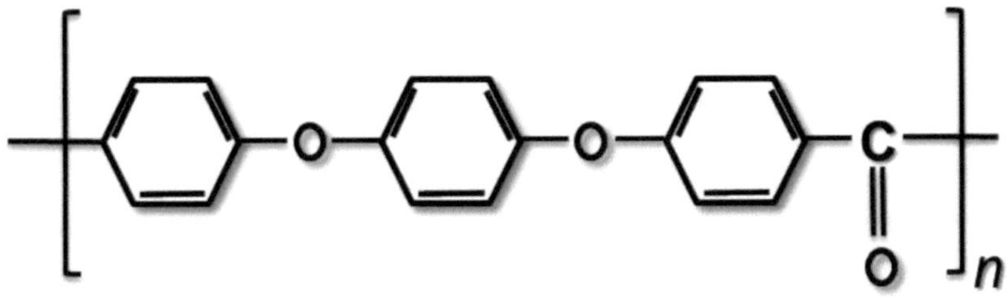

Figure 1. PEEK chemical structure.

In spite of its outstanding properties, PEEK exhibits some drawback. As an example, it has a high friction coefficient in dry sliding, therefore tribology applications are restricted. Nevertheless, several application limits can be overcome adopting some expedient during or after processing, aiming to modify surface nanostructure or creating a thin surface layer, as demonstrated by some examples cited below.

As for nanostructure, excellent properties of PEEK are partly the result of its semi-crystalline structure developed during processing [1–3], therefore a variation of the processing temperature could result in improved mechanical interactions between its surface and surroundings, such as in tribology or anticorrosive applications.

For some biomedical applications, native PEEK film was found to be a very poor substrate of cell cultivation, extremely reluctant to allow cellular adhesion. Improvement of the bioadhesion properties could possibly result from an increase of the surface hydrophilicity, for instance by the introduction of carboxyl groups or by UV irradiation. In such a way, these reactive surface functionalisations should be excellent anchorage points for the immobilization of bioactive molecules [4–6].

Moreover, it is well noted that pure PEEK results is limited in osteoconduction and osseointegration resulting in a poor biocompatibility. To achieve good bone-implant interfaces, the surface properties of permanent implants made of PEEK could be modified, acting on morphology and chemical structures of the surface [7, 8]. In the review [9], different physical modifications of PEEK, used to improve the bone implant interface are described. These modification methods include forming a bioactive layer by coating technology and changing the surface properties by wet treatment.

In general and from all these considerations on PEEK, it is evident that the surface of a bulk material, or better its thin superficial layer, plays a fundamental role in interactions between a material and working ambient. It is worthwhile to note that these modifications are at nanoscale level, but nevertheless they can influence the macroscopic properties of a material. For this reason, it is reasonable to study how to modify a material, and in particular the physical or chemical structure of its surface, to improve some peculiar characteristics. Specifically for PEEK, experimental results demonstrated that its outstanding properties or structures related to surface were improved by adding a secondary phase. The aim of this chapter is to illustrate the interesting results obtained for thin layers of PEEK in terms of some properties related to surface, such as friction coefficient, wear resistance and hydrophobicity.

2. Deposition process

The deposition technique used to obtain thin layers based on PEEK was Electrophoretic Deposition (EPD). It is one of the most outstanding coating techniques based on electrodeposition that does not induce polymer degradation with respect to other deposition techniques such as thermal spraying or printing processes.

Although EPD usually is used to obtain coating some microns thick, it was demonstrated that by performing a post-deposition treatment with opportune parameters, or by using some kind of secondary phase together with PEEK, it was possible to obtain a nanoscale surface layer with modified characteristics able to improve the outstanding properties of bulk material.

EPD is a method traditionally employed to obtain ceramic coatings. Nowadays, both academics and industrialists are more and more interested in it, due to its wide potential in coating processing technology, also applied to polymer and composite coatings. EPD coating process consists in applying an electric field between two electrodes immersed in a suspension where some solid particles or polymer chain are present. The particles or the polymer chains inside the liquid medium of the suspension acquire a surface charge, also for effect of additives or stabilisers added to the suspension. When the electric field is on, it enables a movement of particles towards the electrode with opposite charge. There, the nanoparticles or the polymer chains coagulate and form a thin layer covering the electrode (**Figure 2**).

The main advantages of this deposition technique are high versatility and cost effectiveness. Indeed, EPD can be used with different materials or their combinations and requires simple and cheap equipment. Moreover, it can be used both on a large scale, also when coating objects with a complex geometrical form, and on a small scale, to fabricate composite micro- and nanostructures. New application areas for EPD are the low-cost fabrication of composite

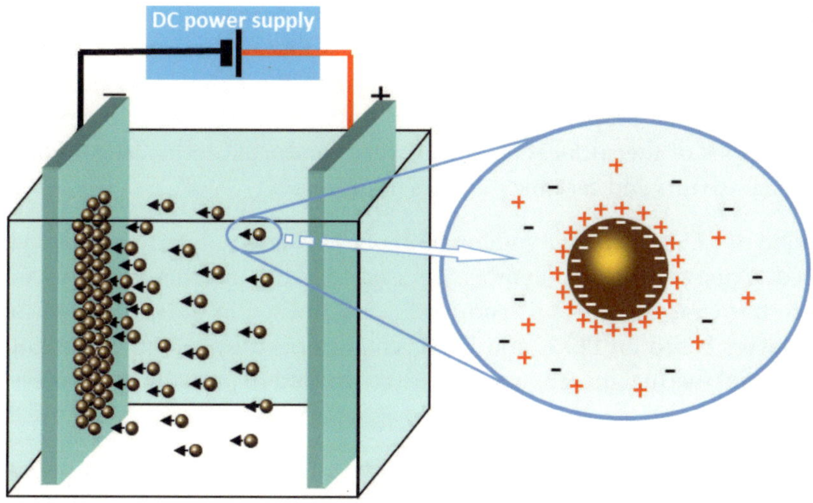

Figure 2. Scheme of an EPD cell.

materials including advanced coatings based on nanocomposites, laminate structures and functional graded materials [10–13].

It is clear that the main requirement to obtain an efficient EPD process is to use suitable suspensions where particles or other components, also at nano scale, are well suspended and dispersed. When a particle is in a liquid medium, it can be charged through four mechanisms [14]:

a. selective adsorption of ions onto the solid particle from liquid,

b. dissociation of ions from solid phase into the liquid,

c. adsorption or orientation of dipolar molecules at the particle surface and

d. electron transfer between the solid and the liquid phase due to the difference in work function.

Usually, the stability of a suspension is evaluated by means of zeta potential measurements, referring to the chemical surroundings of a particle while it is in a suspension; when the zeta potential value is maximised, with positive or negative value indifferently, the dispersion of suspension components is good and a uniform coating can be obtained by EPD.

For the analysis and discussion on charging mechanisms and particles interactions, that are at the base of this processing method, one can refer to the fundamentals of colloid science widely discussed in literature [15].

In this chapter, the experimental data regarding three different composite layers based on PEEK will be reported.

The first kind of composite PEEK coatings contains alumina particles, the most used ceramic material in the field of anti-wear application. These composite layers were prepared in order to obtain coatings with higher wear resistance than the pure PEEK without altering significantly its low friction coefficient. After deposition, a thermal treatment was performed in opportune conditions, so a nanolayer of PEEK covered the alumina particles, providing an improvement in sliding properties of PEEK-alumina coating. Morphological and chemical characterisations of the surface nanolayer were conducted to investigate the interaction between polymer chains and ceramic particles in processing.

As second kind of PEEK coating, a combination of PEEK and polytetrafluoroethylene (PTFE) was prepared. Often these two polymers are used together in order to improve the friction coefficient. In this case, the process conditions were studied in order to obtain superhydrophobic nanolayers based on PEEK and PTFE, characterised by a high crystallinity. For this aim, structural and thermal investigations were conducted to pinpoint the process conditions that resulted optimal.

Finally, the third kind of PEEK based coatings was formed by using an organic additive of vegetal origin together with PEEK. In such a way, these composite coatings showed an improved adhesion to substrate. Moreover, on the surface of coatings, a thin composite layer was formed providing an improved corrosion resistance. Adhesion and corrosion tests, as

http://dx.doi.org/10.5772/67570

well as morphological and chemical characterisations, were fundamental in the investigation of functional properties of PEEK layers.

3. Nanolayers of PEEK on alumina particles

PEEK semicrystalline powder (Victrex® Manufacturing Ltd.) and alumina particles (AES 11 Sumitomo) were used to prepare suspensions suitable for EPD process. The weight ratio between the alumina and PEEK content, hereafter referred as to WR, was varied between 0 and 1 in the suspension, whereas the total amount of solid content was fixed at 1wt%. Ethyl alcohol was used as liquid medium and a solution of triethylamine and citric acid was added as dispersant (10wt% with respect to solid content) [16]. The zeta potential values measured on the suspensions with different composition were quite similar, meaning that the addition of alumina particles to PEEK suspension did not modify the potential efficiency of the EPD process. EPD depositions were performed at constant voltage conditions (25 V/cm applied for 30 sec) and pieces of Si wafer were used as depositing electrode. Since a drawback of the EPD is the low density of the deposit, made of particles put together by the weak van der Waals forces [17], the coatings were submitted to a post-deposition treatment. It was a conventional thermal treatment in a furnace, operated at different temperatures to reach PEEK melting.

Before post-deposition treatments, the thermal behaviour of PEEK was analysed both in absence and in presence of alumina particles. From Thermogravimetric Analysis (TGA) and Differential Scanning Calorimetry (DSC) measurements, reported in **Figure 3**, it is possible to extrapolate that PEEK is a polymer stable up to 580°C, its crystallisation temperature at 250°C, and its melting temperature at 340°C. Moreover, no significant differences in the thermal behaviour of the PEEK powder and the powder of PEEK mixed with alumina particles, in the range of temperature between 0 and 450°C, are appreciable.

For the conventional thermal treatment, two thermal programs were used with different maximum temperature and different time. The operating parameters were reported in **Table 1**. It is worthwhile to note that the samples treated at 340°C took longer time to pass through the crystallisation range (250°C) than those treated at 400°C, as it was impossible to externally control the cooling down step.

The morphology of the coatings thermally treated was very different, as demonstrated by SEM observations. In **Figure 4**, SEM images at high magnification referring to the surface of PEEK-alumina coatings with different content of alumina particles are reported. The different maximum temperature of thermal treatment and the different cooling rate allowed to obtain different nanostructure of thin PEEK layer covering the alumina particles, more rough at 340°C than at 400°C. In fact, as visible in **Figure 4a**, **c**, **e** and **g** a fine structure was visible on the alumina particles surface, more evident when the alumina content was lower. When the maximum temperature was higher (**Figure 4b**, **d**, **f** and **h**), the granular nanostructure on alumina particles is less visible. It is reasonable to suppose that these features are due to PEEK solidified on alumina particles, remaining at solid state, as melting temperature of alumina is much higher than the maximum temperature of thermal treatments. As a comparison, in **Figure 5**, a SEM image of alumina particles as received is reported.

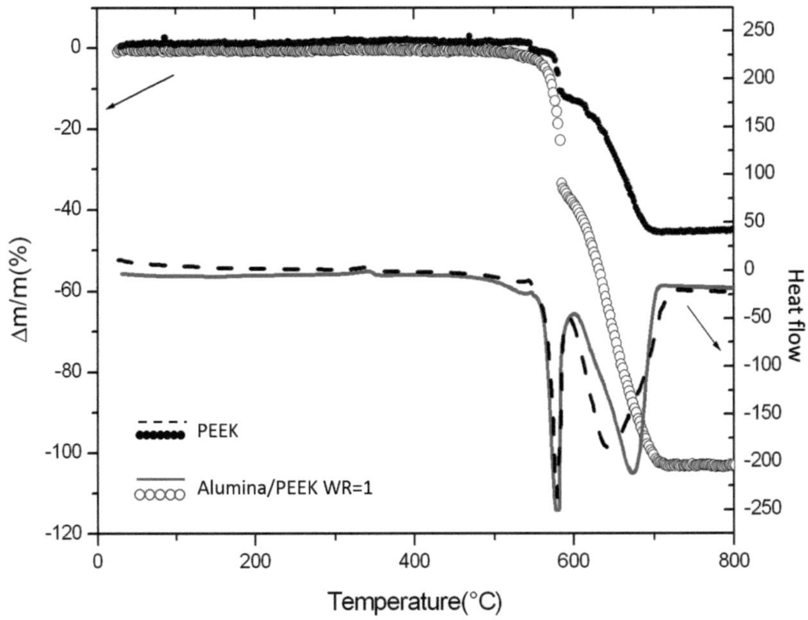

Figure 3. TGA and DSC curves conducted on pure PEEK and on PEEK-alumina powders (WR=1).

	Heating rate	Treatment time
Thermal program #1	10°C/min	30 min @340°C
Thermal program #2	10°C/min	5 min @400°C

Table 1. Operating parameters of the conventional thermal treatment conducted on PEEK-alumina layers.

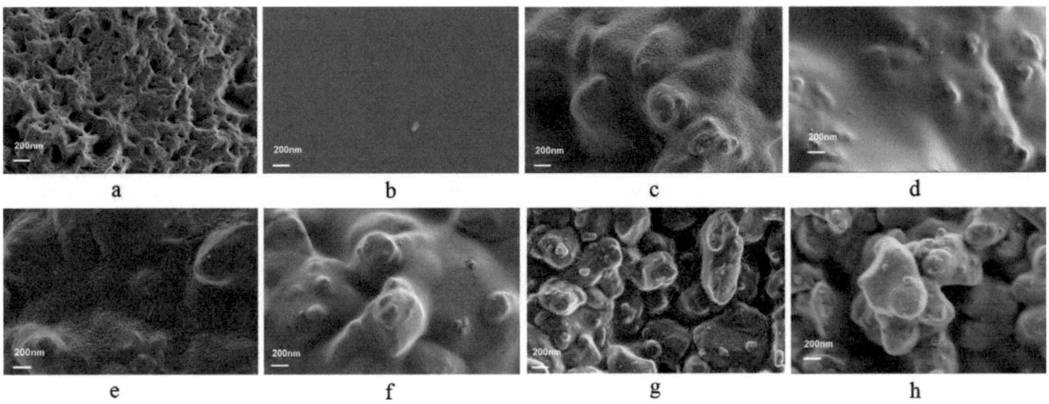

Figure 4. SEM images of fine surface structure of PEEK-alumina coating at different alumina content (a-b: WR = 0, c-d: WR = 0.3, e-f: WR = 0.7, g-h: WR = 1) treated at different temperatures (a, c, e and g @340°C, b, d, f and h @400°C).

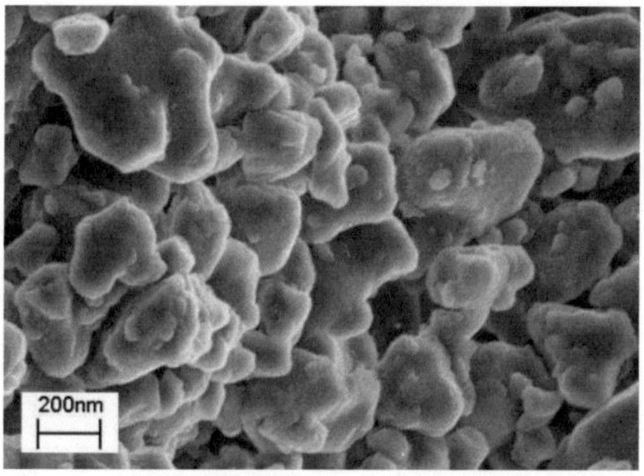

Figure 5. SEM image of alumina particles as received.

XRD measurements acquired in glancing incidence scattering configuration (GIXRD) were performed by keeping the incident angle ω_i (angle between incident beam and sample surface) fixed at 1.5°, in order to analyse surface structures. As a result, in PEEK coatings without alumina particles (WR = 0) treated at 400°C, the PEEK structure is amorphous whereas in the coatings treated at 340°C the characteristic diffraction peaks of PEEK are present (**Figure 6a**). Of course, when alumina particles are present in the coating, also the alumina peaks are visible (**Figure 6b**). Moreover, the peaks of PEEK, whose crystallites are wide 5.3 nm, are more intense when the amount of alumina is lower. In general, it is known that inorganic fillers have two inconsistent influences on the crystallization of semicrystalline polymers. On the one hand, they act as nucleating agents that facilitate the crystallisation of polymers. On the other hand, they hinder the motion of the segments of the polymer chains retarding the crystallisation of polymers. In this case, the hindering effect of alumina particles dominated.

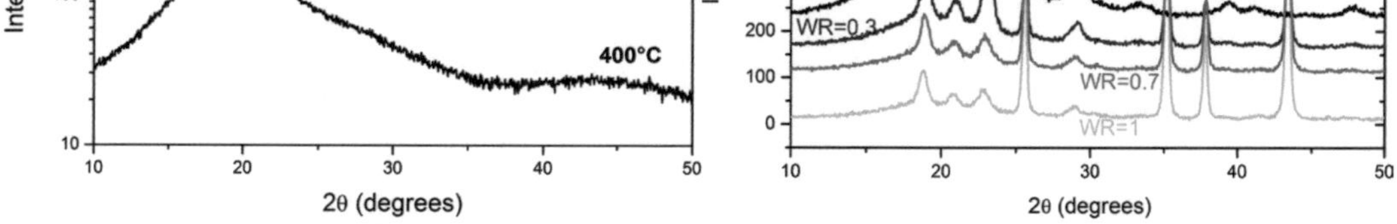

Figure 6. XRD spectra of pure PEEK coating treated at different temperature (a), and PEEK-alumina coating surface with different alumina content, after thermal treatment at 340°C (b).

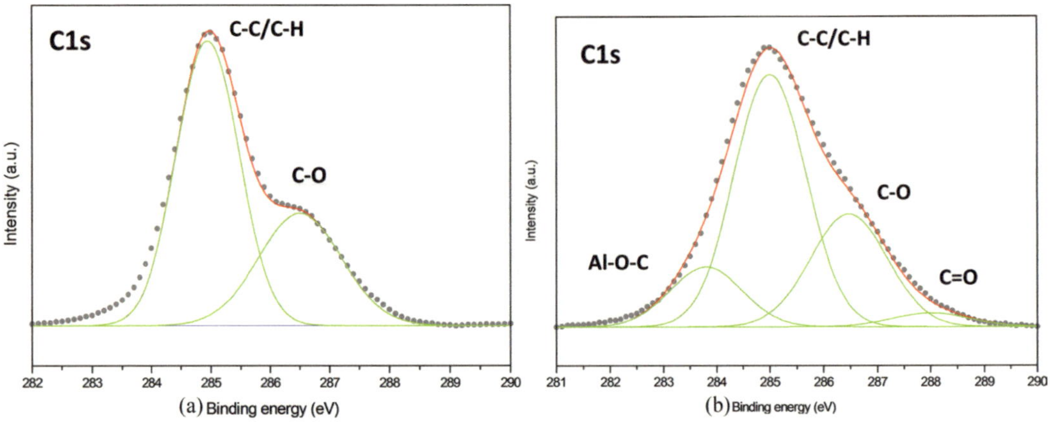

Figure 7. Deconvolution of XPS C1s signal acquired on the surface of pure PEEK coating (a) and of PEEK-alumina coating (WR = 1) (b).

The surface composition of the PEEK nanolayer was analysed by XPS measurements and the C1s peak was deeply studied. On pure PEEK, no important differences were visible in the shape of the C1s peak in layer treated at 340 and 400°C. Moreover, the thermal treatment at 340°C did not induce any significant modification in the O/C ratio with respect to untreated PEEK, but this ratio is lightly increased at 400°C, probably due to some thermal degradation of the PEEK chain.

C1s peaks acquired on surface of PEEK enriched of alumina particles were deconvoluted and different components in the binding energy signal were recognised. No particular differences were observed between C1s peak acquired on coating treated at 340°C and that of coating treated at 400°C. Therefore, the discussion is referred to C1s peak related to 340°C (**Figure 7**). On C1s peak acquired on pure PEEK, at high binding energy the C-O bond component was recognisable, less intense of the component at lower binding energy attributable to C-C and C-H bonds. The deconvolution of C1s peak of the sample PEEK with alumina particles (WR = 1), showed

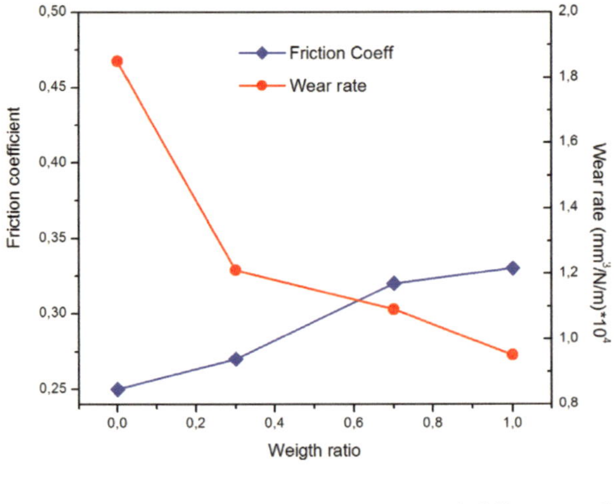

Figure 8. Friction coefficient and wear rate measured on PEEK coatings with different weight ratio between PEEK and alumina content.

the presence of a contribute of C = O and, more significantly, that of the component Al-O-C both at 340 and at 400°C, meaning that a sort of bond between PEEK and alumina was created. This result was confirmed also by the O1s signal (not reported here), that is formed by a larger peak when alumina was present.

The presence of this PEEK nanolayer on alumina particles positively affected macroscopic properties related to the surface that are friction coefficient and sliding wear. As reported in **Figure 8**, the wear rate, evaluated by means of a microtribometer (CSM), decreased when the alumina content increased. Of course, the wear rate minimum is achieved when alumina particles are present in the PEEK coating with a WR = 1. In correlation with this result, the friction coefficient increased lightly.

4. PEEK-PTFE nanolayers

As mentioned before, PEEK exhibits a high friction coefficient in dry sliding, which restricts its tribology applications. In order to reduce the friction coefficient of PEEK and therefore to increase its range of applications, some solid lubricants, such as glass and carbon fibres, graphite or carbon nanotubes [3, 18, 19, 20], as well as polymers with low surface energy, such as PTFE [21, 22], were added to the PEEK matrix. In this section, it was demonstrated that an EPD coating based on PEEK and PTFE treated in opportune condition after deposition had significant tribology properties as well as a remarkable super hydrophobicity with respect the pure PEEK coatings.

Recently, EPD has demonstrated to be a suitable method to obtain polymer coatings based on PEEK [23–26]. As regard to PTFE, to the best of author's knowledge, only few papers referred to EPD of this polymer [27, 28] and always PTFE was mixed to inorganic powders.

As first time, EPD from optimised suspensions containing PTFE and PEEK was widely discussed in reference [29]. After deposition, thermal treatments were performed in order to modify crystallinity of polymeric blend. The coatings used for this aim were indicated in **Table 2**.

DSC is the best analytical technique used for assessing polymer crystallinity, by quantifying the heat associated with melting of the polymer. It is known that, assuming that the heat

Sample	wt% PEEK	wt% PTFE
PEEK 100	100	-
MIX1	80	20
MIX2	50	50
MIX3	20	80
PTFE 100	-	100

Table 2. PEEK and PTFE content in suspension used for the EPD coatings.

capacity of a polymer is substantially independent from the temperature, it is possible to calculate the mass fraction crystallinity (χ) by evaluating the melting enthalpy from the DSC thermogram, according to the following formula:

$$\chi = \frac{H_m}{H_f} \quad (1)$$

Where, H_m is the melting enthalpy, calculated as the area under the crystal melting transition per unit weight, and H_f is the theoretical heat of fusion for a pure crystalline phase, equal to 130 J/g and 82 J/g for PEEK and PTFE, respectively. Therefore by normalizing the recorded heat value of fusion to that of a 100% crystalline sample of the same polymer, the mass fraction crystallinity was obtained.

In order to increase, and then quantify, the crystallinity of PEEK-PTFE blends, a composed DSC thermal program was used. Therefore, the PEEK-PTFE powder mixtures with composition equal to those reported in **Table 2**, were exposed to repeated processes of *heat/cool/heat*, corresponding to a melting/solidification/measuring cycle, respectively. The rates of melting and cooling processes were 1, 5, 20 and 50°C/min, whereas the second heating process was performed at 10°C/min in order to acquire a suitable thermal signal. In this way, the first heating and the low rate cooling were expected to increase polymer crystallinity. As an example, in **Figure 9**, a typical DSC-thermogram of the PEEK 50%-PTFE 50% was reported.

First of all, χ was measured on 100% PEEK and 100% PTFE powder as received, by a single heating at 10°C/min. The crystallinity of these materials was evaluated 58.0 and 94.3% for PEEK and PTFE, respectively. Some mixture of PEK-PTFE powder with the nominal composition of EPD suspensions were treated as described before. For each concentration of polymers different from 100%, the crystallinity was evaluated agreeing Eq. 1 and compared

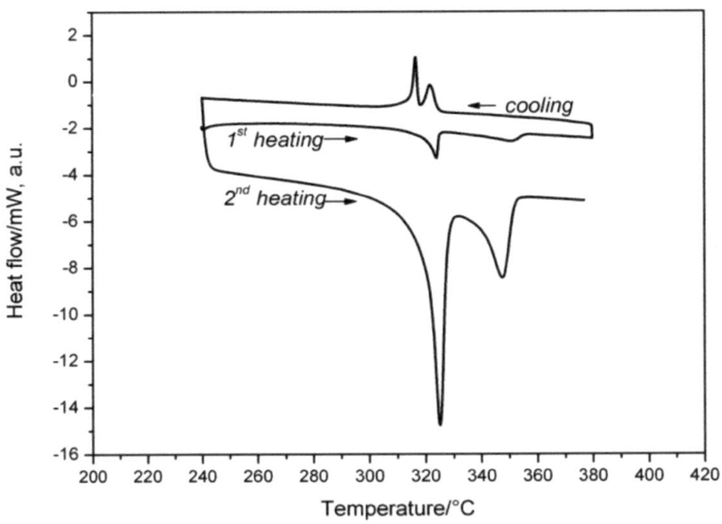

Figure 9. DSC-thermogram of the PEEK 50%-PTFE 50% cycled with 5°C/min (2nd heating = 10°C/min).

http://dx.doi.org/10.5772/67570

with the theoretical value of χ (χ_{th}), given by the ratio between the crystallinity of 100% polymer and the nominal content in the blend. In other words, for a given concentration C of PEEK

$$\chi_{th}(C) = \frac{\chi_{100}}{C} \tag{2}$$

Same evaluations were conducted for PTFE.

As first result, PEEK/PTFE blends thermal treated exhibited a different crystallinity with respect to 'as received' PEEK and PTFE, as an effect of the cooling rate. It was observed that the sample PEEK 100 treated at 1 and 5°C/min has a higher crystallinity fraction than the PEEK 'as received', whereas in presence of PTFE, χ of PEEK is lower than both the theoretical value and that 'as received' (**Figure 10**).

PTFE had a similar behaviour but the theoretical values of crystallinity fraction are significantly lower than the measured χ (**Figure 11**). Moreover, in PTFE 100, the crystallinity was higher than that 'as received' for all the thermal rate. As first conclusion, one can affirm that the presence of PTFE influenced the crystallinity of PEEK and vice versa, at a fixed value of cooling rate.

As, mentioned before, thermal treatments in DSC apparatus and consequent evaluations of fraction crystallinity were performed on mixtures of two polymer as powder, agreeing with the composition reported in **Table 1**. The successive measurements of x-ray diffraction, SEM

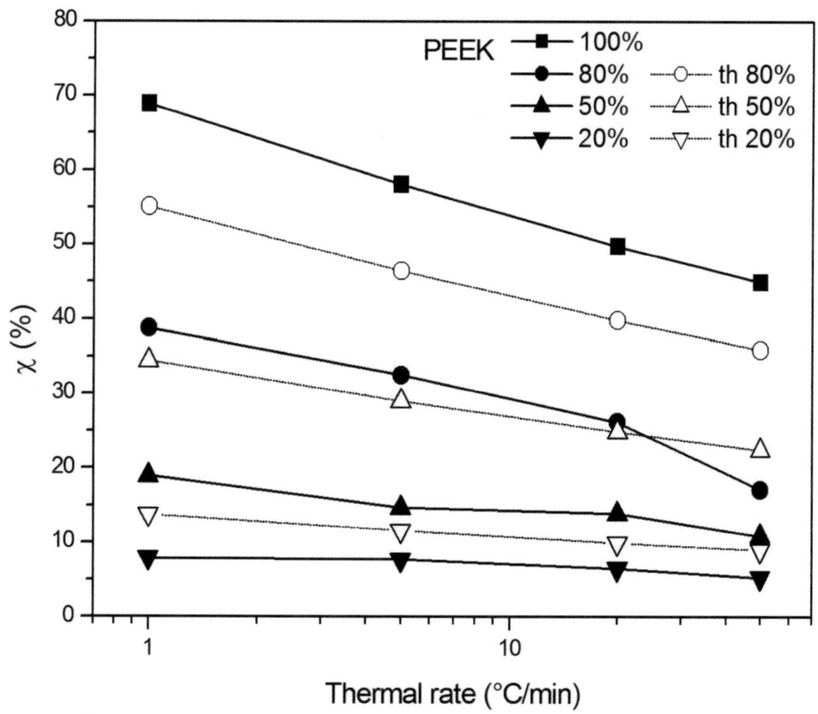

Figure 10. PEEK crystallinity evaluated at different thermal rates for different content in PEEK-PTFE blends.

and TEM observations, contact angle, friction coefficient and wear rate were performed on PEEK-PTFE coatings treated in similar conditions. Nevertheless, result discussion was consistent with DSC conclusion.

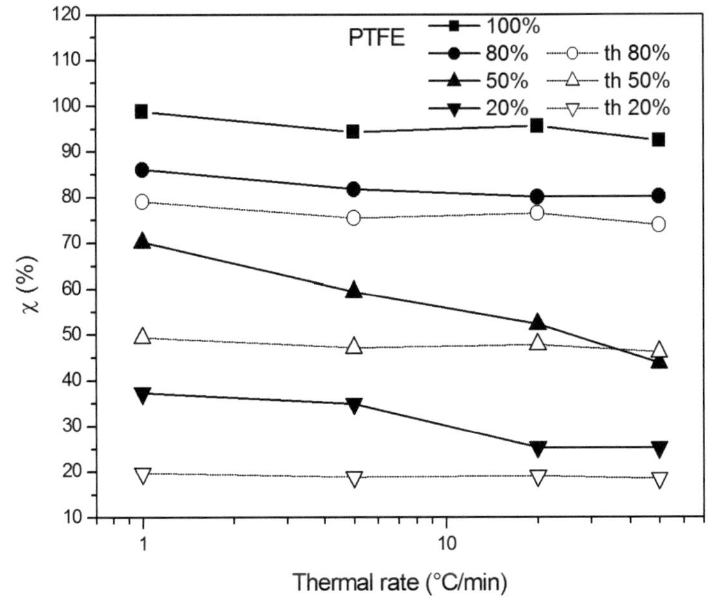

Figure 11. PTFE crystallinity evaluated at different thermal rates for different content in PEEK-PTFE blends.

XRD measurements, performed both in Bragg-Brentano configuration (θ/2θ) and in glancing-incidence configuration (GIXRD), confirmed the crystalline structure of PEEK and PTFE. Moreover, the peak intensities of PTFE were higher than those of PEEK, agreeing with DSC analysis.

SEM images at high magnification highlighted a surface morphology characterised by small granular features, recalling a fine crystalline structure (**Figure 12**). Cross section of these coatings was observed by SEM and TEM. In all coatings, a thin layer about 30 nm thick was revealed in correspondence of the surface (**Figures 13a** and **14**), sometimes detached from underlying material (**Figure 13b**). It is evident that this thin layer had a different structure with respect to the bulk of coating that appears without a structured morphology. This surface peculiarity was attributed to the interaction between first surface layer of material and ambient during cooling step.

The crystalline structure as well as the morphological features at nanoscale of the surface of these coatings conferred a super-hydrophobicity nature to PEEK. In fact, contact angle measured on PEEK-PTFE coatings was remarkably higher than that on pure PEEK surface, comparable to pure PTFE (**Figure 15**). Finally, friction coefficient and wear rate of PEEK-PTFE coatings resulted lowered with respect to pure PEEK and pure PTFE (**Figure 16**), meaning that the nanoscale layer, formed on the surface of these polymer coatings, as a consequence of carefully performed thermal treatment, resulted to have more outstanding properties than the pure PEEK.

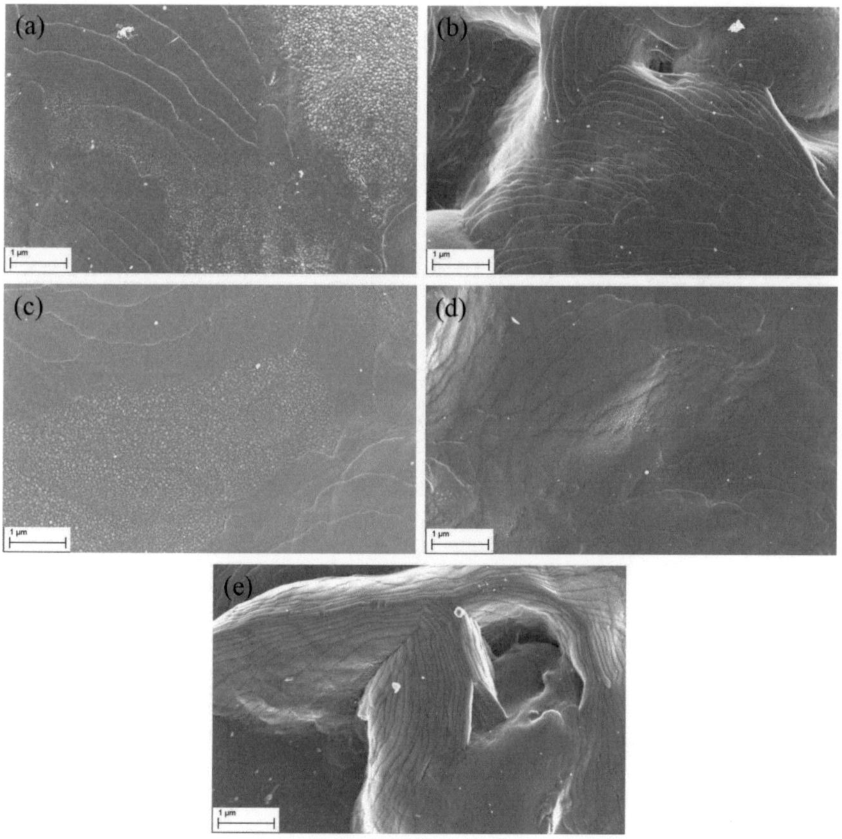

Figure 12. SEM images of fine surface structure on PEEK 100 (a), PTFE 100 (b), MIX1 (c), MIX2 (d), and MIX3 (e) coatings.

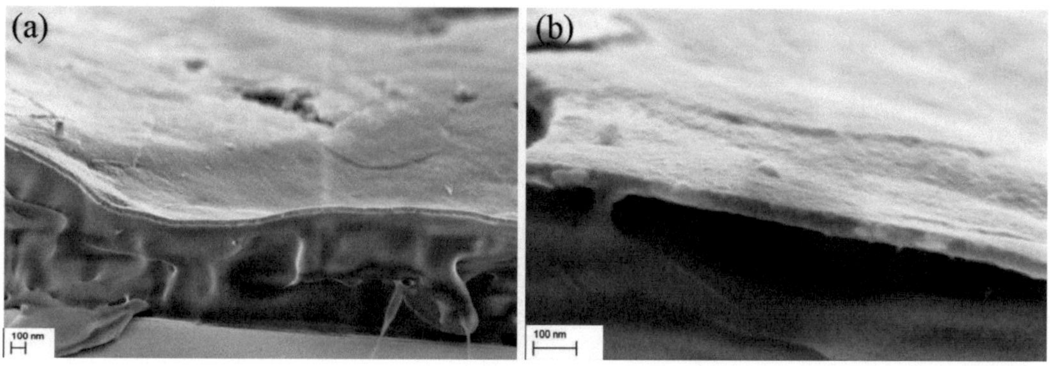

Figure 13. SEM images of cross section of a typical PEEK-PTFE coating with a thin surface layer (a). In (b) the thin layer appears detached from underlying material.

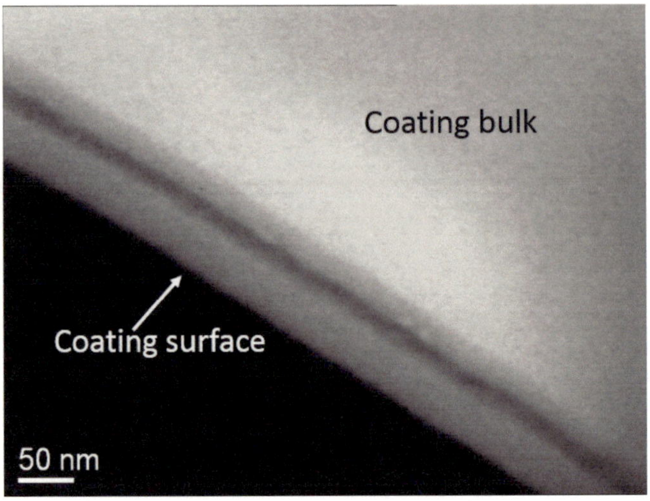

Figure 14. TEM image of thin surface layer of PEEK-PTFE coating in cross section.

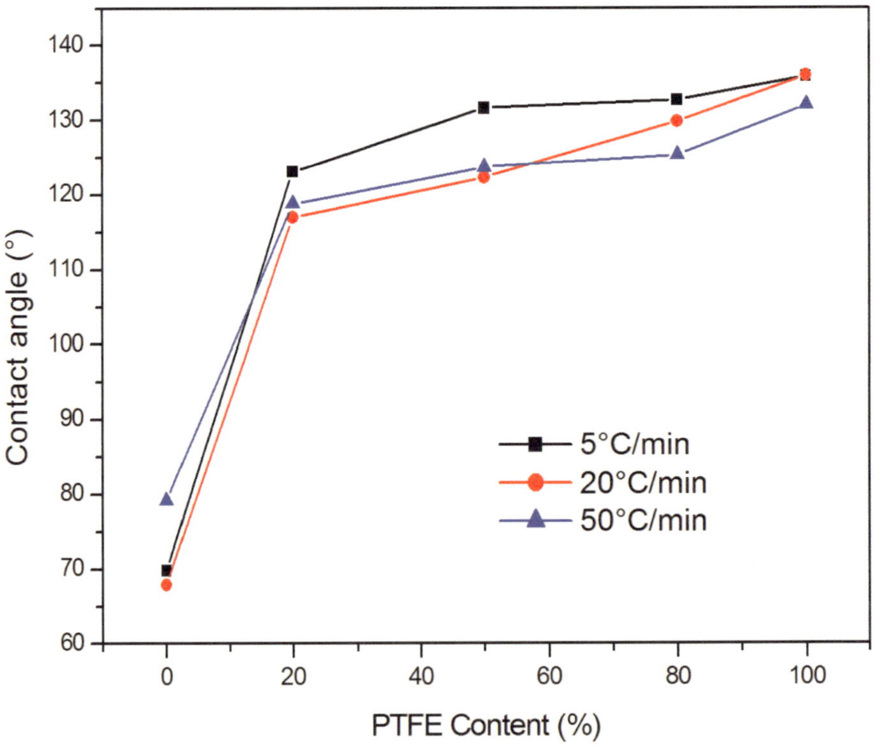

Figure 15. Contact angle measured on PEEK-PTFE coatings.

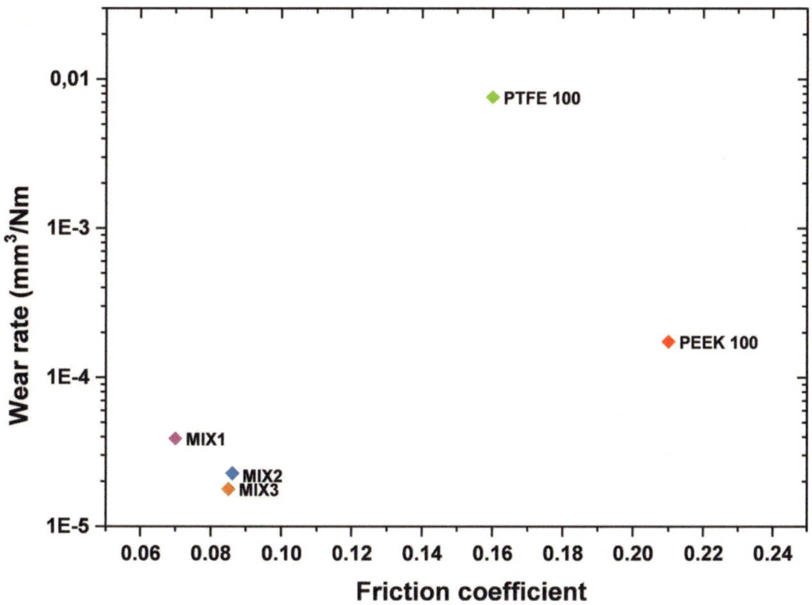

Figure 16. Friction coefficient and wear rate measured on PEEK-PTFE coatings.

5. PEEK-Lignin nanolayers

As described before, EPD has been successfully applied to obtain polymers coatings such as PTFE and PEEK, also in combination with ceramic and biocompatible particles [30]. Sometimes EPD coating adhesion to substrate is critical, principally for applications where interactions between coating surface and working environment are stressed. In order to improve the adhesion of PEEK coatings to conductive substrates, Lignin (hereafter referred as to LGN) was added to EPD PEEK suspensions. Experimental conditions and results were discussed in reference [31]. As a further result, a remarkable corrosion resistance was detected in PEEK-LGN coatings, as demonstrated by electrochemical corrosion measurements.

As an abundant non-toxic amorphous natural polymer, LGN is the second most abundant biopolymer after cellulose. It is a polymer found extensively in the cell walls of all woody plants, and constitutes 25–30% of the total dry weight of trees. LGN, formed by removal of water from sugars to create aromatic structures, resists attack by most microorganisms, and anaerobic processes tend not to attack the aromatic rings at all. Aerobic breakdown of LGN is slow and may take many days.

Due to its very complex structure, LGN has had historically a limited industrial use but the study of its properties has showed its potentiality in new value-added applications. As an example, the functional groups of LGN determine the high polarity of the macromolecular structure and make it an extremely promising material as a chemical component in polymer blends or as an organic filler. Moreover, LGN helps to lower the cost of finished products, since it is an inexpensive by-product of the paper industry. In this specific case, LGN addition could further improve the remarkable properties of PEEK without interfering with its applications.

To investigate anticorrosion properties, some EPD coatings were prepared by using a suspension containing PEEK and LGN dispersed in acetone without further dispersant. The solid content of PEEK and LGN was fixed at 1.6 and 0.8 g/L, respectively.

LGN is a phenolic polymer with a complex structure where several chemical functional groups, including hydroxyl, methoxyl, carbonyl and carboxyl groups, are present in various amounts and proportions, depending on genetic origin and extraction processes (**Figure 17**). Differently from other solvent (deionised water, ethanol and isopropanol), when was dispersed in acetone, LGN conferred a brown colour upon acetone, meaning that some interactions between solid and liquid occurred. In order to investigate dispersion mechanism of LGN, Attenuated Total Reflection Fourier Transform Infrared Spectroscopy (ATR FT-IR) analysis was performed on LGN powder after evaporation of different solvents. By comparing the spectra acquired on as received LGN powder and on LGN residual after removing solvent (**Figure 18**), a new relevant peak at 1700 cm^{-1} appeared in LGN-Acetone, due to stretching of C=O bond. Moreover, other differences are appreciable: (1) an increase on the intensity of the broad peak due to stretching of C-H and O-H in the range 3500–3200 cm^{-1} (not reported here); (2) a general increase of peak intensity in the region 1600–1000 cm^{-1} and (3) a shift towards lower wavenumber in 1425, 1370 and 1220 cm^{-1} peaks attributed to C-H bonds [31]. These results allowed to presume that an interaction between the carbonyl group (C=O) of acetone and methyl (CH_3), methoxyl (CH_3O) or hydroxyl (OH) groups of LGN occurred and therefore LGN resulted functionalised.

After deposition on stainless steel substrates, pure PEEK and PEEK-LGN coatings were thermal treated in air at 300 and 320°C for 10 min. The morphology of the surface was quite different depending on maximum temperature of treatment, as visible in **Figure 19**. Pure PEEK appears quite rough whereas PEEK-LGN coating surface presents some small ridges, more evident at 300°C.

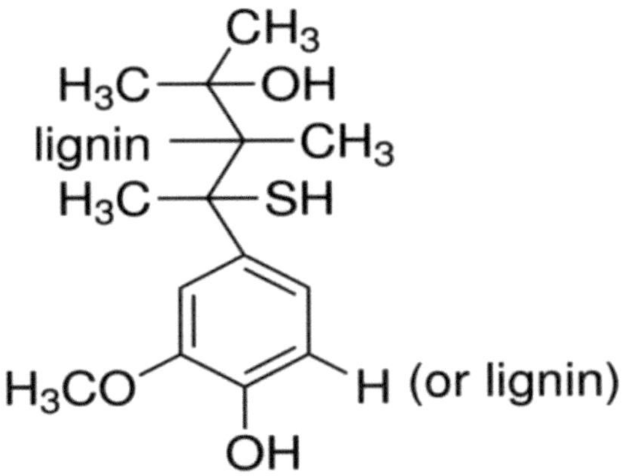

Figure 17. Lignin chemical structure.

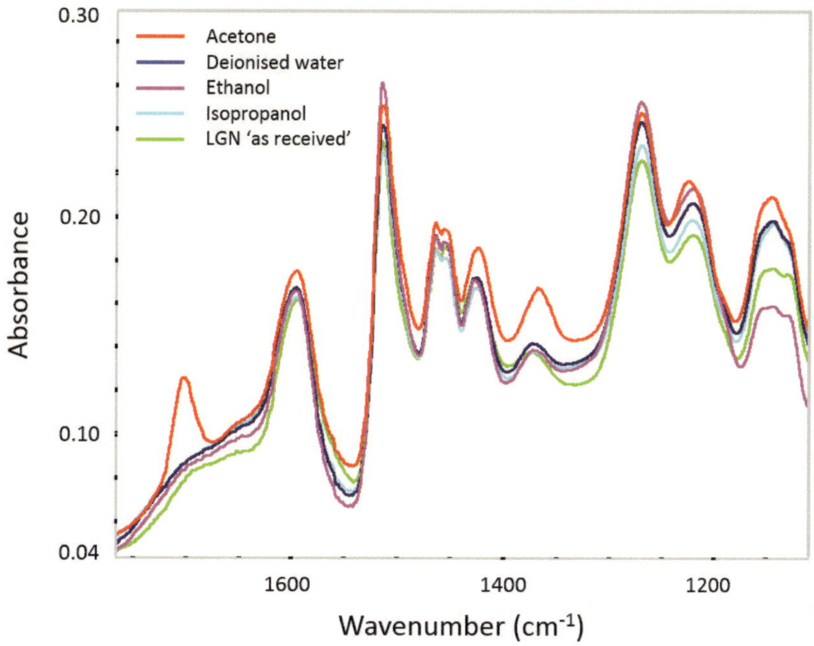

Figure 18. ATR-FTIR spectra acquired on LGN residual powder after evaporation of different solvents, compared with as received LGN.

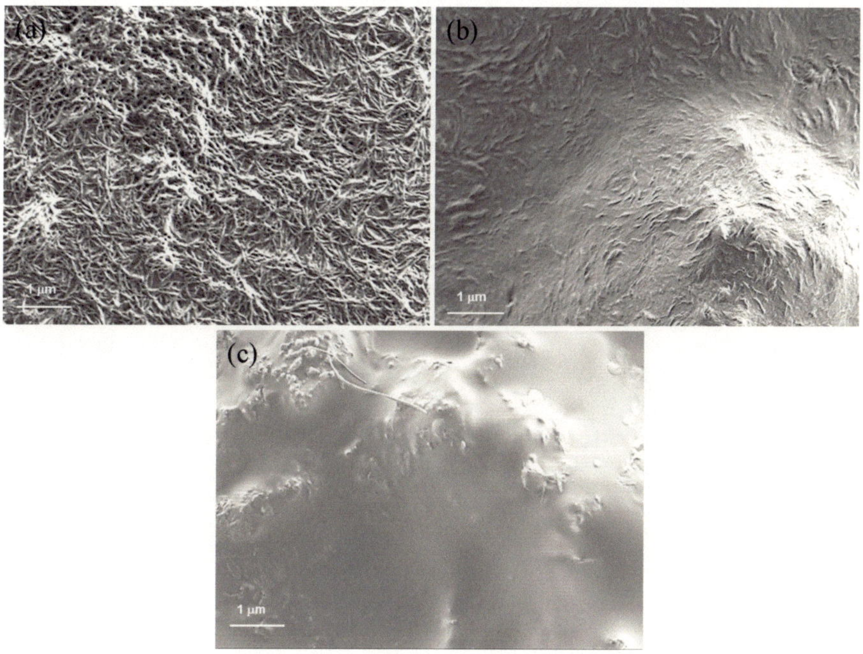

Figure 19. Surface SEM images of PEEK coating treated at 320°C (a), PEEK-LGN coating treated at 300°C (b) and at 320°C (c).

As demonstrated by TG analysis, when LGN is present together with PEEK in the deposited layers, the thermal degradation starts at lower temperature than pure PEEK (**Figure 20**). Therefore, at 320°C the PEEK-LGN layer resulted melted with a lower surface roughness.

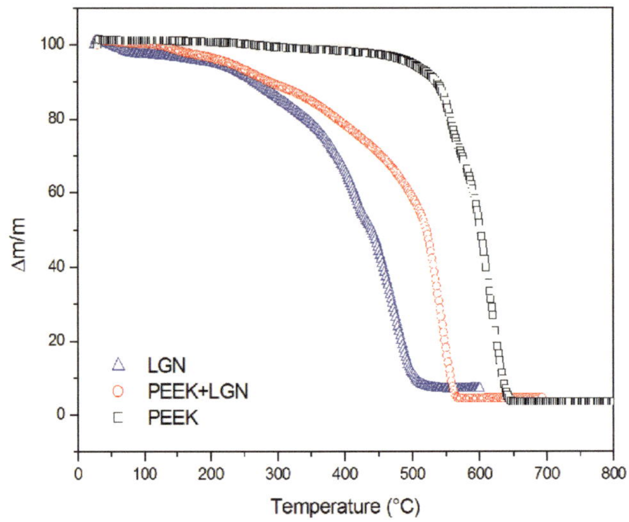

Figure 20. TG curves acquired on pure PEEK (squares), PEEK-LGN (circles) and pure LGN (triangles).

Potentiodynamic curves acquired in NaCl 3.5 g/L solution demonstrated a better anticorrosion behaviour of PEEK+LGN layers than the pure PEEK layer (**Figure 21**). Moreover, corrosion potential was higher and corrosion current was lower when PEEK+LGN layer was treated at 320°C than at 300°C (**Table 3**). This behaviour can be attributed to surface being

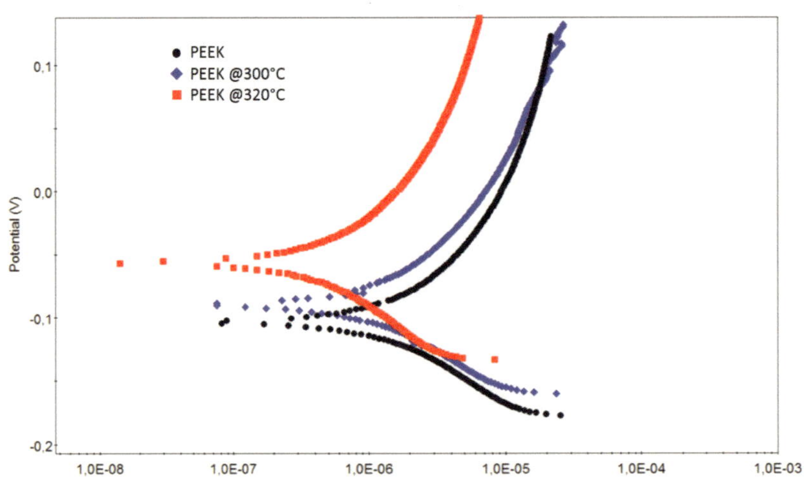

Figure 21. Potentiodynamic curves acquired on PEEK (circles) treated at 320°C and PEEK-LGN coatings in NaCl 3.5 g/L solution treated at 300°C (diamonds) and 320°C (squares) (V vs. Ag/AgCl, 10 mV/s).

characterised by a low porosity. In fact, it is well known that morphological characteristics of a layer influence electrochemical response of its surface to electrolytes interaction when the layer is immersed in an electrolytic solution [32–35].

Coating	Corrosion potential (V vs. Ag/AgCl)	Corrosion current (A/cm²)
PEEK 320°C	−0.10	1.2×10^{-6}
PEEK-LGN 300°C	−0.09	7.2×10^{-7}
PEEK-LGN 320°C	−0.05	3.4×10^{-7}

Table 3. Corrosion potential and corrosion current evaluated for PEEK and PEEK-LGN coatings.

Finally, this result was supported by electrochemical impedance spectroscopy (EIS) measurements. For EIS measurements, a small (usually 10 mV) AC signal is superimposed on the electrochemical system of interest and the system response to this perturbation is measured. The system is composed principally by electrolyte and electrolyte-coating interface and may be represented by an equivalent electrical circuit having a similar electrical behaviour. By studying the complex impedance, it is possible to identify the analysed system features.

Nyquist plot relative to PEEK layer thermally treated at 320°C is reported in **Figure 22**. As discussed in another author's work [36], PEEK layer can be electrically represented as depicted in **Figure 22**, where constant phase element (CPE) component is typical of a porous layer.

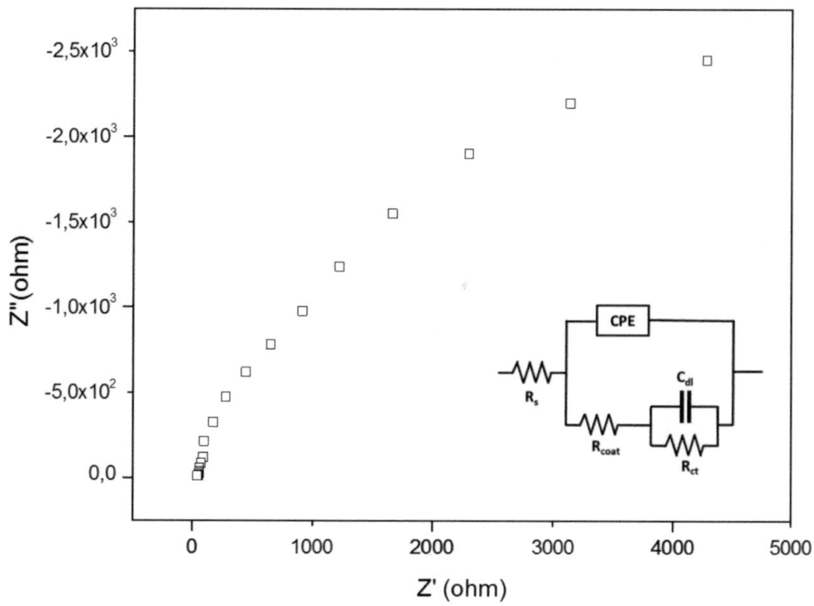

Figure 22. Nyquist plot acquired on PEEK coating treated at 320°C in NaCl 3.5 g/L solution (V vs. Ag/AgCl). In the inset is reported the equivalent circuit representing the coating.

Differently from pure PEEK layer, EIS spectra acquired on PEEK-LGN layers treated at 300 and 320°C (**Figure 23**) corresponded to an electrical circuit composed by two RC loop (inset in **Figure 23**), meaning that two layers could be recognised in the electrochemical behaviour. In other words, the PEEK-LGN-coating surface interacted in a way almost separated from the remaining layer with solution electrolytes.

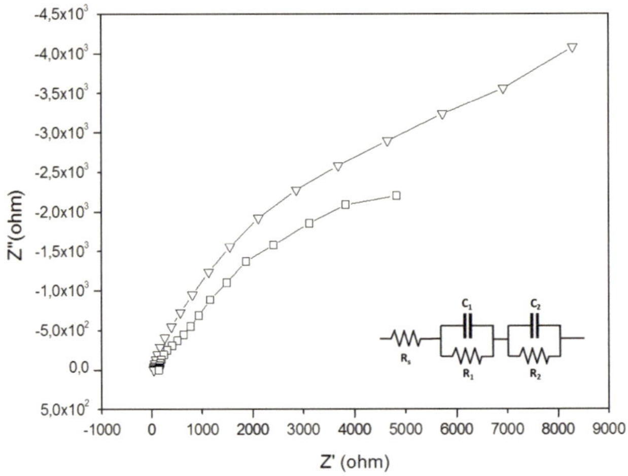

Figure 23. Nyquist plot acquired on PEEK-LGN coating treated at 300 and 320°C in NaCl 3.5 g/L solution (V vs. Ag/AgCl). In the inset is reported the equivalent circuit representing the coating.

The surface of these PEEK based layers was studied by ATR-FTIR measurements, estimating that the thickness interested to infrared analysis was lower than one micron. By comparing spectra acquired on PEEK-LGN coatings before and after thermal treatment (**Figure 24**), it is possible to note for treatments at higher temperature: (1) a lower intensity

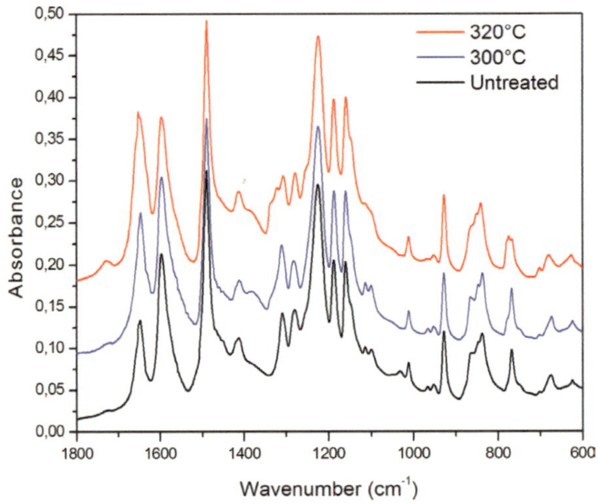

Figure 24. ATR-FTIR spectra acquired on PEEK-LGN coatings before and after thermal treatments at 300°C and at 320°.

http://dx.doi.org/10.5772/67570

of C-H bend stretching of aromatic ring in the range 750–900 cm^{-1}, (2) a decrease of the intensity of C-O ether bend (1300–1350 cm^{-1}), (3) a changed ratio between peak intensities of C-O-R bend stretching at 1600 and 1650 cm^{-1}. All these results, together with the decreased intensity of large band 3200–3600 cm^{-1} (not reported here), mean that an interaction between LGN and PEEK occurred on surface, promoted by temperature, probably through hydrogen bonds between functional groups of LGN and PEEK. It was demonstrated that by adding a certain amount of LGN to PEEK, the thin surface layer of PEEK-LGN coatings showed improved anticorrosion properties, attributed to morphology as well as to chemical resistance of LGN.

6. Conclusions

The outstanding properties of PEEK were further improved by creating a surface nanolayer. This interesting result was obtained by adding a secondary phase to PEEK and thermally treating the composite material based on PEEK. The starting point is EPD technique used to obtain a coating containing PEEK and the secondary phase. After deposition, the coating was submitted to thermal treatment in a conventional furnace, paying particular carefulness to controlling thermal parameters. In such a way, nanostructure or chemical composition of the surface nanolayer was modified with respect to bulk material. As a result, physical and chemical interactions with working ambient produced an improved macroscopic functional behaviour of PEEK.

Author details

Maria Federica De Riccardis

Address all correspondence to: federica.dericcardis@enea.it

ENEA-Italian National Agency for New Technologies, Energy and Sustainable Economic Development, SSPT-PROMAS-MATAS, Brindisi, Italy

References

[1] Liu T, Mo Z, Wang S, Zhang H. Non isothermal melt and cold crystallisation kinetics of poly(aryl-ether ether ketone ketone). Polymer Engineering and Science. 1997;**37**:568–575. doi:10/1002/pen.11700.

[2] Hamdan S, Swallowe G. Crystallinity in PEEK and PEK after mechanical testing and its dependence on strain rate and temperature. Journal of Polymer Science Part B: Polymer Physics.1996;**34**:699–705. doi:10.1002/(SICI)1099-0488(199603)34:4.

[3] Zhang G, Schlarb A K. Correlation of the tribological behaviors with the mechanical properties of poly-ether-ether-ketones (PEEKs) with different molecular weights and their fiber filled composites. Wear. 2009;**266**:337–344. doi:10.1016/j.wear.2008.07.004.

[4] Henneuse C, Goret B, Marchand-Brynaert J. Surface carboxylation of PEEK film by selective wet-chemistry. Polymer. 1998;**39**(4):835–844. doi:10.1016/S0032-3861(97)00362-5.

[5] Henneuse-Boxus C, Dulière E, Marchand-Brynaert J. Surface functionalization of PEEK films using photochemical routes. European Polymer Journal. 2001;**37**:9–18. doi:10.1016/S0014-3057(00)00094-X.

[6] Shi H, Sinke J, Benedictus R. Surface modification of PEEK by UV irradiation for direct co-curing with carbon fibre reinforced epoxy prepregs. International Journal of Adhesion & Adhesives. 2017;**73:**51–57. doi:10.1016/j.ijadhadh.2016.07.017.

[7] Charest J L, Eliason M T, García A J, King W P. Combined microscale mechanical topography and chemical patterns on polymer cell culture substrates. Biomaterials, 2006;**27**(11): 2487–2494. doi:10.1016/j.biomaterials.2005.11.022.

[8] Anselme K, Linez P, Bigerelle M, Noel B, Dufresne E, Judas D, Iost, A, Hardouin P. Qualitative and quantitative study of human osteoblast adhesion on materials with various surface roughnesses. Journal of Biomedical Materials Research.2000;**49**(2): 155–166. doi:10.1002/(SICI)1097-4636(200002)49:2<155::AID-JBM2>3.0.CO;2-J.

[9] Du Y W, Zhang L N, Hou Z T, Ye X, Gu H S, Yan G P, Shang P. Physical modification of polyetheretherketone for orthopedic implants. Frontiers Materials Science. 2014;**8**(4):313–324. doi:10.1007/s11706-014-0266-4.

[10] Besra L, Liu M. A review on fundamental and application of electrophoretic deposition (EPD). Progress in Materials Science. 2007; 52(1):1–61. doi:10.1016/j.pmatsci.2006.07.001.

[11] Boccaccini A R, Roether J A, Thomas B J C, Shaffer M S P, Chavez E, Stoll E, Minay E J. The electrophoretic deposition of inorganic nanoscaled materials. Journal of the Ceramic Society of Japan. 2003; **114**(1):1–14. doi:10.2109/jcersj.114.1.

[12] De Riccardis M F. Ceramic coatings obtained by electrophoretic deposition: Fundamentals, models, post-deposition processes and applications. In: Feng Shi editor. Ceramic coatings: Applications in engineering. Croatia: InTech;2012, pp. 43–68. doi:10.5772/29435.

[13] Corni I, Ryan M P, Boccaccini A R. Electrophoretic deposition: From traditional ceramics to nanotechnology. Journal of the American Ceramic Society. 2008;**28**(7): 1353–1367. doi:10.1016/j.jeurceramsoc.2007.12.011.

[14] Van der Biest O, Vandeperre L J. Electrophoretic deposition of materials. Annual Review Material Science. 1999; **29**:327–352. doi:10.1146/annurev.matsci.29.1.327.

[15] Lewis J A. Colloidal processing of ceramics. Journal of the American Ceramic Society. 2000;**83**: 2341–2359. doi:10.1111/j.1151-2916.2000.tb01560.x.

[16] De Riccardis M F, Carbone D, Rizzo A. A novel method for preparing and characterizing alcoholic EPD suspensions. Journal of Colloid and Interface Science. 2007;307: 109–115. doi:10.1016/j.jcis.2006.10.037.

[17] Sarkar P, Nicholson P S. Electrophoretic deposition (EPD): Mechanisms, kinetics, and application to ceramics. Journal of the American Ceramic Society. 1996;**8**:1987–2002. doi:10.1111/j.1151-2916.1996.tb08929.x.

[18] Voss H, Friedrich K. On the wear behaviour of short-fibre-reinforced PEEK composite. Wear. 1987;**116**:1–18. doi:10.1016/0043-1648(87)90262-6.

[19] Flock J, Friedrich K, Yuan Q. On the friction and wear behaviour of PAN- and pitch-carbon fiber-reinforced PEEK composites. Wear.1999; **229**:304–311. doi:10.1016/S0043-1648(99)00022-8.

[20] Werner P, Altstadt V, Jaskulka R, Jacobs O, Sandler J K W, Shaffer M S P, Windle A H. Tribological behaviour of carbon-nanofibre-reinforced poly(ether ether ketone). Wear. 2004; **257**:1006–1014. doi:10.1016/j.wear.2004.07.010.

[21] Zhang G, Li W-Y, Cherigui M, Zhang C, Liao H, Bordes J M, Coddet C. Structures and tribological performances of PEEK (poly-ether-ether-ketone)-based coatings designed for tribological application. Progress in Organic Coatings. 2007;**60**:39–44. doi:10.1016/j.porgcoat.2007.06.004.

[22] Burris D L, Sawyer W G. Tribological behavior of PEEK components with compositionally graded PEEK/PTFE surfaces. Wear. 2007;**262**:220–224. doi:10.1016/j.wear.2006.03.045.

[23] Wang C, Ma J, Cheng W. Formation of polyetheretherketone polymer coating by electrophoretic deposition method. Surface and Coatings Technology. 2003;**173**:271–275. doi:10.1016/S0257-8972(03)00626-1.

[24] Corni I, Neumann N, Eifler D, Boccaccini A R. Polyetheretherketone (PEEK) coatings on stainless steel by electrophoretic deposition. Advanced Engineering Materials. 2008;**10**: 559–564. doi:10.1002/adem.200800010.

[25] Corni I, Neumann N, Novak S, König K, Veronesi P, Chen Q, Ryan M P, Boccaccini A R. Electrophoretic deposition of PEEK-nano alumina composite coatings on stainless steel. Surface and Coatings Technology. 2009;**203**:1349–1359. doi:10.1016/j.surfcoat.2008.11.005.

[26] Martina V, De Riccardis M F, Carbone D. A chemometric study of alumina/PEEK and hydroxyapatite/PEEK suspensions prepared for electrophoretic deposition of multifunctional coatings. Advances in Science and Technology. 2010;**66**:29–34. doi:10.4028/www.scientific.net/AST.66.29.

[27] Zhang D, Dong G, Chen Y, Zeng Q. Electrophoretic deposition of PTFE particles on porous anodic aluminium oxide film and its tribological properties. Applied Surface Science. 2014;**290**:466–474. doi:10.1016/j.apsucs.2013.11.114.

[28] Jang JH, Machida K, Kim Y, Naoi K. Electrophoretic deposition (EPD) of hydrous ruthenium oxides with PTFE and their supercapacitor performances. Electrochimica Acta. 2006;**52**:1733–1741. doi:10.1016/j.electacta.2006.01.075.

[29] De Riccardis M F, Martina V, Carbone D. Study of polymer particles suspensions for electrophoretic deposition. Journal of Physical Chemistry B. 2013;**117**(6):1592–1599. doi:10.1021/jp3051752.

[30] Boccaccini A R, Keim S, Ma R, Li Y, Zhitomirsky I. Electrophoretic deposition of biomaterials. Journal of the Royal Society Interface. 2010;**7**:S581–S613. doi:10.1098/rsif.2010.0156.focus.

[31] De Riccardis M F, Carbone D, Cuna D. Electrophoretic deposition of lignin reinforced polymer coatings. Key Engineering Materials. 2015;**654**:247–251. doi:10.4028/www.scientific.net/KEM.654.247.

[32] Barsoukov E, Macdonald J R. Impedance spectroscopy: Theory, experiment, and applications. 2nd ed. Hoboken, NJ: Wiley; 2005. ISBN 978-0-471-64749-2.

[33] Orazem M E, Tribollet B. Electrochemical impedance spectroscopy. Hoboken, NJ: John Wiley & Sons; 2008. doi:10.1002/9780470381588.

[34] Lasia A. Electrochemical impedance spectroscopy and its applications. New York: Springer-Verlag; 2014. ISBN 978-1-4614-8933-7.

[35] Bard A J, Faulkner L R. Electrochemical methods: Fundamentals and applications, 2nd ed. Hoboken, NJ: John Wiley & Sons; 2000. ISBN 978-0-471-04372-0.

[36] De Riccardis M F, Carbone D, Capodieci L. Electrophoretically deposited polymeric films: Evidence of correlation between electrochemical impedance spectroscopy measurements and morphological properties. Journal of the Electrochemical Society. 2015;**162:** D3071–D3076. doi:10.1149/2.0201511jes.

Chapter 2

Electroless Deposition of Nanolayered Metallic Coatings

Jothi Sudagar, Rajendraprasad Tamilarasan,
Udaykumar Sanjith, Raj Rajendran and Ravi Kumar

Additional information is available at the end of the chapter

http://dx.doi.org/10.5772/intechopen.68220

Abstract

Electroless metallic coating is referred as the deposition of a substrate material by the process of chemical or autocatalytic reduction of aqueous metal ions deposited to a substrate material without any external supply of power. Electroless nickel alloys are generally considered synonymous to the word *"electroless coating"* as ~90% of productions in industries are of this alloy coating. Rest of the electroless metallic coatings includes gold, copper, palladium, cobalt, silver, etc. These electroless metallic coatings (other than electroless nickel coatings) are also one of the vibrant areas in the field of materials properties and surface engineering research. From the year 2000 to till date, nearly 1000 SCI indexed research papers were published on this topic. However, no comprehensive studies about the recent progress on this topic were reported elsewhere so far. In this context, the present chapter aims to give a complete overview on various aspects of the rest of the electroless metallic nanocoatings/layer as a whole. More importance will be on the recent developments of the nanocharacteristics and future scopes.

Keywords: electroless deposition, metallic coating, electroless nanocoating, electroless gold, electroless copper, electroless palladium, electroless cobalt, electroless silver

1. Introduction

The term "Electroless coating" is referred to as the reduction of aqueous metal ions plated to a substrate by autocatalytic or chemical means, in the absence of external current [1–3], and it disregards the technique used to perform coating in the absence of current such as immersion plating (deposition of copper on steel dipped in copper sulfate solution or nickel on steel dipped in chloride and boric acid bath) and the process of homogenous chemical reduction (silvering). The mechanism of autocatalytically deposited electroless metallic coatings differs in a way completely from coatings obtained by immersion plating or silvering, wherein in the latter case, nature of base material itself will behave as a reducing agent and does not necessitate any external reducing agent to initiate any metal ion to metal reduction. However, these processes have not gained wide acceptance, as it has poor adherence and non-productive behavior. Electroless process has received widespread acceptance in the market due to their exceptional anti-corrosion [4–6], wear resistance properties [3, 5–7] and also good for soldering and brazing applications [8]. Many metals like nickel, copper, gold, silver, palladium and cobalt are being deposited employing this technique [3, 9–12]. Rest of electroless coatings includes gold, copper, palladium, cobalt and silver [13–15]. These metallic coating/ layers (other than electroless nickel) are also one of the vibrant areas in the field of materials properties and surface engineering research. This review chapter provides an insight into the electroless metallic gold, copper, palladium, cobalt, silver, aluminum and the new processing technology in deposition mechanism and nanocharacteristics. Besides, more stress has been laid to understand the recent developments and its future scopes.

1.1. Historical overview

The term "electroless plating" also defined as "autocatalytic" is nearly as old as electroplating. Von Liebig first described it in 1835 with the reduction of silver salts by reducing aldehydes. In 1988, metallic nickel deposition by electroless technique from an aqueous solution in the existence of hypophosphite (reducing agent) was first stated as a chemical tragedy by Wurtz [3, 16]. Roux, in the year 1911, positively reported that there was clear precipitation of metal in the form of powder. However, these workings were not practically used for any applications. Development in this field did not receive much of patronage until World War II. The method established by Brenner and Riddell for coating a layer of nickel-tungsten alloy in the inner parts of tubes by means of a citrate based bath containing an insoluble anode yielded the hypophosphite with uncommon reducing properties. The Patent Office of US declared that this patent, issued in 1950, is distinct and varied from the earlier patent owned by Roux, in which the process was unstructured and complete. In contrast, Brenner and Riddell's method was described as a controlled catalytic process as deposition process took place only on catalytically active surfaces dipped in the plating bath. Dr. Brenner later manifested that his patent was an unintended outcome comparable to the efforts of Wurtz and Roux. But he added that the patent was taken in order to safeguard US government rights. Actually, in 1963, a publicized US Army based technical report was printed that pronounced mostly about the work of Wurtz and Roux and gave much of the finding credit to Brenner and also covered by patent in 1956 [17]. This phenomenon ascribed to the action of chemical reduction of nickel ions was later identified.

After initiation of electroless nickel deposit, rest of electroless coatings was followed. In 1970, semiconductors and circuit boards were plated with thick, pure soft gold after the development of the first electroless gold coating bath at Bell Laboratories [14]. Narcus first reported the electroless copper deposit, and the first profitable application was realized by Cahill and Zeblisky et al. [18] in tartarate baths, where the reducing agent used was formaldehyde. Improved formulations of electroless copper [19] recorded a higher rate of plating with exceptionally stable conditions under diverse operating conditions. Baum et al. developed a process for selective deposition of copper by first selectively depositing palladium seeds in

1986. Palladium and alloys of palladium have been identified as a pecuniary alternative to applications involving gold plating. Electroless cobalt discovered along with nickel deposition by Brenner and Riddell [1–3]. This technique of electroless cobalt deposition has been exploited frequently for the preparation of magnetic films. In particular, thin films of cobalt have found application as recording media on account of their thinness, high coercivity, and high remanence. The commercial application of electroless silver was developed early in 1970 [20]. Electroless aluminum [21], electroless platinum [22], electroless ruthenium, and electroless rhodium were developed. In the early 1984, materials like plastics; ceramics, polymers and other non-conducting materials are coated by electroless technique before even being subjected to electroplating. This phenomenon led to the development of the rest of electroless metallic deposits. Meanwhile, the field of electroless coating chemistry has materialized as one of the promising and key areas of the surface engineering and metal finishing research, etc.

1.2. Primary aspects of electroless metallic coating

The basic elements of any electroless bath and their role are briefly reported in **Table 1**. A typical electroless gold, copper, palladium and cobalt plating baths available in literature are illustrated in **Table 2** [3, 23]

Components	Functions
Metal ions	Source of metal
Reducing agents	Source of electrons to reduce the metal ions
Complexants	Stop excess of free metal ions concentration
Accelerators	Quicken the reducing agent and increase the deposition
Stabilizers	Mitigate the bath from decomposition by protecting catalytically active deposition
Buffers	Endure the pH for long time
pH regulators	pH tuning
Temperature	Energy to the bath deposition

Table 1. Electroless bath components and their functions.

Electroless bath	Gold	Copper	Palladium	Cobalt
pH, temperature and deposition rate	10–13, 65–85°C and 2–5 μm/h	9–13, 25–70°C and 1–5 μm/h	8–12, 45–75°C and 2–5 μm/h	9–14, 85–95°C and 2–10 μm/h
Metal salt(s) or Source	Gold chloride, Gold cyanide, Potassium aurate.	Copper sulfate, Copper carbonate, Copper formate, Copper acetate, Copper nitrate.	Palladium chloride, Palladium bromide.	Cobalt sulfate, Cobalt chloride.
Reducing agent(s)	Dimethylamine (DMAB), Sodium hypophosphite, potassium borohydride, Potassium cynoborohydride.	Formate, Formaldehyde, DMAB, Sodium hypophosphite.	Sodium hypophosphite, DMAB, Triethylamine borane.	DMAB, Sodium hypophosphite.

Electroless bath	Gold	Copper	Palladium	Cobalt
Complexing agent(s)	Potassium citrate, sodium phosphate, sodium borate, Potassium tartrate.	Rochelle salt, Ammonium hydroxide, pyridium-3-sulfonic acid, Potassium tartrate, Quadrol, EDTA.	Ammonia, Methylamine, EDTA.	Sodium citrate, Succinic acid. Ammonium chloride, Citric acid.
Stabilizer(s)	Alkali metal cyanide, Alkali hydrogen, fluoride, Acetyl acetone.	Thiodiglycolic, Thiourea, Sodium cyanide, Vanadium pentoxide, Potassium ferrocyanide, MBT.	Thioorganic compound. Thiourea, Organic cyanide, Thiocyanates.	Urea, Thioorganic compounds.
pH adjuster(s)	Potassium hydroxide, Phosphoric acid, Sulfuric acid.	Hydrochloric, Sulfuric acids, Sodium and Potassium hydroxide.	Ammonium hydroxide, Hydrochloric acid.	Ammonium hydroxide, Sodium hydroxide.
Major applications	Integrated circuit, chips, connector, and semi-conductor devices in electronic Industry.	Conductive layer prior to deposit on plastics or non-metallic, Printed circuit board and decorative purpose.	Printed circuit board, electronic switch contact, and it are alternative for electroless gold.	Magnetic memory disc and storage devices in electronic industry.

Table 2. Plating bath composition of typical electroless gold, copper, palladium, and cobalt with their major applications.

In the process of electroless deposition, the reduction of metal ions to metal takes place by the action of reducing agents, which are basically donors of electrons. The metal ions act as acceptors of electron and are subject to chemical reaction with the electron donors. This process is autocatalytic in which the acceleration of electroless chemical reaction subjects the reducing agent used to oxidize. The experimental apparatus shows the basic illustration of the setup usually used in experiments of electroless coatings (**Figure 1**). In addition to the basic setup, ultrasound improved the benefits in the electroless nickel, copper and cobalt plating [24].

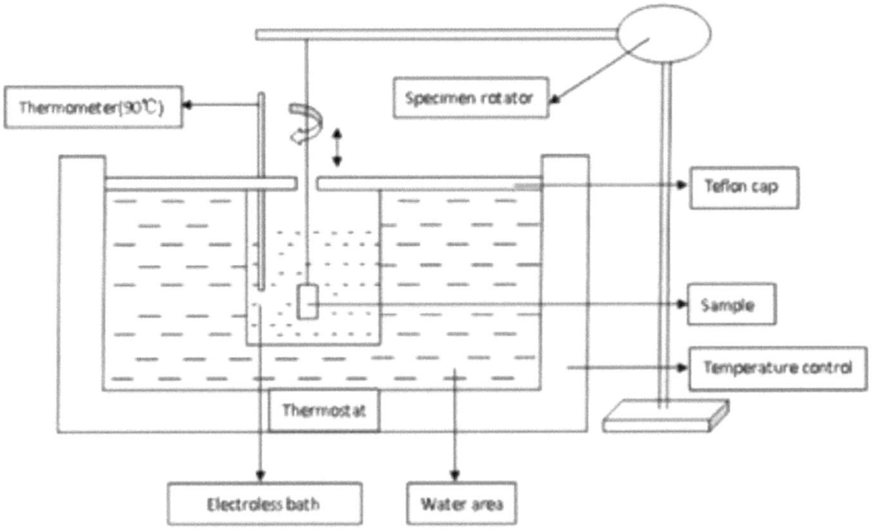

Figure 1. Basic illustration of the apparatus used in electroless coating experiments.

2. Rest of the electroless metallic coatings

2.1. Electroless gold

In most of the existing practices of electroless gold, coatings are produced initially by depositing a thin layer of immersion gold, followed by electroless gold plating [14, 15]. However, there are several shortcomings which depict electroless gold processes: (a) low deposition rates, (b) necessity to carefully regulate plating conditions, (c) substrates should be meticulously cleaned, (d) plating baths are likely to have relatively short lives, (e) stirring. Electroless gold deposition has been most successfully accomplished using a gold sulfite and ammonium salts of sulfite. Those salts of gold comprise of potassium gold chloride, gold cyanide and gold sulfite among which the most commonly used source is potassium gold cyanide. The reducing agents essential for electroless plating are as varied as the potential gold salts. A wide variety of reducing agents required for electroless are available as various potential gold salts such as sodium hypophosphite, dimethylamine borane, sodium borohydride, and hydrazine. It has been serving as the primary application for electronic industry to perform selective plating, thereby conserving plating costs and improve circuit design in integrated circuits [25]. There are some realistic cases of autocatalytic processes of coating gold with the ability of having achieved 99.99% purity. **Table 3** gives a typical bath composition for the electroless gold plating. Electroless plating of gold salts on germanium results in morphologically complex nanostructure metallic films [26]. Deposition simply takes place by means of galvanic displacement without the presence of pH adjusters, fluoride, complexing agents, or other external reducing agents. This facile method gives good control over deposition rate and surface morphology by proper variation of coating parameters such as temperature, immersion time and metal ion concentration. The preparation of Au nanowires of high aspect ratio by employing electroless reduction of gold in the hexagonally ordered, thiol-modified nanosized channels [27]. The outcomes evidently display that the development of the Au nanowires is templated by the channel structure of the base substrate. Ecofriendly electroless plating bath was developed by chloroauric acid ($HAuCl_4$) and hydrogen peroxide (H_2O_2) for depositing a film of gold onto (3-aminopropyl)-trimethoxysilane (APTMS). It could prove as a feasible replacement of using hazardous source of potassium gold cyanide [28].

Electroless gold bath composition	
Gold hydrochloride trihydrate	0.01 M
Sodium potassium tartrate	0.014 M
Dimethylamine borane	0.013 M
Sodium cyanide	400.0 mg/L
pH (adjusted with NaOH)	13
Temperature	60°C

Table 3. A typical electroless gold bath.

The various characteristics of electroless gold coatings attained from baths of borohydride have been abridged by Feldstein [29], and it is given in **Table 4**. This table shows mainly the physical properties of the gold plating (i.e., adhesion, density, porosity, and resistivity)

2.1.1. Recent developments of electroless gold

Electroless gold layer was developed on hydrogen-terminated Si substrate from aqueous hydrofluoric solutions [30]. The Au deposition is kinetically inadequate by diffusion at first, but then development of Au clusters is kinetically restricted by a surface reaction containing a fluoride species. This possibly necessitates for Au to be coated in a relatively mobile state initially which is only moderately discharged. Hou et al. reported about the preparation of gold films and affirmed the possibility for them to subsequently be used as base materials to aid the formation of self-assembled monolayers (SAMs) from alkanethiols [31]. The formation of SAMs on the films of electroless gold can be ascribed to two primary reasons. Firstly, any type of wet-chemical laboratory can be used to prepare electroless gold. In addition, electroless gold can be coated on intricate and complicated shapes of substrate, wherein no technique for the evaporation process could be recommended. The densely packed SAMs (prepared from hexadecanethiol), present on the surface of electroless gold, are only deposited as a thin film of evaporated gold. Electroless coating promoted the epitaxial growth of Au (111) on a seed layer of evaporated gold that makes it highly beneficial for microfabrication applications [32]. Wang et al. [33] have recently found that electroless Au microelectrodes could be fabricated on polycarbonate microfluidic chips with SAM, after cleaning the microelectrodes with plasma. The system comprised of a polycarbonate microfluidic chip with an electrochemical detector, a gravity pump, and an automatic sample loading and injection unit. Lei et al. [34] reported the analysis of prepared gold films with a surface plasmon resonance (SPR) device to detect the rapid and label-free detection of the white spot syndrome virus (WSSV). This method, key issues in the field of pisciculture and environmental toxicology, has been addressed and supplemented the expansion in the range of applications of the SPR technology. Schwarz et al. have recently [35] deposited a thin layer of soft gold onto polypyrrole and copper-coated paraaramid

Property	Value
Adhesion	Excellent on metals
Appearance	Matt yellow
Density	Bulk gold (19.3 g/cm^3)
Hardness	Soft (Knoop 60–80)
Porosity	~Zero for deposits, ≥ 1 μm on uniform substrate
Resistivity	Bulk gold (0.03 ohm/square at 1 μm)
Purity	99.90%
Thermo compression bond ability	Excellent

Table 4. Properties of electroless gold deposits.

yarns. Copper ions go into the solution as less noble copper dissolves, and consequently, there is a reduction of gold ions in the solution thereby getting deposited on the yarn surface. Once a thick layer of gold has been deposited, further dissolution of copper cannot happen and the process of charge exchange gets stopped. As gold layer is now formed on the yarn surface, autocatalytic chemical reduction takes place, and deposition continues further. Result revealed that the electroless coated yarns exhibited improved mechanical properties plus excellent electrical conductivity and considerable resistance to washing. Also, electrochemical results displayed that the gold coated yarns can be used to measure biomedical signals as they are promising electrode materials. An effective and simple electroless deposition technique was proven to deposit mixed SnO_2-Pd-Au film for sensing hydrogen. Electrical conductivity of the film was improved by the co-deposited gold. The sensor was easy-to-use and can be fabricated easily. This sensor can be used in areas like hydrogen gas alarm in commercial or domestic security [36]. Conducting electroless gold pads on light-emitting diodes chips are fabricated which exhibit no color difference, reliable wire bonding ability, and high values of electrical conductivity. The hardness of pads formed by electroless plating is three times softer than those formed by evaporation and the force gauges [37]. The industrial application of this plating is feasible.

2.1.2. Nanocharacteristics of electroless gold

The procedure for fabricating high-yield integrated nano-gap electrodes concurrently on a single sample was developed, which contains iodine solution (usually known as tincture (a medical liquid)) and ascorbic acid [38]. Results revealed that methods like nanogap fabrication and the fabricated gold nano-gap electrodes are beneficial for realizing applications like mono-electron and molecular nanodevices. Ding et al. have [39] produced Au nanoparticles using conducting polymer nanoparticles through electroless collecting $[AuCl_4]^-$ from solutions. A considerable improvement of the recovery capability of Au was realized by the polypyrrole nanoparticle when compared to the film or polypyrrole powder, due to the high specific surface area. Researcher [40] investigated in wielding electroless plated Au as a provision for carbon nanotube (CNT) electrodes. Further, they stressed the need for developing electroless techniques to meet the requirements of creating a more heterogeneous, uniform layer of gold to reduce desorption of cysteamine monolayer. An innovative nanotemplating technique was established to produce spherical gold nanoparticles (NPs) or regular arrays of silane rings on silicon substrates through selective electroless plating on layers of particle-lithographed silane [41]. This novel nanotemplating method can synthesize even and smooth metal NP arrays over huge areas to enhance the potential of improved spectral features in opto-plasmonic devices in spite of disregarding the requisites of large and expensive lithography and metal coating equipment. Research [42] defined a new electroless technique of depositing single-crystalline Au-NPs on and inside an organic single-matrix (SM) confined with both a stabilizer and a reducing agent. This process is appropriate for direct deposition of mechanically stable and optically transparent Au-NPs on and inside a SM. The development of palladium and platinum nanoparticles was made possible only with this method when the actual reaction was carried out on the surface of the SM in the presence of some gold nanocrystals.

2.2. Electroless copper

Electroless copper deposits have found their greatest applications for imparting a conductive layer for non-conductors before being coated by electrolytically. Electroless copper solutions resembling today's technology were first reported in 1957 by Cahill with the report of alkaline copper tartrate baths using formaldehyde as reducing agent. The pH range of 12.0–13.0 is generally optimal in formaldehyde-reduced baths. Formaldehyde-reduced electroless copper bath is given in **Table 5**. In printed-circuit boards industry, copper is deposited on the inner wall of the insulating hole that connects the two sides of the substrate by employing plating-through-hole process, one of the techniques of electroless copper deposition. Electroless Cu plating is becoming attractive because of its favorable nature of electrical & thermal conductivity, ornamental surface and so on [15].

Among the various chelating agents in electroless Cu plating (ethylenediamine tetraacetic, triethanolamine, and ethylenediamine), studies recognized ethylenediamine as an outstanding grain-refining agent owing to its strong adsorption on the Cu surface [43]. Formaldehyde is the reducing agent used in the great majority of commercial electroless Cu baths and other reducing agents have also been used successfully. The copper deposit increases the surface roughness by using pyridine-2,6-dicarboxylic and 4-hydroxypyridine-2,6-dicarboxylic acids as Cu (II) ions ligands in formaldehyde having alkaline electroless Cu plating baths [44]. The surface roughness factor (i.e., ratio of real surface area (nano-scale roughness) to geometrical surface area) was found to be the highest for pyridine-2,6-dicarboxylic acid having solutions at pH 13.

2.2.1. *Recent developments of electroless copper*

One of the most important applications of electroless copper is the electronic industry. The electroless Cu/Ni/Au deposits had the attributes of excellent adhesion in addition to having a low sheet resistivity, which are prerequisites for low attenuation at microwave frequencies. The above coating has been recommended as a cost effective means for microwave components because the thickness of gold in the proposed system is much less than that of conventional chromium/gold. Researcher studied the effective acceleration of deposition of copper at a temperature range of 20–30°C by a factor of 2–4, even for lesser additions of

Electroless copper bath composition	
Copper salt as Cu^{2+}	1.8 g/L
Rochelle salt	25 g/L
Formaldehyde as HCHO	10 g/L
Sodium hydroxide	5 g/L
2-Mercaptobenzothiazole (MBT)	<2 g/L
pH	12
Temperature	RT

Table 5. A typical electroless copper bath.

ammonia (1–3 mM) [45]. Acidic electroless Cu on aluminum-seeded ABS plastics was an alternative option for plating plastics, [46] and this was critically criticized because of its very slow rate. Hanna et al. [47] studied the role of various organic additives (cytosine, pyridine, benzotriazole and 2-mercaptobenzothiozole) which stabilized electroless copper baths in addition to enhancing the rate of plating. The bath stability increased 20 times to that of the aeration lacking-bath, as an effect of mild air agitation. This is an unusual method of non-isothermal deposition to enhance the deposition rate without affecting the plating bath. The deposition rate increased with increasing temperature of the substrate; nevertheless, the influence of bath temperature cannot be ignored. Result showed that this non-isothermal technique allows for deposition rates ~12 μmh^{-1} at certain conditions, which is 3 times than conventional isothermal method [48]. Recently, found long-term stability of Cu/Pd nanoparticles by using poly-vinyl-pyrrodione (PVP). In his study, acceptable activity and superior stability have been exhibited by the newly developed Cu/Pd colloidal system thus, indicating its capacity as an auxiliary for the prevailing Pd/Sn-based activator in PCBs [49]. The reduction of Cu(II) autocatalytically by formaldehyde from solutions having saccharose as the ligand initiates at a level of pH beyond 12, increases with an additional increase in pH, reaches a maximum value at pH 12.75, and then decelerates at higher pH values [50]. Tamayo-Ariztondo et al. [51] prepared electroless Cu deposits preferably on the surfaces parallel to graphene layers, due to the exposure of π bonds in the outside surface, and deposition was subdued on surfaces perpendicular to the graphene layers. Result revealed that there is maximum coverage of electroless Cu along the plane of graphene, and it is reduced at the edges of the plane. Garcia et al. [52] narrated two methods based on the ligand-induced electroless plating (LIEP) process to obtain patterns of Cu onto flexible polymer substrates. The LIEP process permits plating of copper selectively with stable electrical properties onto flexible polymer substrates. Therefore, the LIEP process collaborated with any of those patterning methods could perhaps be a better auxiliary for the classical processes of cost-effective fabrication of large-area plastic electronic devices and to endure substantial mechanical deformation with only a negligible loss in performance. Researcher developed activation for the dielectric surface by the cobalt compounds. This palladium-free activation looks hopeful for the practical application due to its stability and can be potential compare to many of the patented Pd-free activation solutions [53].

2.2.2. *Nanocharacteristics of electroless copper*

A research thrust has been initiated [54] that sub-100 nm copper films of low resistivity be deposited by electroless means on SAM of 3-aminopropyltrimethoxysilane and activated by 5 nm gold nano-particles. The resistivity achieved in this process is fairly relatable to the effective resistivity expected by the International Technology Roadmap for Semiconductors (ITRS) for realization in 2010–2011 years for 45 nm ULSI metallization. Noteworthy developments in wetting of aluminum can be attained by applying an electroless Cu on Al_2O_3 and SiC (Al/Cu-Al_2O_3 and Al/Cu-SiC) ceramics [55]. The copper coating present in the interface hindered the reactivity of SiC toward Al thus causing a clean interface. Recently, [56] a new method has been proposed to coat the surface of fly ash particles with conducting metal Cu by electroless, where titania/ultraviolet radiation/metal catalyst-system has been used instead of conventional Pd/Sn-based activator. More work has to be done, whether the proposed one will be suit for the rest. Organic protection coatings act as an interlayer for highly reactive metals (magnesium) to coat electroless Cu and Ni or its alloys, and the process is criticized for its long time process. Daoush et al. studied on enhancing the strength of interfacial bonding between Cu and CNT by acid treatment and electroless copper coating of multiwall CNT to produce CNT/Cu nanocomposite powder with different CNT volume fractions. The electrical conductivity decreased, and the hardness increased with increase in volume fraction of CNT. The yield strength of the sintered materials had enhanced by increasing the volume fraction of CNT except in case of 20 vol. % CNT/Cu composite where the material fractured even before yielding. In addition, the increase in volume fraction of CNT in copper matrix witnessed an increase in Young's modulus and a subsequent decrease in elongation [57]. Wear and mechanical behavior of reinforced carbon nanofibers improved by electroless Cu alloy-based composites [58]. A new method proposed for plating a Cu layer onto an aramid film with a strong bonding by adhesion [59]. Deposition of electroless copper nanofilms on silicon and on polycrystalline germanium substrates seemed to cover an extensive area of the base material compared to other noble metals such as Au, Pt, Pd, and Ag [60]. Bruning et al. developed an in situ X-ray diffraction measurement for electroless Cu films. The strain evolution of films was determined based on three types of electrolyte. Within the experimental ambiguity, the correlation among stress and strain for the Cu films approves with the properties of bulk polycrystalline Cu [61]. The coating of Cu on particles of B_4C is required in order to synthesize metal-ceramic composites with improved sinterability and dispersability. A surface pretreatment similar to that of acid and alkali treated particles was carried out for B_4C particles. There was uniformity in the observed copper coating in alkali-treated particles at a level of pH-12, when matched to others. This is due to the effective elimination of impurities during the processes of production and processing of commercially existing B_4C [62]. The electroless Cu-P-SiC composite coating on carbon steel improved the anti-corrodibility behavior of deposits of electroless Cu [63].

Zangmeister and van Zee [64] observed the possibility of Cu to be deposited by the reducing action of formaldehyde on Cu^{2+} ions in 4-mercaptobenzoic acid SAMs, but inferred that the same cannot be performed on octadecanethiolate or 3-mercaptobenzoic acid SAMs. The formation of a surface-bound Cu complex when reduced by formaldehyde leads to the deposition of Cu metal. Garno et al. [65] observed another prospect for Cu to be deposited on COOH-terminated SAMs. Moreover, it was observed that smaller amounts of Cu could be deposited on CH_3-terminated SAMs, while narrowing likelihood of depositing Cu on OH-terminated SAMs. Recently, two main requirements for selective electroless deposition of Cu to take place were presented. Firstly, in cases where more than one type of functional group is present in the SAM, Cu^{2+} ions must interact with any one specific terminal group, if there is more than one group. Secondly, the temperature needs to be maintained adequately high to avert adsorption of non-specific Cu on non-interacting SAMs. It was significant, however, to maintain the reaction temperature in such a way, so that it was not too high to damage the SAM. The penetration of Cu into the Au/S interface through the monolayer was observed in addition to the deposition of Cu at the SAM/air interface. Moreover, the penetrating action through the SAM was noticed not to be ceasing even after plating was stopped [66]. This advises that electroless deposition of Cu taking place parallel with Cu evaporation may not be an appropriate technique to develop Cu/SAM/metal or Cu/SAM/semiconductor junctions in molecular architectures based applications. This is because of the gradual shifting of upper Cu metal contact with respect to time. Alternative conditions of electroless deposition and metal types (e.g., Au, Pt, and Pd,) should be encouraged and moreover be explored to find the scope for formation of more stable metal over-layers with slight or without any metal atom penetration. Electroless Cu plating developed on nanofibers with the attributes of [67] high modulus and high strength, and particularly high electric conductivity on the surface of poly (p-) phenylene benzobisoxazole. A simple electroless Cu-coated prepared on glass nanofiber with excellent conductivity. Actually, this method is simple, low-cost, and large production and can be stretched to make other metal coated glass fibers by distinctive conductivity [68].

2.3. Electroless palladium

Palladium and palladium alloys can be deposited by the electroless mechanism using hypophosphite or hydrazine reducing agent. Palladium deposits find application in electrical contacts and connectors and serve as a diffusion barrier between metals such as copper and gold. It has been established as an economic substitute to gold plating. It has also been used as a best replacement for rhodium for wear application [69]. **Table 6** gives the constituents and composition details for an electroless palladium (hypophosphite-reduced) bath. The deposit can be hardened or be bonded to electroless nickel or be acquired with desired coating characteristics by means of varying specific bath components or their composition. For example, deposits with greater bond strength than the actual tensile strength of the palladium plate itself also can be attained. The plating can be direct in metals like stainless steel and nickel, whereas copper, brass, and other copper alloys would require an electroless nickel preplate.

The catalytic properties in Palladium have marked the metal notable in serving several applications of chemical and automotive industries. The usage of the metal is, however, been limited owing to high cost and difficulty to put it to realistic processing for other commercial applications. It is, yet, studied and proposed that thin films of palladium plated by electroless means onto ceramic sponge or other support materials may well be used effectually as reaction or auto emission catalysts. The metal and its alloys have been established as a profitable

Electroless copper palladium composition	
Palladium chloride	10 g/L
Rochelle salt	19 g/L
Ethylenediamine	25.6 g/L
Cool solution to 20°C (68°F) and then add:	
Sodium hypophosphite	4.1 g/L
pH (adjusted with HC1)	8.5 g/L
Temperature	68–73°C

Table 6. A typical electroless palladium bath.

substitute to gold plating. In spite of the differences in some of the properties of Pd and Au such as melting point and deposit hardness (**Table 7**), one character the metals share in common is the superior oxidation resistance.

2.3.1. Recent developments and nanocharacteristics of electroless palladium

It is often proposed that Pd/Sn-based activator is used to activate the substrate surface to deposit electroless Cu. In some cases, electroless Pd acts as an activator [70]. Recently, electroless Pd was coated on Iridium (Ir) and tungsten (W) substrates. The thickness of Pd was 20 and 30 Å on Ir and W substrates, respectively. A very strong adhesion of the electroless Cu to Ir and W was observed, when Pd was used as a catalytic layer [71]. In practical application point of view, modification of Pd by electroless on ZrO_2-TiO_2 selective layer produced membranes, which detached hydrogen and nitrogen gases in the Knudsen diffusion domain (H_2:N_2~3.75). In addition, using γ-aminopropyltriethoxysilane (γ-APTES) will ultimately lead to the proficient electroless deposition of amorphous and porous layers of Pd [72]. A method of vacuum electroless plating for synthesizing of thin dense Pd membranes on porous alumina tubes was developed. The diverse application of these membranes was ascribed to their high permeation performance, good thermal durability, and favorable chemical stability. Furthermore, these membranes had withstood tests for thermal durability over 470 h under H_2 or Ar atmosphere for both cycles of temperature cycles and gas-exchange. Additionally, these membranes offered a strong resistance to fluctuations in the chemical stability under various H_2 mixtures over a varied pressure and temperature range for 2000 h. No significant

Property	Electroless palladium	Electroless gold
Color	Silver/White	Yellow
Crystal	FCC	FCC
Resistivity	10.5×10^8 ohm-m	2.2×10^8 ohm-m
Magnetic	Paramagnetic	Diamagnetic
Density	12.0 g/cc	19.3 g/cc
Molecular weight	106	197
Melting point	1555°C	1063°C
Boiling point	3140°C	2660°C
Hardness	50–200 HV	20–150 HV
Elongation	24%	44%
Tensile strength	24 Ksi	18 Ksi
Co-efficient of expansion	6.5 μ-in./in./°F	7.9 μ-in./in./°F
Maximum thickness	Unlimited	Unlimited
Reacts with	HNO_3; HF	H_2SO_4; KCN; aqua regia

Table 7. A comparison between electroless palladium and gold.

changes in hydrogen permeation performance were observed, and sustained hydrogen permeation for the acute temperature fluctuations was exhibited. Another application oriented in electroless Pd membranes onto cordierite mini-channel network was developed recently and testified hydrogen separation [73]. Initially, the cordierite channels were coated in its interiors with alumina layers as supporting layers, consisting of a layer of micropowder alumina that covered a structure of bare cordierites structure of high porosity, and followed by a layer of nanopowder alumina resulting in an even surface free of defects for electroless plating of defect-free palladium films. The perm-selective palladium films fabricated in this new support structure also permits for extraction of hydrogen from hydrocarbon fuels in this design of integrated membrane reformers [74]. In some cases, the function of electroless Pd film on stainless steel is to increase the electrode potential and boost the rate of passivation of steel to endure strong corrosive environments [69]. Strukova et al. developed Pd-Au, Pd-Ag, Pd-Ni, Pd-Pb, and ternary Pd-Au-Ni alloy system onto different metallic substrates. The coatings of Pd-Au, Pd-Ag, and Pd-Ni have a solid solution structure, whereas Pd-Pb is intermetallic compound whose films after deposition comprised of nanocrystalline grains with sizes in the range of 11–35 nm [75]. Self-supported thin Pd alloys [76] membranes without defects were prepared for hydrogen permeation performance. The subsequent self-supported Pd membranes, less than 10 μm in thickness, demonstrate exceptional performance of hydrogen permeation and an extensive selectivity. In addition, a new type of non-alloy Ru/Pd composite membrane fabricated for hydrogen separation [77].

2.4. Electroless cobalt

Cobalt deposits are mainly produced from alkaline-hypophosphite baths. These coatings are produced from sodium hypophosphite-based solutions at a slightly alkaline pH range at elevated temperatures. The electroless cobalt deposition by hypophosphite is always supplemented by the co-deposition of phosphorus. The fabrication of magnetic storage devices with high area recording density generally requires magnetic material that is soft in nature. Its application includes large variety of magnetic properties and has found their major applications for switching and memory storage devices [78]. **Table 8** gives the information on constituents and composition of a typical electroless cobalt bath. Thin electroless Co deposits have their applications in the electronics industry (magnetic memory discs and storage devices) exclusively for

Electroless cobalt palladium composition	
Cobalt chloride	30 g/L
Sodium hypophosphite	20 g/L
Sodium citrate	35 g/L
Ammonium chloride	50 g/L
pH	9.5
Temperature	95°C

Table 8. A typical electroless cobalt bath.

their magnetic properties. In ternary Ni-Co-P alloy coatings, the Co content displayed a linear dependency with both pH of bath and temperature and was also influenced by the composition of Co sulfate in the electrolyte. It showed a spherical nodular structure with finer, compact grains and a homogeneous crystal structure. In addition, the maximum hardness was 804 HV_{50} for the deposits having 19% cobalt [79]. The maleic and succinic acid maintained the stability, long plating of the bath, and morphology of the as-plated deposits [80]. The deposition rate was more in the bath with succinic acid addition. These results are, though, inadequate for reporting evidences of electroless Co-based alloys with magnetic properties due to lack of systematic study on the conditions of preparation and intrinsic magnetic properties of the material.

2.4.1. Recent developments and nanocharacteristics of electroless cobalt

In Ni-Co-B ternary alloy coatings, the saturation magnetic moment was found to be increasing with rise in content of cobalt in the deposit and with prolonged annealing of the deposit [81]. SiC (Co-P-SiC) entrapment was more favored within the cobalt matrix composite coatings than Ni (Ni-P-SiC) matrix. The magnetic properties of the electroless Co-Fe-P films formed out of a stable sulfate bath [82]. In high pH bath with heat treatment films of Co-Fe-P, showed good soft magnetic properties. The magnetic properties of Co-W-P films have been studied by many authors [83–84]. Magagnin et al. [85] have proposed that electroless cobalt-phosphorus acts as a metallization barrier for copper in lead-free soldering. However, electroless Co-P/Au finish with about 4 wt. % P content sturdily restricts inter diffusion and formation of intermetallic compounds when paralleled to Ni-P/Au finish with Sn-Pb and Sn-Ag-Cu solder alloys. Furthermore, the shear test results suggested that Co-P/Au deposit having higher joint strength than Ni-P/Au deposit. Jiang et al. [86] have prepared SiC-W/Co nanocomposite particles by electroless cobalt on SiC whiskers. The bonding between the substrate and the cobalt coating is so weak due to which the thermal stability of SiC-W/Co composite is low. Hence, a research thrust has been initiated toward encouraging further investigation on the thermal stability of metal coating plated on whiskers. The electroless Co composites have not seemed to have received widespread attention and application; however, future prospects will be very attractive. The Co-Zn-P thin film coated nano-diamond materials were the basis for providing a feasible solution over the expensive cobalt material in the synthesis of this type of magnetic thin film materials. These magnetic film materials have been witnessing myriad prospective applications in the field of Magnetic Abrasive Lapping Materials in the near future [87]. CNT were decorated with FeCO using one-step polymer-stabilization activation step and low-cost electroless deposition. The approach recommends a viable method that can be followed for preparing nanoparticles of FeCo in application of cancer thermotherapy [88].

2.5. Electroless silver

Silver can be deposited from dimethylamine borane-based baths. The major application is to coat in the interior of waveguides. Many authors have reported and illustrated evidences of depositing electroless silver on different substrates. Abbott et al reported studies carried out about Ag coating on copper substrates using an ionic liquid [89]. A super hydrophobic surface formed by developing of a monolayer of polyfluoroalkyl thiol layer on copper or zinc substrate followed by electroless deposition of silver was reported by Larmour et al [90]. Various methods for depositing silver on silicon substrates of definite patterns and plain Si substrate have also been recently explored [91].

2.5.1. Recent developments and nanocharacteristics of electroless silver

Electroless Ag nanoparticles were effectively deposited on ZnO nanorod surfaces for the purpose of decreasing the infrared emissivity values, due to its high reflectance and will lead to innovative options for producing materials of low infrared emissivity by doping metal to semiconductor materials. To impart electrical conductivity to non-conducting glass particles, possibility to deposit electroless Ag over the glass was reported to be feasible and efficient [92]. Electroless Ag was deposited on calcite and was first reported by Srikanth and Jeevanandam [93]. In their report, lower concentration of silver ions (e.g., 0.01 M $AgNO_3$) and shorter deposition times (e.g., 30 min) led to the formation of silver nanoparticles on calcite. Sun et al. [94] coated a uniform silver film about 50-nm thick on a graphite nanosheet surface by an enhanced electroless plating using 3-aminepropyltrimethoxysilane. This silver-coated graphite nanosheet exhibited high conductivity that was equivalent to that of the silver powder. There is still a need for new and simple methods for electroless deposition of silver metal on different substrates. A protocol was developed for a solid templating mask, which is utilized for the electroless modification of sulfate-terminated polystyrene spheres with caps comprised of silver nanoparticles. Miyoshi et al. developed an electroconductive Ag nanoarray on a Si wafer. The fabrication of nano-interconnections in electric circuits, nanowire grid polarizers, molecular sensors, and other functional devices has been attained with promising attributes of pattern and material variety on the scale of ≤50 nm for the above technique. Ag nanoparticles on hydrogenated SiN_x:H layers for photovoltaic applications. A novel activation procedure was developed, via electroless Ag deposition and comparable to the wet Sn-Ag activation. The investigated approach may find applications in the fabrication of metal microstructures and nanostructures on various substrates and is projected to have numerous applications in catalysis, plasmonic devices, sensors, and many other fields. Radke et al. successfully fabricated 3D metallic bichiral crystals via direct laser writing and electroless silver plating and this method exposes a route toward very complex 3D plasmonic structures in the optical range, for example, toroidal structures with completely unusual and novel types of optical resonances. Electroless Ag coating on tetraethoxy silane-bridged fiber glass has lowest electrical resistance of 1.56×10^3 Ω/cm² and good mechanical stability. Kim et al. developed uniform compact silver layer by ecofriendly electroless method on a Fe/TiO_2/Ag core-shell structure [95–98].

2.6. Miscellaneous electroless coatings

2.6.1. Electroless aluminum

Electroless aluminum is capable of becoming one of the beneficial methods to develop thin films of Al and aluminum wiring at a very low cost. It is very difficult to perform electro deposition of aluminum in an aqueous solution because aluminum is not a very noble metal. In spite of being able to deposit Al from a room-temperature ionic liquid [99], there have been no established techniques well in virtue. A method used for the electroless plating of Al based on using $AlCl_3$-1-ethyl-3-methylimidazolium chloride ($AlCl_3$-EMIC) ionic liquid as the electrolyte and lithium hydride (LiH) as the solid reducing agent. It is criticized for its bath composition (contained LiH) as it was challenging to control the bath condition and stability. The main reasons for the limited use of LiH-based baths are that LiH solubility in the plating bath is very low and usually supplemented with excessive temperature during the deposition reaction. Recently, the same group has further investigated electroless aluminum plating based on using $AlCl_3$-EMIC ionic liquid with di-isobutyl aluminum hydride (DIBAH) as a liquid reducing agent [100]. The DIBAH-based baths were easier to control and regulate for stability than that containing LiH. However, the reports on this topic such as film composition, plating condition, and reaction mechanism are still scarce. If the plating technique is established, it would widen the likelihood of obtaining thin and thick film coatings on the substrates of insulating material and intricate structures without electricity.

2.6.2. Electroless platinum and its alloy

Electroless bath and method of coating platinum and platinum alloys contain up to about 20% rhodium, up to about 10% iridium, and up to about 10% ruthenium on an active surface, wherein the bath is an alkaline solution containing about 2 to about 20 g/L of platinum, an alkali metal hydroxide to give a minimum bath pH of about 8, up to about 1 g/L of hydrazine. Electroless platinum deposits in the absence of the stabilizer have catalytic properties, whereas platinum and platinum alloy deposits in the presence of the stabilizer are bright. Electroless coating of platinum [101] group metals has reasonable descriptions on techniques for preparing solutions and setting up conditions of plating. However, all the information on individual processes was obtained only from the patent literature, and the basic interpretation of those processes is not well known. Besides, in a few exceptional cases, there are still lacunae about the detailed information on process characteristics and deposit properties.

2.6.3. Electroless ruthenium

Electroless ruthenium developed in a patent by using Ru-nitosylammine complexes with hyrazine. Hydroxylamine added to the bath acts as a stabilizer as it is generally done to the similar baths of electroless platinum. The baths contain both $[Ru(No)(OH)(NH_3)_4]^{2+}$ and $[Ru(No)(NH_3)_3]^{3+}$. The active ruthenium species are either added as their chloride salts or produced in-situ from other ruthenium salts, such as $RuCl_3$ or $K_2[Ru(No)Cl_3]$ with $NaNo_2$ and NH_4OH. The inventor found that the low operating temperature brands this process of ruthenium deposition highly suitable for materials that are mercurial at high temperatures [102].

2.6.4. Electroless rhodium

Electroless rhodium deposits were developed by Strejcek [103] by using hydrazine as the reducing agent. Rhodium bath solution: 0.1 g $RhCl_3.3H_2O$ + 100 ml water and a large excess of $NaNo_2$ (10 g). After heating in the range of 95–98°C for about 30 min, the color of the solution changes from red to pale yellow. After cooling, 5 ml of conc. $(NH_3)_n$ Rh $(No_2)_4$ a copper wire (with Al foil contacting) and a nickel sheet was immersed in this solution. With continuous agitation and heating, a 2% solution of $N_2H_4.H_2O$ was added drop wise. At 60°C, a bright deposit of rhodium was deposited.

3. Conclusions and future perspectives

The literature briefs and demonstrates the numerous attempts prepared to identify the interdependence of the parameters which influence the performance of electroless nanolayered metallic coatings (concentration, the problems, easier for the impending user to prepare a bath). There is scope for further research and empirical analysis to be done toward formulating a rest electroless plating bath that would be reliable for extensive application than the baths existing at present.

Simple work is needed to replace of using hazardous source of potassium gold cyanide in electroless Au layer coatings. Long-term stability of Cu/Pd nanoparticles requires lot of research, and more work has to be done, whether the proposed one will be suit for the rest. The thermal stability of SiC-W/Co composite layer remains less because of the weak bonding between the substrate and deposit. This Co composite seems to be inadequate, future prospects will be very attractive. This silver-coated graphite nanosheets exhibited excellent conductivity, equivalent to silver powder. Finding of new and simple approaches is needed for the deposition of silver metal on numerous substrates. If the electroless Al plating technique is established, it will be able to achieve the thin and thick film coating on the substrates of shielding material and complicated structures. The reports on this topic are still scarce. It was difficult to make critical judgment on the practical usefulness of electroless platinum and its platinum alloy. Nevertheless, future work is to improve the existing process as well as to develop new process useful for today's application.

It is expected that this review, together with the ideas proposed by the authors, will be helpful toward the development of newer practical applications. These studies have highlighted commercial viability for rest of electroless processes.

Acknowledgements

We would like to thank Indian Institute of Technology Madras and B.S. Abdur Rahman University, Chennai, India, for their kind permission and support. We confirm that there are no conflicts of interest.

Author details

Jothi Sudagar[1, 2*], Rajendraprasad Tamilarasan[3], Udaykumar Sanjith[3], Raj Rajendran[3] and Ravi Kumar[1]

*Address all correspondence to: sendme2sudagar@gmail.com

1 Department of Metallurgical & Materials Engineering, Indian Institute of Technology Madras, Chennai, Tamil Nadu, India

2 Department of Physics, Vellore Institute of Technology-Amaravati, Andhrapradesh, India

3 Department of Mechanical Engineering, School of Mechanical Sciences, B.S. Abdur Rahman University, Chennai, Tamil Nadu, India

References

[1] Brenner A, Riddell GE. Deposition of nickel and cobalt by chemical reduction. Journal of Research of the National Bureau of Standards. 1947;**39**:385–395.

[2] Brenner A, Riddell GE. Electroless plating comes of age. Metal Fishing. 1954;**52**:68–76.

[3] Sudagar J, Lian JS, Sha W. Electroless nickel, alloy, composite and nano coatings—A critical review. Journal of Alloys and Compounds. 2013;**571**:183–204.

[4] Liu J, Zhu X, Sudagar J, Diao W, Yu S. Increased corrosion resistance of closed-cell aluminum foams by electroless Ni-P coatings. Materials Transactions. 2011;**52**:2282–2284.

[5] Sudagar J, Bi G, Jiang Z, Li G, Jiang Q, Lian J. Electrochemical polarization behaviour of electroless Ni-P deposits with different chromium-free pre-treatment on magnesium alloy. International Journal of Electrochemical Science. 2011;**6**:2767–2788.

[6] Sudagar J, Venkateswarlu K, Lian J. Dry sliding wear properties of a 7075-T6 aluminum alloy coated with Ni-P (h) in different pretreatment conditions. Journal of Materials Engineering and Performance. 2010;**19**:810–818.

[7] Tamilarasan TR, Rajendran R, Sanjith U, Rajagopal G, Sudagar J. Wear and scratch behaviour of electroless Ni-P-nano-TiO_2: Effect of surfactants. Wear. 2016; **346**: 148–157.

[8] Krishnan KH, John S, Srinivasan K, Praveen J, Ganesan M, Kavimani P. An overall aspect of electroless Ni-P depositions—A review article. Metallurgical and Materials Transactions A. 2006;**37**:1917–1926.

[9] Agarwala R, Agarwala V. Electroless alloy/composite coatings: A review. Sadhana. 2003;**28**:475–493.

[10] Ruan DW, Sudagar J, Liang YQ, Elansezhian R, Lian JS. Effects of surfactants on corrosion behavior of chromium-free electroless nickel deposit on magnesium alloy. Journal of Jilin University (Engineering and Technology Edition). 2013;**43**:363–367.

[11] Selvakumar A, Perumalraj R, Sudagar J, Mohan S. Nickel-multiwalled carbon nanotube composite coating on aluminum alloy rotor for textile industries. Proceedings of the Institution of Mechanical Engineers, Part L: Journal of Materials: Design and Applications. 2016;**230**:319–327.

[12] Balaraju J, Sankara Narayanan T, Seshadri S: Electroless Ni-P composite coatings. Journal of Applied Electrochemistry. 2003;**33**:807–816.

[13] Baudrand D, Bengston J. Electroless plating processes: Developing technologies for electroless nickel, palladium, and gold. Metal Fishing. 1995;**93**:55–57.

[14] Ali HO, Christie IRA. A review of electroless gold deposition processes. Gold Bulletin. 1984;**17**:118–127.

[15] Henry JR. Electroless (autocatalytic) plating. Metal Finishing. 2003;**101**:345–356.

[16] Wurtz A. Formation of a cuprous hydride, by the action of hypophosphorus acid on a cupric salt solution. Annales de Chimie et de Physique. 1844;**11**:250–252.

[17] Gutzeit G. Processes of chemical nickel plating. 1956, US Patent 2,766,138.

[18] Zeblisky RJ, McCormack JF, Williamson JD, Shneble FW. Electroless copper plating. 1963, U.S. Patent 3,095,309.

[19] Shipley CR. Historical highlights of electroless plating. Plat Surf Finish. 1984;**71**:92–99.

[20] Sivertz C, Soltys JF. Deposition of silver. 1973, U.S. Patent 3,776,740.

[21] Schultz PB, Yarkosky EF. Electroless nickel plating of aluminium. 1989, U.S. Patent 4,840,820.

[22] Rhoda RN, Vines RF. Bath and process for platinum and platinum alloys. 1969, US Patent, 3,486,928.

[23] Tamilarasan TR, Rajendran R, Rajagopal G, Sudagar J. Effect of surfactants on the coating properties and corrosion behaviour of Ni-P-nano-TiO_2 coatings. Surface and Coatings Technology. 2015;**276**:320–326.

[24] Cobley AJ, Mason TJ, Saez V. Review of effect of ultrasound on electroless plating processes. Transactions of the Institute of Metal Finishing. 2011;**89**:303–309.

[25] Sone T, Wachi H. Plating method by means of an electroless gold plating solution and system therefore. 1997, U.S. Patent 5,691,003.

[26] Porter Jr. LA, Choi HC, Ribbe AE, Buriak JM. Controlled electroless deposition of noble metal nanoparticle films on germanium surfaces. Nano Letters. 2002;**2**:1067–1071.

[27] Petkov N, Stock N, Bein T. Gold electroless reduction in nanosized channels of thiol-modified SBA-15 material. The Journal of Physical Chemistry B. 2005;**109**:10737–10743.

[28] Hu J, Li W, Chen J, Zhang X, Zhao X: Novel plating solution for electroless deposition of gold film onto glass surface. Surface and Coatings Technology. 2008;**202**:2922–2926.

[29] Feldstein N. Electroless plating in the electronic industry. Plating. 1974;61:146–153.

[30] Srinivasan R, Suni II. Electroless deposition of Au onto Si (111) studied by surface second harmonic generation. Surface Sciences. 1998;**408**:698–702.

[31] Hou Z, Abbott NL, Stroeve P. Electroless gold as a substrate for self-assembled monolayers. Langmuir. 1998;**14**:3287–3297.

[32] Hou Z, Dante S, Abbott NL, Stroeve P. Self-assembled monolayers on (111) textured electroless gold. Langmuir. 1999;**15**:3011–3014.

[33] Wang Y, Luo J, Chen H, He Q, Gan N, Li T. A microchip-based flow injection-amperometry system with mercaptopropionic acid modified electroless gold microelectrode for the selective determination of dopamine. Analytica Chimica Acta. 2008;**652**:180–182.

[34] Lei Y, Chen H, Dai H, Zeng Z, Lin Y, Zhou F, Pang D. Electroless-plated gold films for sensitive surface plasmon resonance detection of white spot syndrome virus. Biosensors and Bioelectronics. 2008;**23**:1200–1207.

[35] Schwarz A, Hakuzimana J, Kaczynska A, Banaszczyk J, Westbroek P, McAdams E, Moody G, Chronis Y, Priniotakis G, De Mey G. Gold coated para-aramid yarns through electroless deposition. Surface and Coatings Technology. 2010;**204**:1412–1418.

[36] Wen X, Wang M, Wang C, Jiang J. Electroless plated SnO_2-Pd-Au composite thin film for room temperature H_2 detection. Electrochimica Acta. 2011;**56**:6524–6529.

[37] Chen P, Wong H, Hu S, Huang C, Liu R. Functional electroless gold Ohmic contacts in light emitting diodes. Applied Physics Letters. 2011;**99**:63511–63513.

[38] Yasutake Y, Kono K, Kanehara M, Teranishi T, Buitelaar MR, Smith CG, Majima Y. Simultaneous fabrication of nanogap gold electrodes by electroless gold plating using a common medical liquid. Applied Physics Letters. 2007;**91**:203107.

[39] Ding J, Wang H, Lin T, Lee B. Electroless synthesis of nano-structured gold particles using conducting polymer nanoparticles. Synthetic Metals. 2008;**158**:585–589.

[40] Flavel BS, Yu J, Ellis AV, Shapter JG. Electroless plated gold as a support for carbon nanotube electrodes. Electrochimica Acta. 2009;**54**:3191–3198.

[41] Ahn W, Roper DK. Periodic nanotemplating by selective deposition of electroless gold island films on particle-lithographed dimethyldichlorosilane layers. ACS Nano. 2010;**4**:4181–4189.

[42] Banu K, Shimura T. A novel electroless method for the deposition of single-crystalline gold nanocrystals on and inside an organic solid-matrix. New Journal of Chemistry. 2011;**35**:1031–1041.

[43] Lin YM, Yen SC. Effects of additives and chelating agents on electroless copper plating. Applied Surface Science. 2001;**178**:116–126.

[44] Norkus E, Vaskelis A, Jaciauskiene J, Stalnioniene I, Stalnionis G: Obtaining of high surface roughness copper deposits by electroless plating technique. Electrochimica Acta. 2006;**51**:3495–3499.

[45] Vaskelis A, Jaciauskiene J, Stalnioniene I, Norkus E. Accelerating effect of ammonia on electroless copper deposition in alkaline formaldehyde-containing solutions. Journal of Electroanalytical Chemistry. 2007;**600**:6–12.

[46] Li D, Yang CL: Acidic electroless copper deposition on aluminum-seeded ABS plastics. Surface and Coatings Technology. 2009;**203**:3559–3568.

[47] Hanna F, Hamid ZA, Aal AA. Controlling factors affecting the stability and rate of electroless copper plating. Materials Letters. 2004;**58**:104–109.

[48] Sung Y, Chou Y, Hwu W, Fan Y, Cheng J, Ger M. Electroless copper deposition by non-isothermal deposition technology. Materials Chemistry and Physics. 2009;**113**:303–308.

[49] Lo S, Wang YY, Wan CC. Long-term stability of Cu/Pd nanoparticles and their feasibility for electroless copper deposition. Electrochimica Acta. 2008;**54**:727–732.

[50] Norkus E, Prusinskas K, Vaskelis A, Jaciauskiene J, Stalnioniene I, Macalady DL. Application of saccharose as copper (II) ligand for electroless copper plating solutions. Carbohydrate Research. 2007;**342**:71–78.

[51] Tamayo-Ariztondo J, Córdoba J, Odén M, Molina-Aldareguia J, Elizalde M: Effect of heat treatment of carbon nanofibres on electroless copper deposition. Composites Science and Technology. 2010;**70**:2269–2275.

[52] Garcia A, Polesel Maris J, Viel P, Palacin S, Berthelot T. Localized ligand induced electroless plating (LIEP) process for the fabrication of copper patterns onto flexible polymer substrates. Advanced Functional Materials. 2011;**21**:2096–2102.

[53] Naruškevičius L, Tamašauskaitė-Tamašiūnaitė L, Žielienė A, Jasulaitienė V. A Co-based surface activator for electroless copper deposition. Surface and Coatings Technology. 2012;**206**:2967–2971.

[54] Asher T, Inberg A, Glickman E, Fishelson N, Shacham-Diamand Y. Formation and characterization of low resistivity sub-100 nm copper films deposited by electroless on SAM. Electrochimica Acta. 2009;**54**:6053–6057.

[55] Leon C, Mendoza-Suarez G, Drew R. Wettability and spreading kinetics of molten aluminum on copper-coated ceramics. Journal of Materials Science. 2006;**41**:5081–5087.

[56] Shijitha T, Baiju K, Shukla S, Patil K, Warrier K. Novel electroless process for copper coating of flyash using titania/ultraviolet-radiation/metal catalyst-system. Applied Surface Science. 2009;**255**:6696–6704.

[57] Daoush WM, Lim BK, Mo CB, Nam DH, Hong SH. Electrical and mechanical properties of carbon nanotube reinforced copper nanocomposites fabricated by electroless deposition process. Materials Science and Engineering: A. 2009;**513**:247–253.

[58] Xia L, Jia B, Zeng J, Xu J. Wear and mechanical properties of carbon fiber reinforced copper alloy composites. Materials Characterization. 2009;**60**:363–369.

[59] Belmas M, Tabata I, Hisada K, Hori T. Supercritical fluid-assisted electroless metal plating onto aramid films: The influence of thermal treatment. Journal of Applied Polymer Science. 2011;**119**:2283–2291.

[60] Scudiero L, Fasasi A, Griffiths PR. Characterization of a controlled electroless deposition of copper thin film on germanium and silicon surfaces. Applied Surface Science. 2011;**257**:4422–4427.

[61] Bruning R, Muir B, McCalla E, Lempereur É, Bruning F, Etzkorn J. Strain in electroless copper films monitored by X-ray diffraction during and after deposition and its dependence on bath chemistry. Thin Solid Films. 2011;**519**:4377–4383.

[62] Deepa J, Resmi V, Rajan T, Pavithran C, Pai B. Studies on the influence of surface pretreatments on electroless copper coating of boron carbide particles. Applied Surface Science. 2011;**257**:7466–7474.

[63] Faraji S, Rahim AA, Mohamed N, Sipaut CS, Raja B. The influence of SiC particles on the corrosion resistance of electroless, Cu-P composite coating in 1 M HCl. Materials Chemistry and Physics. 2011;**129**:1063–1070.

[64] Zangmeister CD, Van zee RD. Electroless deposition of copper onto 4-mercaptobenzoic acid self-assembled on gold. Langmuir. 2003;**19**:8065–8068.

[65] Garno JC, Zangmeister CD, Batteas JD. Directed electroless growth of metal nanostructures on patterned self-assembled monolayers. Langmuir. 2007;**23**:7874–7879.

[66] Lu P, Walker AV. Investigation of the mechanism of electroless deposition of copper on functionalized alkane thiolate self-assembled monolayers adsorbed on gold. Langmuir. 2007;**23**:12577–12582.

[67] Matsuo M, Ishikawa H, Xi Y, Bin Y. High modulus and high strength fibers with high electric conductivity prepared by copper electroless plating on the surface of poly (p-phenylenebenzobisoxazole) (PBO). Polymer Journal. 2007;**39**:389–396.

[68] Xu C, Liu G, Chen H, Zhou R, Liu Y. Fabrication of conductive copper-coated glass fibers through electroless plating process. Journal of Materials Science: Materials in Electronics. 2014;**25**:2611–2617.

[69] Zuo Y, Tang J, Fan C, Tang Y, Xiong J. An electroless plating film of palladium on 304 stainless steel and its excellent corrosion resistance. Thin Solid Films. 2008;**516**: 7565–7570.

[70] Ma Z, Tan K, Kang E. Electroless plating of palladium and copper on polyaniline films. Synthetic Metals. 2000;**114**:17–25.

[71] Kim YS, Shin J, Cho JH, Ten Eyck GA, Liu DL, Pimanpang S, Lu TM, Senkevich JJ, Shin HS. Surface characterization of copper electroless deposition on atomic layer deposited palladium on iridium and tungsten. Surface and Coatings Technology. 2006;**200**:5760–5766.

[72] Williams M, Pineda-Vargas C, Khataibe E, Bladergroen B, Nechaev A, Linkov V. Surface functionalization of porous ZrO_2-TiO_2 membranes using [gamma]-aminopropyltriethoxysilane in palladium electroless deposition. Applied Surface Science. 2008;**254**:3211–3219.

[73] Zhang X, Xiong G, Yang W. A modified electroless plating technique for thin dense palladium composite membranes with enhanced stability. Journal of Membrane Science. 2008;**314**:226–237.

[74] Kim D, Kellogg A, Livaich E, Wilhite BA. Towards an integrated ceramic micro-membrane network: Electroless-plated palladium membranes in cordierite supports. Journal of Membrane Science. 2009;**340**:109–116.

[75] Strukova G, Strukov G, Batov I, Sakharov M, Kudrenko E, Mazilkin A. Studies of nanocrystalline Pd alloy films coated by electroless deposition. Materials Chemistry and Physics. 2010;**119**:377–383.

[76] Mukaida M, Takahashi N, Hisamatsu K, Ishitsuka M, Hara S, Suda H, Haraya K. Preparation for defect-free self-supported Pd membranes by an electroless plating method. Journal of Membrane Science. 2010;**365**:378–381.

[77] Ryi SK, Li A, Lim CJ, Grace JR. Novel non-alloy Ru/Pd composite membrane fabricated by electroless plating for hydrogen separation. International Journal of Hydrogen Energy. 2011;**36**:9335–9340.

[78] Paunovic M, Jahnes C. Production of electroless Co (P) with designed coercivity. 2001, US Patent, 6,197,364.

[79] Aly I, Younan M, Nageeb M. Autocatalytic (electroless) deposition of ternary nickel-cobalt-phosphorus alloy. Metal Finishing. 2003;**101**:37–42.

[80] Rudnik E, Gorgosz J. The influence of maleic acid on the Co-P electroless deposition. Surface and Coatings Technology. 2007;**201**:6953–6959.

[81] Sankara Narayanan T, Stephan A, Guruskanthan S. Electroless Ni–Co–B ternary alloy deposits: preparation and characteristics. Surface and Coatings Technology. 2004;**179**: 56–62.

[82] Huang WQ, Huang GF, Liang B, Xie CL. Magnetic properties of CoFeP films prepared by electroless deposition. Journal of Magnetism and Magnetic Materials. 2009;**321**:117–1181.

[83] Ge H, Chen Q, Li L, Wu Q, Jin H, Wei G, Chu N, Wang X. Magnetic properties of CoWP film by electroless deposition. Journal of Alloys and Compounds. 2009;**471**:515–518.

[84] Tsai T, Wu S, Liu W, Hsieh S, Chen W. Electroless CoWP as a diffusion barrier between electroless copper and silicon. Journal of Electronic Materials. 2007;**36**:1408–1414.

[85] Magagnin L, Sirtori V, Seregni S, Origo A, Cavallotti P. Electroless Co-P for diffusion barrier in Pb-free soldering. Electrochimica Acta. 2005;**50**:4621–4625.

[86] Jiang J, Zhen L, Xu C., Shao W. Microstructure evolution of cobalt coating electroless plated on SiC whisker during electroless plating and heat treatment. Surface and Coatings Technology. 2007;**201**:6059–6062.

[87] Zheng ZX, Wang R, Wang CM. Electroless plating of Co-Zn-P thin film onto nanodiamond cores. Current Applied Physics. 2011;**11**:227–230.

[88] Wilson L, Kalu EE, Martin L, McHenry ME. Decoration of carbon nanotubes with iron–cobalt (FeCo) alloy using polymer-stabilization and electroless deposition techniques for thermotherapy applications. Journal of Materials Chemistry. 2012;**22**:595–601.

[89] Abbott AP, Griffith J, Nandhra S, O'Connor C, Postlethwaite S, Ryder KS, Smith EL. Sustained electroless deposition of metallic silver from a choline chloride-based ionic liquid. Surface and Coatings Technology. 2008;**202**:2033–2039.

[90] Larmour IA, Bell SEJ, Saunders GC. Remarkably simple fabrication of superhydrophobic surfaces using electroless galvanic deposition. Angewandte Chemie. 2007;**119**: 1740–1742.

[91] Liu FM, Green M. Efficient SERS substrates made by electroless silver deposition into patterned silicon structures. Journal of Materials Chemistry. 2004;**14**:1526–1532.

[92] Moon JH, Kim KH, Choi HW, Lee SW, Park SJ. Electroless silver coating of rod-like glass particles. Ultramicroscopy. 2008;**108**:1307–1310.

[93] Srikanth CK, Jeevanandam P. Electroless deposition of silver on synthesized calcite via surface modification. Applied Surface Science. 2009;**255**:7153–7157.

[94] Sun W, Chen G, Zheng L. Electroless deposition of silver particles on graphite nanosheets. Scripta Materialia. 2008;**59**:1031–1034.

[95] Byeon JH, Kim JW. Fabrication of a pure, uniform electroless silver film using ultrafine silver aerosol particles. Langmuir. 2010;**26**:11928–11933.

[96] Radke A, Gissibl T, Klotzbücher T, Braun PV, Giessen H. Three-dimensional bichiral plasmonic crystals fabricated by direct laser writing and electroless silver plating. Advanced Materials. 2011;**23**:3018–3021.

[97] Lien W-F, Huang P-C, Tseng S-C, Cheng C-H, Lai S-M, Liaw W-C. Electroless silver plating on tetraethoxy silane-bridged fiber glass. Applied Surface Science. 2012;**258**: 2246–2254.

[98] Kim S-D, Choe W-G, Jeong J-R. Environmentally friendly electroless plating for Ag/TiO_2-coated core–shell magnetic particles using ultrasonic treatment. Ultrasonics Sonochemistry. 2013;**20**:1456–1462.

[99] Zhao Y, VanderNoot T. Electrodeposition of aluminium from nonaqueous organic electrolytic systems and room temperature molten salts. Electrochimica Acta. 1997;**42**:3–13.

[100] Shitanda I, Sato A, Itagaki M, Watanabe K, Koura N. Electroless plating of aluminum using disobutyl aluminum hydride as liquid reducing agent in room-temperature ionic liquid. Electrochimica Acta. 2009;**54**:5889–5893.

[101] Rhoda RN, Vines RF. Bath and process for platinum and platinum alloys. 1969, US Patent, 3,486,928.

[102] Torikai E, Kawami Y, Takenaka K. Bath and process for ruthenium alloys. 1984, Japanese patent (Kokai tokkya koho) 84-80766.

[103] Strejcek J. Deposition of rhodium alloys. 1977, Ger. Offen patent. DE 2607988.

Chapter 3

Laser Prepared Thin Films for Optoelectronic Applications

Marcela Socol, Gabriel Socol, Nicoleta Preda,
Anca Stanculescu and Florin Stanculescu

Additional information is available at the end of the chapter

http://dx.doi.org/10.5772/67659

Abstract

Laser techniques such as pulsed laser deposition, combinatorial pulsed laser deposition, and matrix-assisted pulsed laser evaporation were used to deposit thin films for optoelectronic applications. High-quality transparent conductor oxide films ITO, AZO, and IZO were deposited on polyethylene terephthalate by PLD, an important experimental parameter being the target-substrate distance. The TCO films present a high transparency (>95%) and a reduced electrical resistivity (5×10^{-4} Ωcm) characteristics very useful for their integration in the flexible electronics. $In_xZn_{1-x}O$ films with a compositional library were obtained by CPLD. These films are featured by a high optical transmission (>95%), the lowest resistivity (8.6×10^{-4} Ωcm) being observed for an indium content of about 44–49 at.%. Organic heterostructures based on arylenevinylene oligomers (P78 and P13) or arylene polymers (AMC16 and AMC22) were obtained by MAPLE. In the case of ITO/P78/Alq3/Al heterostructures, a higher current value is obtained when the film thickness increases. Also, a photovoltaic effect was observed for heterostructures based on AMC16 or AMC22 deposited on ITO covered by a thin layer of PEDOT:PSS. Due to their optical and electrical properties, such organic heterostructures can be interesting for the organic photovoltaic cells (OPV) applications.

Keywords: PLD, CPLD, TCO, MAPLE, organic thin films

1. Introduction

During the time, the deposition methods have been developed or/and adapted to process materials with special properties as thin films. It is well known that between the deposition methods and the quality of the obtained layers exist a strong correlation.

From the actual deposition techniques, those based on laser prove their great potential in the field of the thin films processing for different applications. Moreover, the methods based on pulsed laser have been widely implied in the preparation of transparent conductor oxide (TCO), organic films, nitrides, biomaterials, etc. [1–4].

TCO are materials integrated in applications such as organic photovoltaic cells (OPV), organic light-emitting devices (OLED), organic field-effect transistors (OFET), or smart windows [5–8]. Generally, in order to be used in optoelectronic applications, these materials must satisfy special requirements such as high optical transmittance in visible parts of the spectrum, significant reflectance in IR, and a reduced electrical resistivity [9, 10]. The most used TCO is indium tin oxide (ITO) based on indium, a rare and expensive material, deposited on glass or flexible substrates [1, 5]. Beside ITO, other TCO material used is ZnO, especially ZnO doped with elements from the III group as aluminum, indium, or gallium for improving its conductivity [11–13]. It was reported that using an aluminum-doped zinc oxide (AZO) electrode, which has absorption under 380 nm, can avoid the degradation of the cell due to the UV irradiation in the OPV applications [14]. Indium zinc oxide (IZO) films are characterized by a high electrical mobility and increased carrier density [15, 16].

Organics are a class of materials containing a large variety of compounds as small molecules, oligomers, and polymers that, lately, have been intensively studied in order to replace the inorganic materials in different domains. The most important applications of the organic semiconductors are OPV, OLED, and OFET [17–19]. Thus, Heliatek reported that a higher efficiency (13.2%) can be obtained in the field of the organic materials using a combination of the three oligomers [20]. Organic light-emitting devices are already implemented in applications as displays (in TV sets or in mobile phone) and lighting sources. OFET are used in the sensor applications [21]. A great advantage of these organic materials comes from the facts that they are ecofriendly and at low cost fabricated by a large variety of deposition techniques.

In this chapter, some of our contributions in the field of the TCO layers and organic thin films deposited by laser techniques (pulsed laser deposition (PLD), combinatorial pulsed laser deposition (CPLD), and matrix-assisted pulsed laser evaporation (MAPLE)) are presented. All these materials have been studied to be further integrated into OPV applications: TCO (ITO, AZO, and IZO) as transparent electrodes (anode) and organic semiconductors as active layers. The TCO materials obtained by PLD and CPLD present suitable optical and electrical properties as high transparency and reduced electrical resistivity. The MAPLE prepared organic films are characterized by a large absorption domain and adequate electrical properties.

2. Insights into the laser-based techniques for deposition of thin films—basic principles and experimental set-ups

Various deposition methods based on chemical or physical processes are used to obtain different materials in thin films form. The technique is chosen to obtain coatings with the expected

http://dx.doi.org/10.5772/67659

properties for targeted applications, by tuning the parameters implied in the deposition process. In the field of OPV applications, the requested TCO can be obtained using methods such as RF magnetron sputtering, oxygen ion beam-assisted deposition, spray chemical vapor deposition (CVD), PLD, and spray pyrolysis [22–26].

For the preparation of the organic active layer, methods such as vacuum thermal evaporation, spin-coating, Langmuir-Blodgett, inkjet printing, and MAPLE can be used [27–31].

In the following are briefly summarized the laser methods used in our work for the preparation of TCO and organic layers, to understand their basic principles, and the way in which the appropriate technique can be chosen in order to obtain layers with adequate properties.

2.1. Pulsed laser deposition (PLD)

PLD is a deposition technique widely used in the preparations of the thin films based on material or on combination of materials. High-quality coatings with special properties can be performed by PLD. Materials with complicated composition can be transferred by PLD on substrate without changing their stoichiometry [3]. During the material transfer process, a high-intensity laser source falls on a solid target containing materials used for deposition inside a vacuum chamber or filled with inert gas as nitrogen (N_2) or reactive gas as oxygen (O_2). Over a particular value of the incident laser intensity, the target elements are heated above their evaporation temperature (evaporation threshold). The materials are ejected from the target forming the plasma plume and moved toward the substrate. The plasma species that have sufficient energy condenses on the substrate producing the nucleation and the thin film grow up [32]. The target is rotated in order to prevent the local deterioration, which can affect the uniformity and quality of the obtained coating. A typical PLD experimental set-up is presented in **Figure 1**. Into a PLD process, the most important parameters are laser fluence, deposition rate, substrate temperature, target-substrate distance, and number of laser pulses [33].

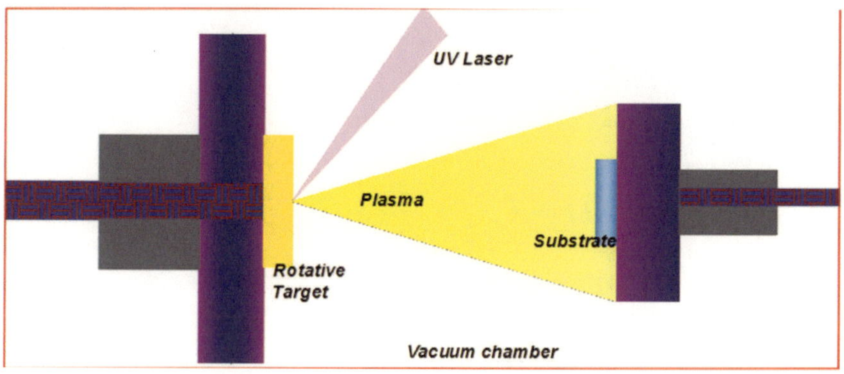

Figure 1. Schematic representation of PLD set-up.

2.2. Combinatorial pulsed laser deposition (CPLD)

In the beginning, the CPLD was introduced, in chemistry, in the fields of drugs from the necessity to develop new active molecules [34, 35], but this technique can be applied for a variety of materials: metals, semiconductors, polymers [36–38], etc. This deposition method is a proper tool to obtain doped materials like TCO layers.

The great advantage of the CPLD over PLD is the possibility to perform in a single experiment samples with different composition. By comparison, in order to find the sample with the best properties using PLD technique, it is necessary to carry out a lot of samples with different compositions; a time-consuming process using CPLD is obtained, a so-called composition library. Practically, along the deposition substrate in each point, the concentration of the thin films is different. In **Figure 2**, a CPLD deposition set-up is illustrated. In comparison with the PLD, in the CPLD deposition process is involved targets with different composition, situated at certain distance to each other [39]. An optical beam-splitter is used to split the laser beam into two beams. The targets are simultaneously ablated, generating intermixed films [40].

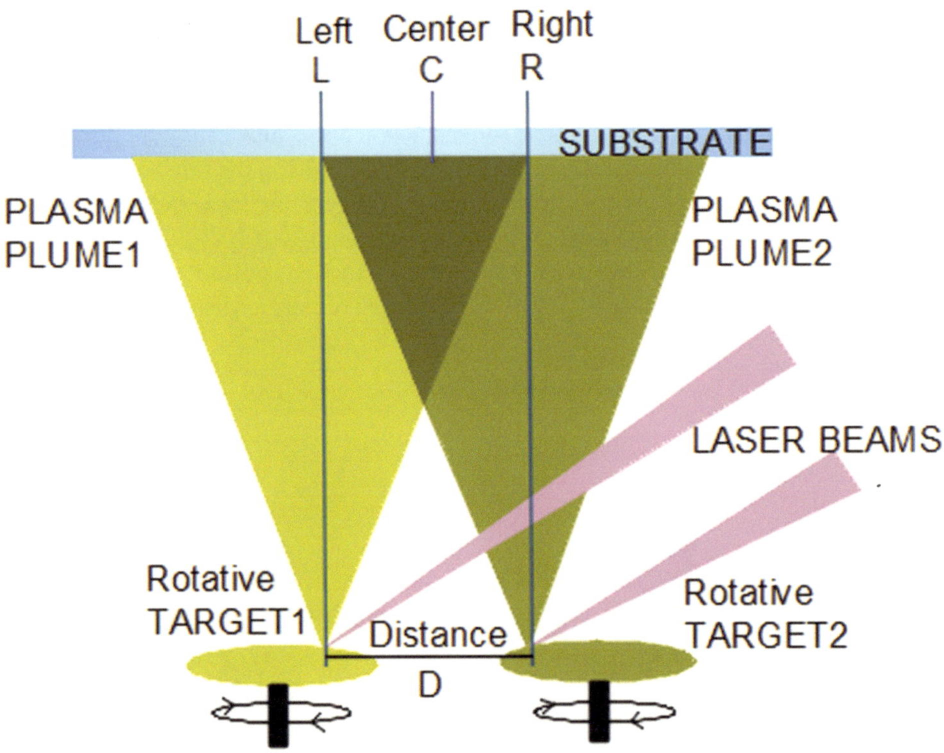

Figure 2. Schematic representation of CPLD set-up.

2.3. Matrix-assisted pulsed laser evaporation (MAPLE)

This technique is also derived from PLD method; Piqué et al. [41] mentioned that this technique was first introduced by Epstein in 1997 [42]. MAPLE has the advantage that it can process soft materials (organics) that could not be transferred by other techniques because there is the risk that takes place—a decomposition of the materials. In MAPLE, the target is formed from the materials (one or more) that must be deposited, and an adequate solvent is used as matrix [43, 44]. The solvent is chosen to obtain a homogeneous mixture (concentration usually below 3%) and to be compatible with the used laser wavelength. The formed mixture (organic material and solvent) is subsequently frozen in liquid nitrogen to form a solid target. During the deposition, the laser energy is absorbed by the solvent and transformed into thermal energy, enabling the evaporation of the solvent, and this being pumped away by the vacuum system while the material of interest reaches the substrate [45–47].

In MAPLE, smaller fluence are used (under 0.5 J/cm^2) in order to prevent the deterioration of the materials [44]. Another great advantage compared with methods used for the deposition of the organic materials is the possibility to obtain stacked layers without deterioration of the preliminary deposited layer [31]. **Figure 3** presents a schematic representation of the experimental set-up used in MAPLE deposition.

In **Table 1**, are presented comparatively the pulsed laser techniques described before which can be used to prepare thin films from different materials with thickness from nano to micrometers.

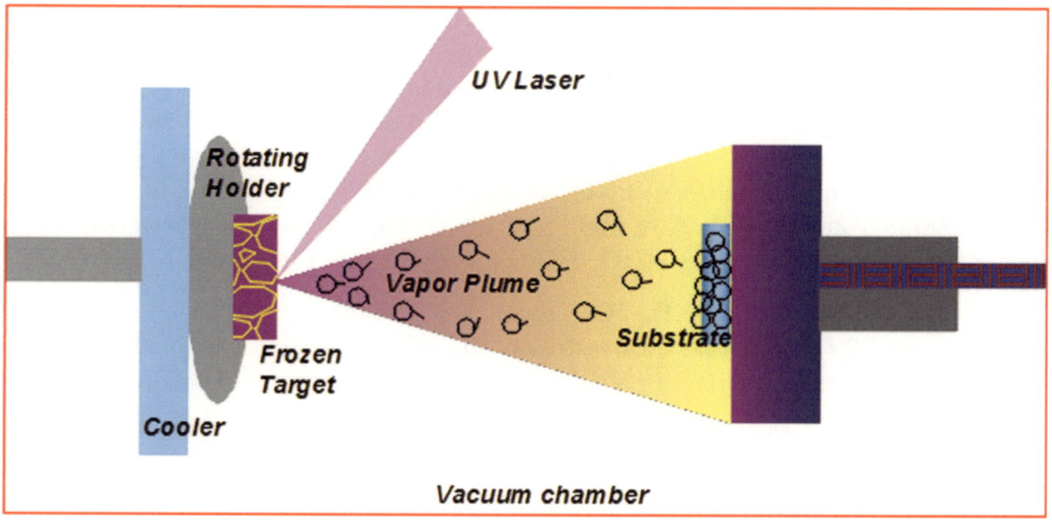

Figure 3. Schematic representation of MAPLE set-up used for the deposition of organic films.

Method	PLD	CPLD	MAPLE
Class of materials	Suitable for any inorganic material with complex and very complex stoichiometry	Combination of any inorganic materials	Suitable for deposition of organic compounds and laser transfer of inorganic or organic nanostructures (e.g., nanoparticles, nanotubes, nano-sheets, etc.)
Typical laser fluences (J/cm^2)	1–10	1–10	0.05–0.5
Principle of the laser transfer	Laser ablation of solid targets	Laser ablation of solid targets with different composition	Laser evaporation of frozen composite targets under ablation threshold
Deposition rate (nm/pulse)	0.01–0.5	0.01–0.5	0.1–0.5
Typical film thickness (nm)	1–5000	10–1000	10–5000

Table 1. Advantages of pulsed laser techniques for deposition of thin films.

3. Transparent conductive oxide (TCO) thin films deposited by PLD or CPLD—influence of the deposition conditions on their structural, morphological, optical, and electrical properties

The great interest in the field of TCO is proved by the high number of the research articles, reviews, books, or chapter books existent in the literature regarding this topic [48–56]. Various attempts were made to find the TCO with high optical and electrical properties and the best method to obtain these properties. The first TCO preparation is attributed to Badeker and dates from 1907, when CdO was obtained by the thermal oxidation of a Cd film deposited by sputtering [57]. ITO was frequently used in different applications, being one of the most-studied materials. In time, materials such as In_2O_3, CdO, ZnO, and SnO_2 were also addressed as TCO. n type semiconductors can be doped in order to improve their electrical conductivity, obtaining materials such as In_2O_3-ZnO, In_2O_3-SnO_2, Ga_2O_3-In_2O_3, In_2O_3-$MgInO_4$, $ZnSnO_3$-$ZnIn_2O_5$, $ZnIn_2O_5$-$GaInO_3$ or $MgIn_2O_4$-$Zn_2In_2O_5$ [58]. Several attempts were made to prepare p type semiconductors: $CuGaO_2$, $SrCu_2O_2$, $CuAl_2O_2$, and $CuCrO_2$ [59–62].

There are many studies regarding the TCO prepared by PLD, the best result being reported for ITO thin films (7.2×10^{-5} Ωcm electrical resistivity and ~90% transparency [63]). Also, by PLD, ITO layers with a smooth surface (root mean square (RMS) ~ 4.5 Å) were obtained [64]. The performances of the ITO deposited by PLD are superior to that presented by commercially available ITO deposited by sputtering.

The TCO film represents a key element in all applications, including OPV, due to the fact that through this electrode passes the light depending on its optical transmittance [65].

Our results regarding the TCO films obtained by PLD and CPLD and their complex characterization by techniques such as X-ray diffraction (XRD), ultraviolet-visible spectroscopy (UV-VIS) and atomic force microscopy (AFM) are presented in the next section.

3.1. Indium tin oxide (ITO)

This material was the subject of many research studies and is the most used electrode in OPV applications because it seems to present the best electrical conductivity and optical properties. It is used as hole injector material, having a work function between 4.5 and 5 eV [33].

Because of the technological progress in the field of flexible electronics, light-weight and cheap TCO materials that are compatible with plastic substrates are necessary.

A PLD experimental set-up like that presented in **Figure 1** was used to prepare ITO thin films on polyethylene terephthalate (PET) substrate. An excimer laser with KrF* (model COMPex-Pro 205, Lambda Physics-Coherent), λ = 248 nm, and τ_{FWHM} = 25 nm was used to irradiate the ITO target (SCI engineered materials) in an ultrahigh vacuum chamber [1]. The depositions were made at room temperature, at 2 J/cm^2 laser fluence, and with 10 Hz repetition rate. The oxygen pressure was between 1 and 1.5 Pa. We have selected different deposition parameters: target-substrate distance (4, 6, or 8 cm) and the number of laser pulses (6000, 9000, or 12,000). The sample was labeled as ITO1 (4 cm distance and 6000 pulses), ITO2 (6 cm distance and 9000 pulses), and ITO3 (8 cm distance and 12,000 pulses).

Analyzing the ITO samples from the structural point of view (**Figure 4a**), it was found that the films present a lower crystallinity. The ITO1 film deposited at the lower target-substrate distance appears as a small and large peak at 35.2° corresponding to the (400) diffraction plane. An increase in the diffraction peak from 35.2° is remarked for the ITO2. A supplementary peak at ~30.2° (a reduced one) that corresponds to (222) diffraction plane of ITO [66] is also disclosed. The film ITO3 deposited at higher target-substrate distance is amorphous. This means that there is an optimum for the deposition condition for assuring an increased crystallinity of the ITO layer.

The UV-VIS spectra for ITO samples (**Figure 4b**) record subtracting the contribution of the flexible substrate and show the differences in the transmission degree. The sample ITO1 presents the lowest transparency (70–75%) that is being attributed to the presence of some defects (cracks) generated during a deposition process at this distance and the defects that scatter the light.

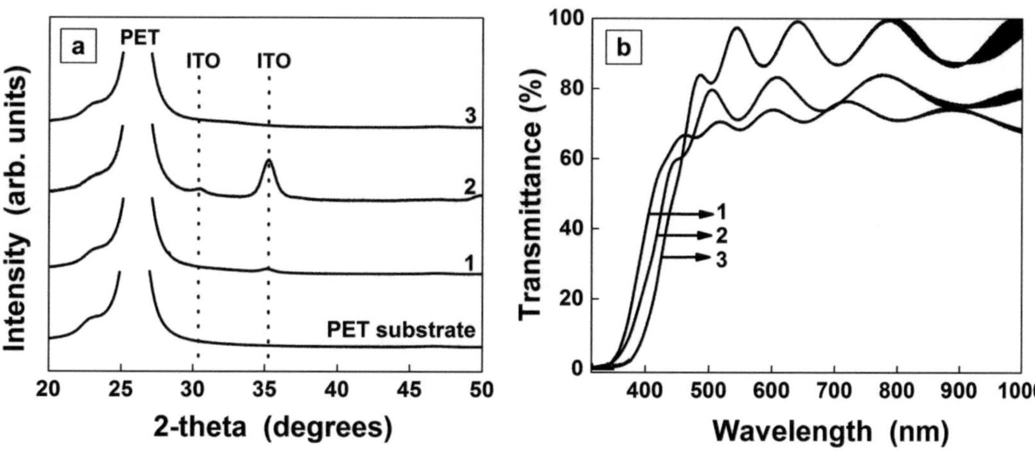

Figure 4. XRD patterns (a) and transmission spectra (b) of the ITO1, ITO2, and ITO3.

In PLD, the deposition appears as species with an increased kinetic energy that can affect the deposition substrate and the way in which this takes place is the nucleation process [1]. Defects can appear due to the energy transferred through the collision of the energetic items such as atoms, ions, and molecules, with the substrate atoms or with the atoms previously deposited. The adatoms must have enough time and mobility to form films with a good adherence at reduced substrate temperature [67].

The transmittance of the film increases with the increase of the target-substrate distance, reaching up to ~90% for the film deposited at a higher distance (ITO3).

Figure 5 shows the topographic images obtained by AFM. The appearance of the cracks in the ITO1 film can be observed. Also for the second sample, it was obtained morphologically with cracks. These cracks are characteristic to ITO deposited on PET substrates in some experimental conditions, which were being reported by the other authors also [68]. The smallest RMS value (2.4 nm) was obtained for the ITO3 film, compared to ITO1 (RMS = 15.5 nm) and to ITO2 (RMS = 31.5 nm), the ITO3 being the film showing the highest transmittance. This aspect of the sample is different from the other two, no cracks being observed on it.

In terms of the electrical resistivity, the sample ITO2 was measured with a good resistivity (5.9×10^{-4} Ωcm). Taking into account that this is the sample with the increased crystallinity, it can be concluded that there is a correlation between its crystallinity and its electrical properties. The sample ITO3 that is amorphous is featured by an increased resistivity (9.7×10^{-4} Ωcm). Higher value for the films is obtained by PLD at room temperature and had been reported by other authors [69]. The electrical properties of the TCO layers can be improved by heating the substrate is mentioned in the literature [70].

3.2. Aluminum-doped zinc oxide (AZO)

Another studied TCO with n type conduction is AZO, which is a nontoxic candidate having properties close to ITO. AZO samples were prepared by PLD on PET substrate in the same condition as ITO using a target with 2% Al content (SCI engineered materials). The samples prepared in different geometrical configuration were also labeled as AZO1 (4 cm), AZO2 (6 cm), and AZO3 (8 cm).

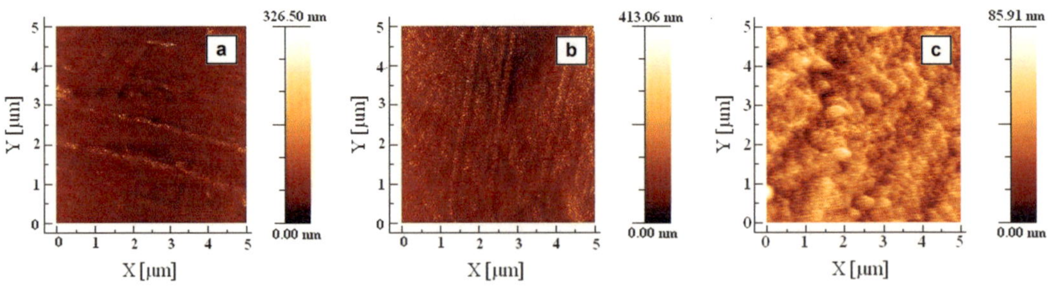

Figure 5. AFM images of the ITO1 (a), ITO2 (b), and ITO3 (c).

As for ITO, defects like cracks were remarked on the AZO layers deposited at 4 cm target-substrate distance (AZO1) and at 6 cm (AZO2). Films with crystalline quality (**Figure 6a**) were obtained at 6 and 8 cm. All AZO samples show a diffraction peak situated at ~34°, corresponding to (002) plane of ZnO [71]. The broadening of this peak from 34° for AZO1 is attributed to the presence of the lattice strain in this film, resulting in a peeling effect of the flayer.

The UV-VIS spectra in **Figure 6b** can be observed has a higher transmittance (over 90%) obtained for AZO3 coating compared with AZO1 and AZO2. A high optical quality is achieved for the film obtained at 8 cm target-substrate distance because, after target ejection, probably the energetic species have enough time to be thermalized in collision with the oxygen molecule when the distance is increased between target and PET substrate.

The AZO films present a morphology with long aggregates (**Figure** 7). The AZO3 has the bigger aggregates (with size between 0.4 and 1 μm) that are oriented in the same direction. This is in concordance with the higher transmittance observed for this layer due to the reduction of the light scattering on the grain boundaries. The AZO2 film presents the morphology similarly to AZO3 but with smaller aggregates, the film having lower RMS value (5.2 nm) compared with AZO1 (RMS = 12.4 nm) and AZO3 (RMS = 8.5 nm).

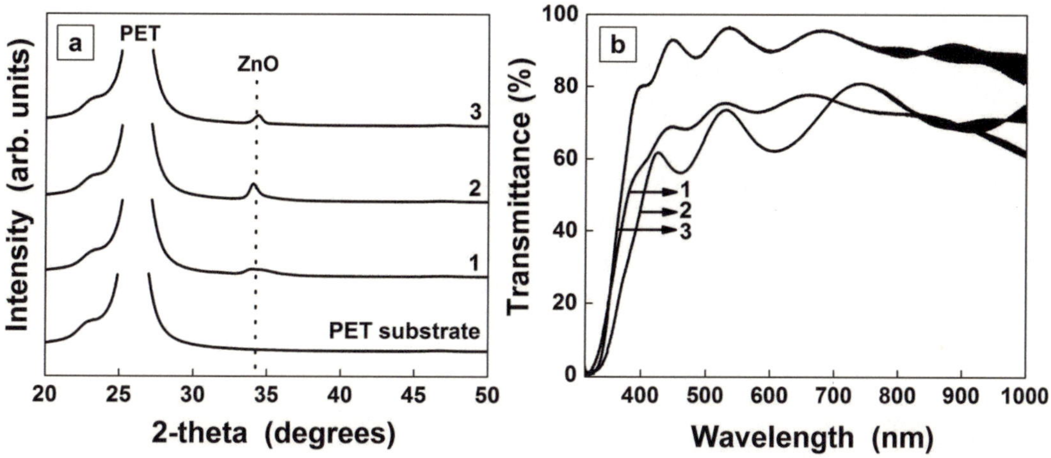

Figure 6. XRD patterns (a) and transmission spectra (b) of the AZO1, AZO2, and AZO3.

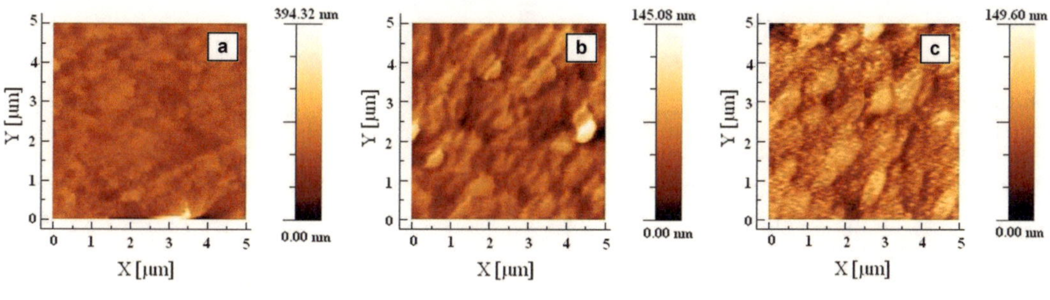

Figure 7. AFM images of the AZO1 (a), AZO2 (b), and AZO3 (c).

The best electrical resistivity (6.8 × 10^{-4} Ωcm) was found for the AZO2 film, with the AZO3 sample having increased in value (2.1 × 10^{-3} Ωcm). Due to the plasma expansion between target and substrate at a distance of 8 cm, it could appear that some oxidation effects affect the ZnO stoichiometry. Collisions and recombination can appear due to the oxygen background. Nevertheless, the resistivity values are better than other reported values (1.1 × 10^{-3} Ωcm) for AZO deposited on PET by RF magnetron sputtering at 100°C [72].

3.3. Indium zinc oxide (IZO)

Competitive optical and electrical properties were found for IZO electrode in comparison with that reported for ITO. IZO has a work function of ~5.2 eV being framed also as an n type semiconductor. The same deposition parameters used before were preserved to deposit IZO by PLD on PET from a solid target formed with In in atomic concentration, In/(In + Zn), of 70% by mixing In_2O_3 and ZnO powders (both from Aldrich). Also, in this case, the samples were labeled in accordance with the target-substrate distance: IZO1 (4 cm), IZO2 (6 cm), and IZO3 (8 cm).

As is reported in the literature [25], generally, IZO layers deposited at room temperature are amorphous (**Figure 8a**). A broad peak appears between 31 and 34° only for the IZO1 film. This peak contains two contributions one attributed to (111) diffraction plane of In_2O_3, which is usually situated ~31° and another to (101) diffraction plane of ZnO, which appears at 34° [25]. For the IZO1 and IZO2 layers, no diffraction peaks were observed.

If for the ITO and AZO, we have observed the appearance of a significant number of defects on the samples deposited at 4 and 6 cm, in the case of IZO, defects were evidenced just for the samples performed at the lower target-substrate distance (IZO1). For IZO layers, the transmittance is reduced compared with ITO and AZO (**Figure 8b**). No noticeable differences appear in the transmission spectra of the IZO films deposited at different target-substrate distances, the transmittance having between 75 and 88%.

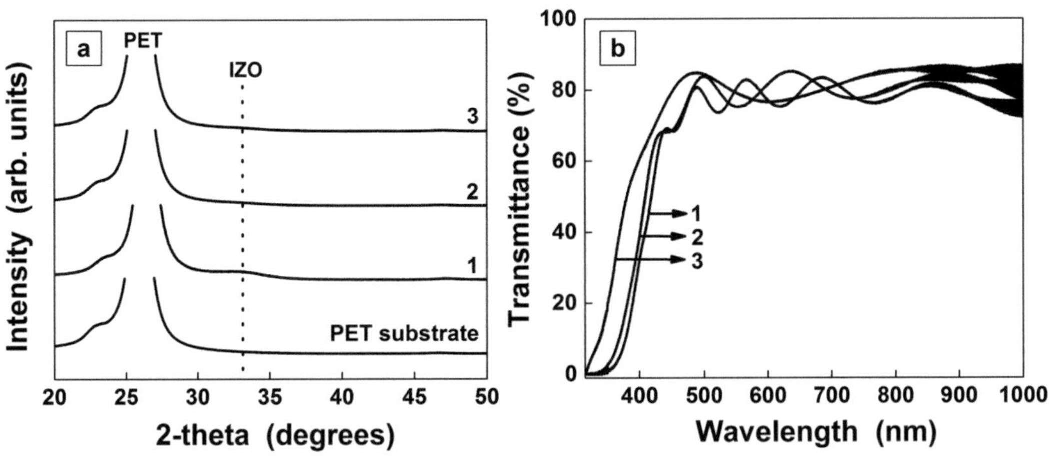

Figure 8. XRD patterns (a) and transmission spectra (b) of the IZO1, IZXO2, and IZO3.

The RMS values of the IZO films were evaluated by AFM (**Figure 9**). Thus, for the IZO3 sample, low RMS value (2.1 nm) was interpolated from AFM measurements. The other two films (IZO1 and IZO2) present higher RMS values (47.4 and 21.2 nm). The great value obtained for the IZO1 can be attributed to the presence of the cracks in this film. The best value for IZO3 is comparable with that obtained for ITO1. In order to be adequate for device applications, in addition to possess good optical and electrical properties, the films must be characterized by a smooth surface [33].

The resistivity values obtained for IZO samples were much better, compared with ITO and AZO layers, being ranged in 5.4–6.7 × 10^{-4} Ωcm domain. These values are in agreement with other results obtained for coatings made on glass substrate [25].

Considering all the investigated properties for ITO, AZO, and IZO deposited by PLD on the flexible substrate, it can be concluded that an important parameter in PLD process is the target-substrate distance. At a higher distance (8 cm) smooth films are obtained characterized by increased transmittances compared with the films obtained at 4 or 6 cm distance. Films with high transparency (over 95%) can be obtained by this laser technique.

As was already pointed, another technique to obtain TCO films is CPLD; this method allows to obtain doped films with a compositional gradient along the deposition substrate.

Using the same laser beam and a combinatorial geometry as that presented in **Figure 2**, IZO films at 3 J/cm^2 laser fluence were deposited. Two targets with atomic In concentration, In/(In + Zn), of 28 and 56 at.% or 42 and 70 at.% were made by mixing In_2O_3 and ZnO powders [73]. The combinatorial samples were obtained in 1 Pa oxygen atmosphere and at room temperature. The laser repetition rate was 10 HZ, and the number of pulses was 3000. Substrates for deposition were 26 × 76 mm microscope glass slides. The distance between the targets and substrate was 5 cm. In order to make a relevant comparison, samples from each target, but just by PLD, were also performed. The samples obtained by CPLD were IZOCMB1 (with 28% and 56%) and IZOCMB2 (with 42% and 70%). The sample produced by PLD are labeled as IZO1A (28%), IZO1B (56%), IZO2A(42%), and IZO2B(70%).

The UV-VIS spectra of the films deposited by PLD are presented in **Figure 10a**. The UV-VIS spectra of the combinatorial samples, IZOCMB1 presented in **Figure 10b** and IZOCMB2 presented in **Figure 10c**, were collected in three points corresponding to L, C,

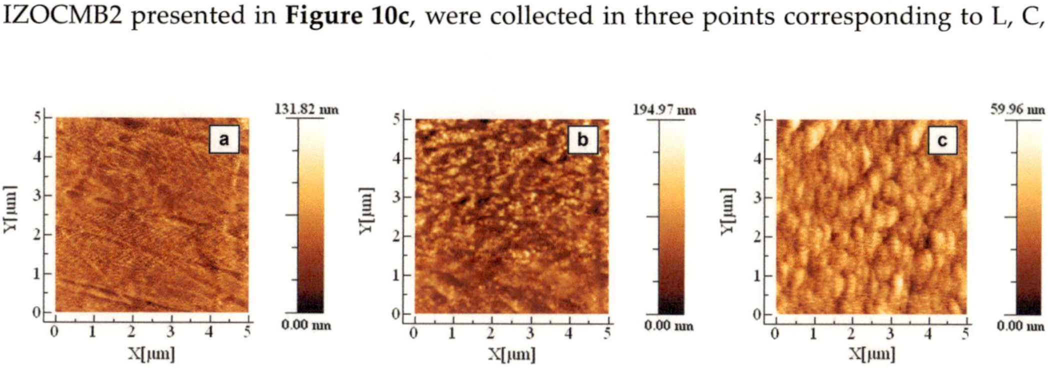

Figure 9. AFM images of the films deposited on PET substrate IZO1 (a), IZO2 (b), and IZO3 (c).

and R position (see **Figure 2**). The spectra are presented subtracting the glass contribution. Independently of the method used for deposition, PLD or CPLD, the samples show a high transmittance (up to 95%) in 500–1000 nm domain, which is a good premise for this TCO material.

For the films obtained by CPLD, a nonlinear variation in the composition of the films has been revealed by Energy dispersive X-ray spectroscopy (EDS) measurements (**Figure 11**). Data from 2.5 mm consecutive section along L-R line were mediated in order to establish the In, In/(In + Zn) content. Moreover, values from four areas were used to obtain the average In content from the target 1(TG1) and target 2 (TG2). We have obtained 28 and 56 at.% In, In/(In + Zn) atomic concentration for the first CPLD target and 42 and 70 at.% In, In/(In + Zn) for the second CPLD target. The films IZOCMB1 and IZOCMB2 have atomic concentration in In between L and R position, of 27–33 and 36–52 at.%, respectively. The lower In content is attributed to the ZnO preferential nucleation on glass substrate.

The AFM images obtained on the combinatorial films (in different areas between L and R points) are presented in **Figure 12**. The films are smooth, and the RMS roughness values are ranged between 7.1 and 26.2 nm for the IZOCMB1 film and 1.0 and 7.3 nm for the IZOCMB2 film.

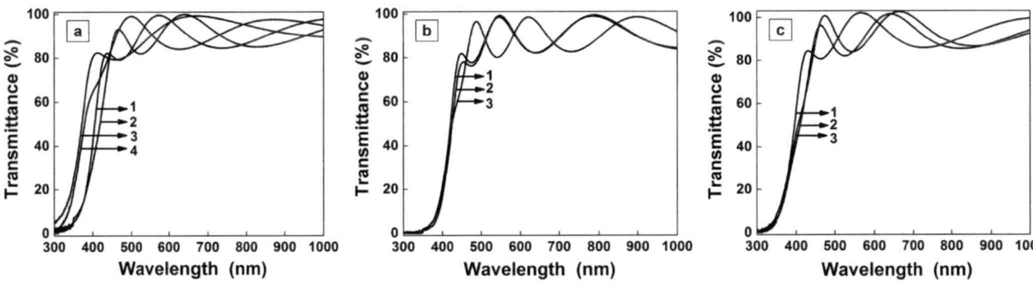

Figure 10. Transmission spectra of the IZO1A(curvel), IZO1B(curve2), IZO2A(curve3), IZO2B(curve4) (a) of the IZOCMB1 (b) and IZOCMB2 (c) in different areas corresponding to positions L, C, and R.

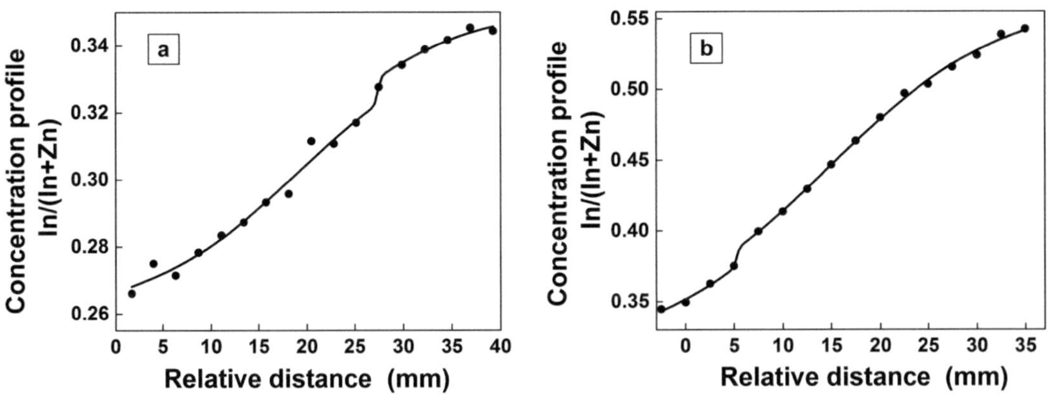

Figure 11. Elemental composition profiles of the IZO films deposited on glass substrate by CPLD using two targets with various In atomic concentration: (a) –28% (TG1) and 56% (TG2) and (b) –42% (TG1) and 70% (TG2).

The variation of RMS as a function by distance is given in **Figure 13**. The RMS mean value from three neighbor areas was used to plot each point in **Figure 13**. A decrease in the RMS value was observed with the increase in the In content between L and R points. This observation is in agreement with other results presented in the literature [74].

The modification of the electrical resistivity was also investigated along L-R direction of the films. The IZOCMB1 film presents a lower resistivity (2.3×10^{-3} Ωcm) for 28.8–29.5 at.% In content (**Figure 14a**), and IZOCMB2 film has a lower resistivity (8.6×10^{-4} Ωcm) in 44–49 at.% of In content region (**Figure 14b**). The literature reported similar data for the minimum resistivity obtained for samples with similar In atomic concentration deposited by magnetron sputtering [75, 76].

In conclusion, $In_xZn_{1-x}O$ ($27 \leq x \leq 52$) systems were performed by CPLD. The best resistivity value obtained was 8.6×10^{-4} Ωcm that correspond to 44–46 at.% In content domain. All investigated films present a high optical transparency (~95%). This technique is useful for preparing TCO films with different composition and adequate optical and electrical properties.

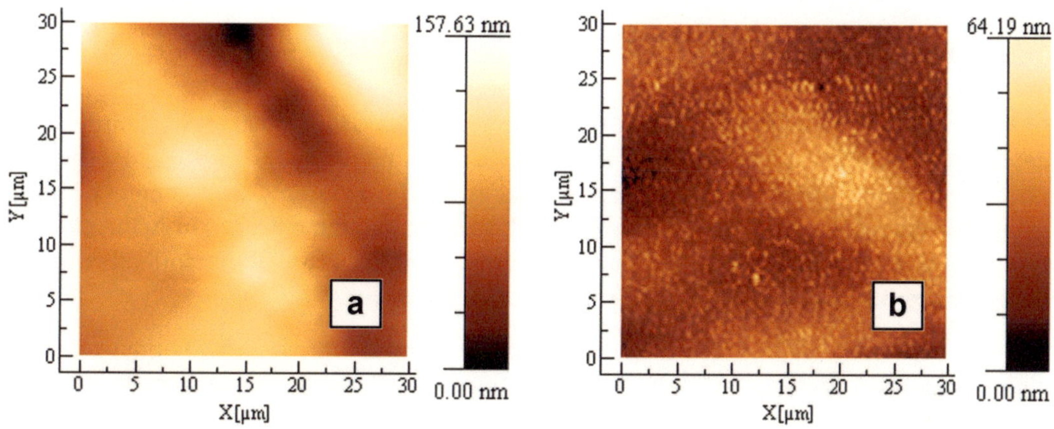

Figure 12. AFM images of IZOCMB1 (a) and IZOCMB2 (b).

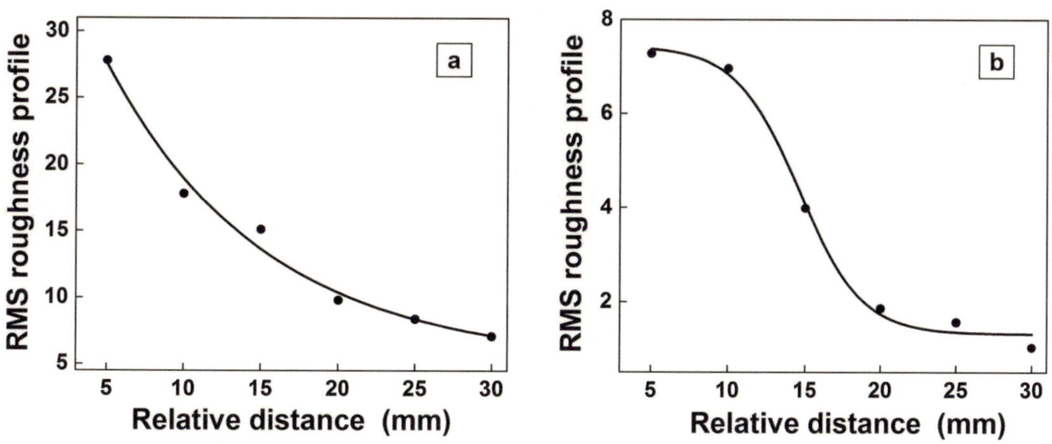

Figure 13. RMS roughness profiles of the IZOCMB1 (a) and IZOCMB2 (b) between L and R positions.

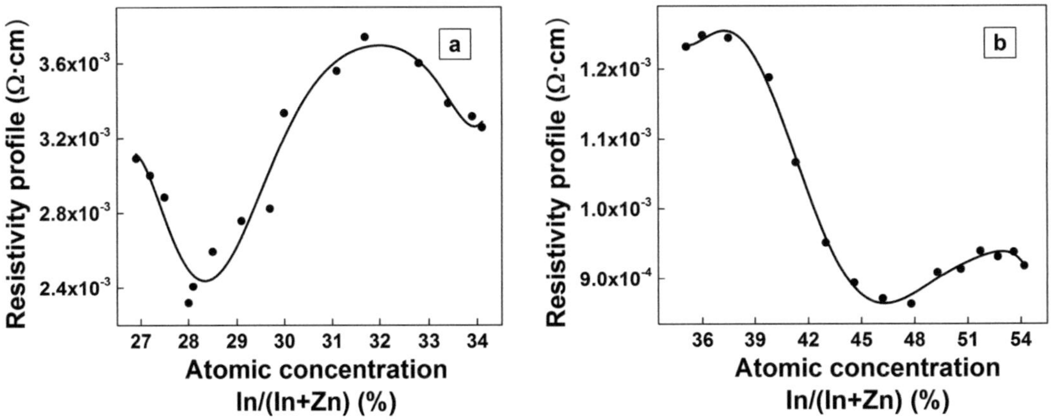

Figure 14. Electrical resistivity profiles of the IZOCMB1 (a) and IZOCMB2 (b).

4. Organic thin films deposited by MAPLE—properties and applications

At the same time with the development of the organic materials domain, a large number of deposition methods had been adapted to obtain them as thin films. The most used technique to deposit organic layers was vacuum evaporation, but for materials as oligomers or polymers, more suitable are techniques involving solution because the risk to destroy the molecular chains during the deposition is reduced. The spin-coating method was frequently used as deposition technique for the polymers thin films preparation.

In the last years, the MAPLE method was introduced in order to process such organic materials, taking into account that MAPLE allows their transfer with the preservation of their chemical composition.

Subsequently are presented the organic thin layers prepared by MAPLE. These were investigated from morphological (AFM) and optical (UV-VIS, photoluminescence spectroscopy—PL, Fourier transform infrared spectroscopy (FTIR), and from electrical (current-voltage characteristics) point of view.

4.1. Oligomers based on arylenevinylene compounds

Arylenevinylene oligomers, 1,4-*bis* [4-(N,N-diphenylamino)phenylvinyl] benzene (P78) and 3,3-*bis* (N-hexylcarbazole)vinylbenzene (P13), with electron-donating groups (triphenylamine and N-alkylcarbazole), are used in combination with *tris*(8-hydroxyquinolinato)aluminum salt (Alq_3) to prepare by MAPLE organic heterostructures with one or two layers. The molecular formula of the oligomers used in this study was presented in **Figure 15**. The concentration of the organic material in the dimethyl sulfoxide (DMSO), used as a solvent, was 2.5% (w/v). The depositions were made at 5 Hz laser frequency and at lower fluence (250 mJ/cm^2). In the heterostructures with two layers, the first deposited layer was the oligomer (P 13 or P78) and the second layer was Alq3 [77]. The number of laser pulses were 80,000 for 1P13, 1P78, and 1Alq3 samples and 160,000 for 2P13, 2P78, and 2Alq_3 samples.

http://dx.doi.org/10.5772/67659

The FTIR spectra (**Figure 16**) of the MAPLE-deposited layers have been analyzed and was concluded that no materials decomposition appear during the laser transfer. The peaks characteristic to P13 and P78 compounds were identified. The peak situated at 960 cm^{-1} is characteristic to HC=HC trans-out-of-plane bending vibration [78], whereas the δ (C-N) stretching vibration appears at 1154 cm^{-1} in P13 layer and at 1328 cm^{-1} in P78. The band from 1589 cm^{-1} in P78 is assigned to the vibration of the C-C phenyl group. The peaks from 1598 cm^{-1}, 1491 cm^{-1}, and 1477 cm^{-1} are due to the ν (C-C) vibration in monosubstituted benzene [78]. The Alq_3 layer presents vibration characteristic to the following chromophoric groups: between 600 and 900 cm^{-1} =C-H stretching, at 1475 cm^{-1} aromatic C=C stretching, at 1390 cm^{-1} C=N bond, at 1600 cm^{-1} the quinolinic ring [79].

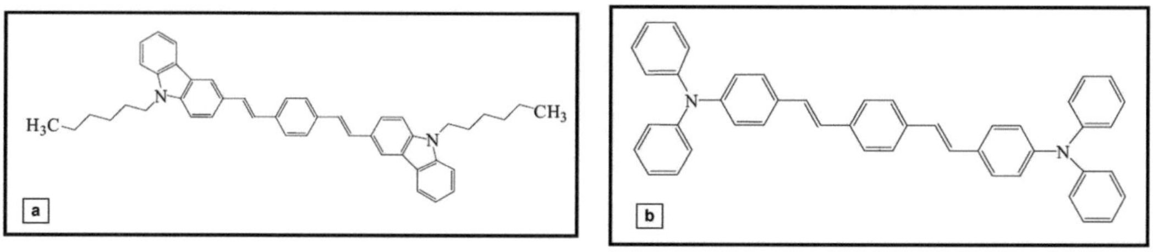

Figure 15. Chemical structures of the arylenevinylene oligomers: P13 (a) and P78 (b).

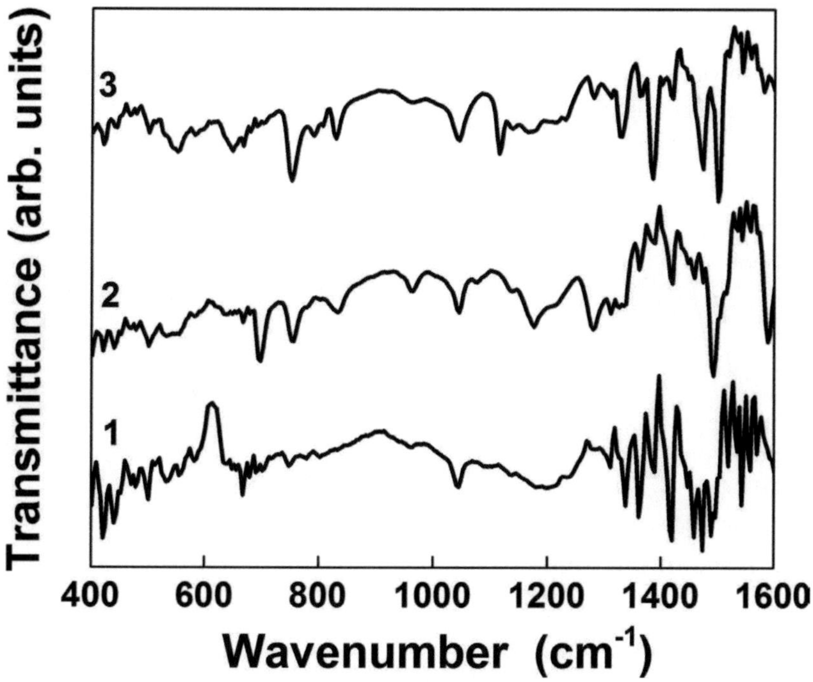

Figure 16. FTIR spectra of the organic thin films deposited on Si substrate by MAPLE: P13 (curve 1), P78 (curve 2), and Alq3 (curve 3).

The UV-VIS spectra of the P13 and P78 layers deposited by MAPLE on ITO and with an additional Alq_3 layer are presented in **Figure 17(a)** and **(b)**. The structures realized with two organic materials are featured by a great transparency, 60% for $\lambda > 550$ nm. Absorption maxima attributed to the electronic π-π^* transitions of the conjugated backbone [80] were evidenced. The absorption maxima characteristic to these materials and for the Alq_3, situated at low wavelength are hidden by the ITO substrate [81].

The Alq_3 is frequently used in OLED applications due to its emission properties. The photoluminescence spectra of oligomers and Alq_3, obtained at 350 nm excitation wavelength are given in **Figure 17(c)** and **(d)**. The emission band with the maximum situated at 523 nm is attributed to the presence of the Alq_3 meridional stereoisomer [80]. P13 oligomer showed two emission maxima (490 and 525 nm), while P78 discloses a raw maxima at 500 nm [80], the prepared films preserving the emission properties of the start materials.

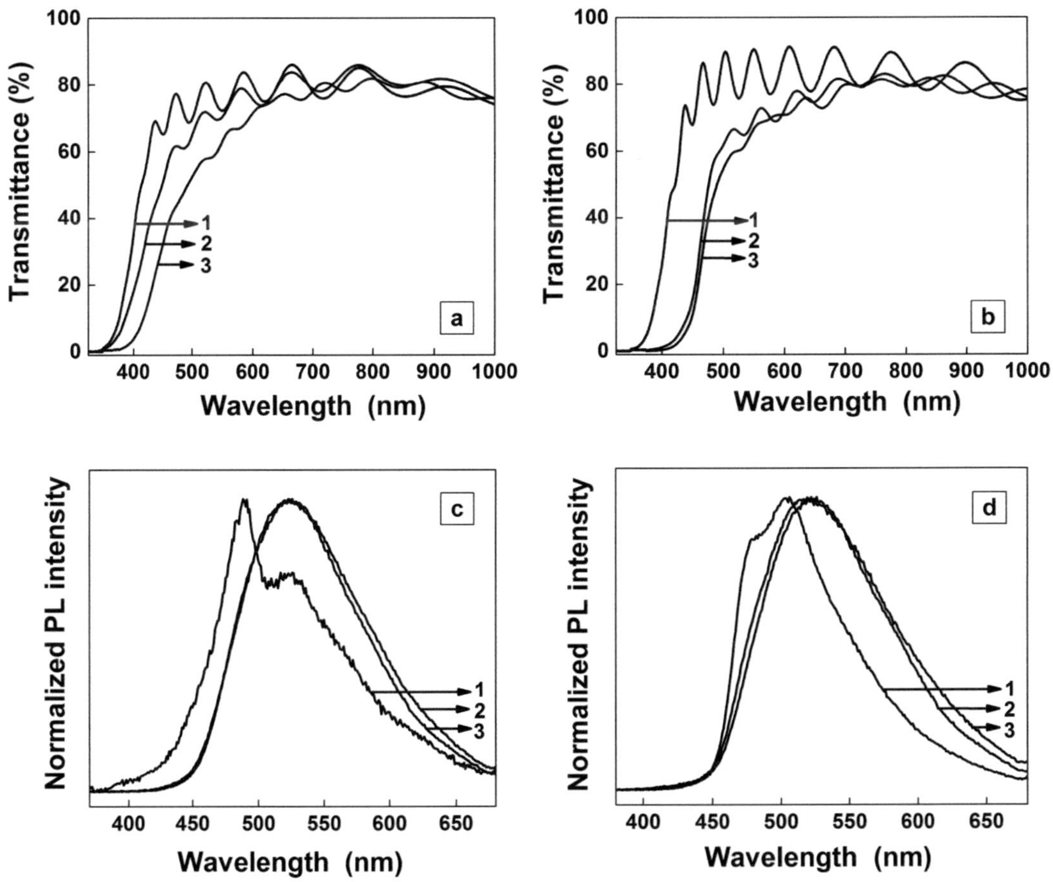

Figure 17. Transmission spectra (a, b) and photoluminescence spectra (c, d) of the organic thin films based on arylenevinylene oligomer (P13 (a, c) and P78 (b, d)) deposited on glass/ITO substrate by MAPLE. For transmission spectra: glass/ITO (curve 1), glass/ITO/arylenevinylene oligomer (curve 2), and glass/ITO/arylenevinylene oligomer/Alq3 (curve 3). For photoluminescence spectra: glass/ITO/arylenevinylene oligomer (curve 1), glass/ITO/Alq3 (curve 2), and glass/ITO/arylenevinylene oligomer/Alq3 (curve 3).

Different topographies of the films deposited by MAPLE are presented in **Figure 18**. The globular morphology is characteristic to MAPLE process [2, 11]. The RMS values obtained by interpolation of a single layer deposited on ITO evidenced that the P78 film (24.6 nm) presents a higher roughness compared with P13 (8.2 nm) (**Table 2**).

The samples made with a doubled number of pulses present RMS values comparable with that obtained for the samples realized with 80,000 laser pulses. A decrease in the RMS values was observed for the samples based on P78 and Alq_3 double layers, while an increase was seen for the sample containing P13 and Alq3 (**Table 2**). The higher roughness of P78 seems to favor a better fit of Alq_3 molecules.

The I-V characteristics recorded in 0–10 V domain in the dark are presented in **Figure 19**. No rectifying properties were observed for the investigated heterostructures. A good current value (8×10^{-6} A at 1 V) was evidenced in the sample based on P78 and Alq3 with thicker layers. The lowest current value was presented by the sample with Alq_3 layer realized at 160,000 laser pulses.

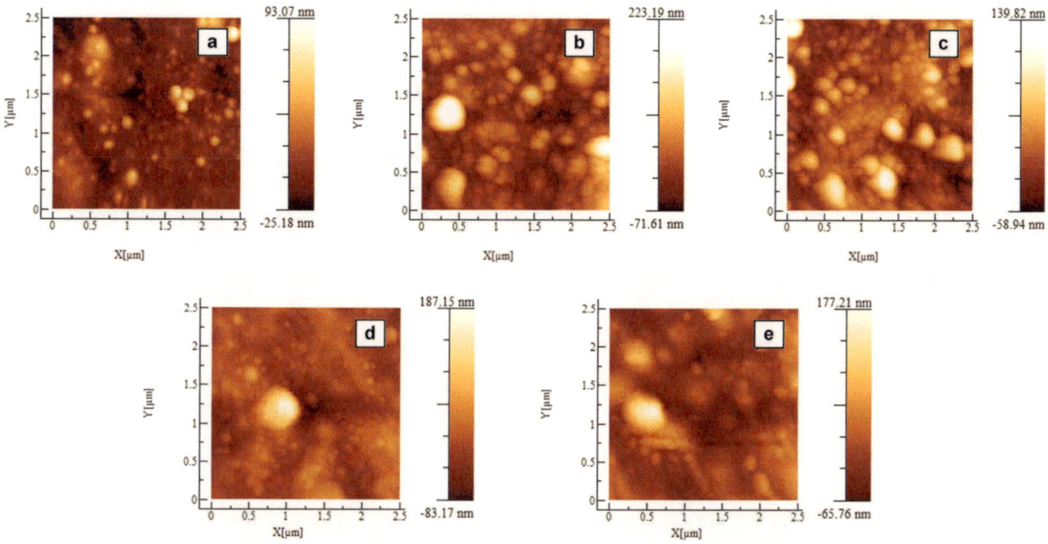

Figure 18. AFM images of the organic thin films deposited on glass/ITO substrate by MAPLE: glass/ITO/1P13 (a), glass/ITO/1P78 (b), glass/ITO/1Alq3 (c), glass/ITO/1P13/1Alq3 (d), and glass/ITO/1P78/1Alq3 (e).

Sample	RMS (nm)	Sample	RMS (nm)
1P78/ITO	35.3	1Alq3/1P13/ITO	26.2
1Alq3/1P78/ITO	31.8	2P13/ITO	12.6
2P78/ITO	36.2	2Alq3/2P13/ITO	28.5
2Alq3/2P78/ITO	29.1	1Alq3/ITO	26.0
1P13/ITO	10.4	2Alq3/ITO	30.1

Table 2. The RMS values obtained from the AFM on the organic thin films.

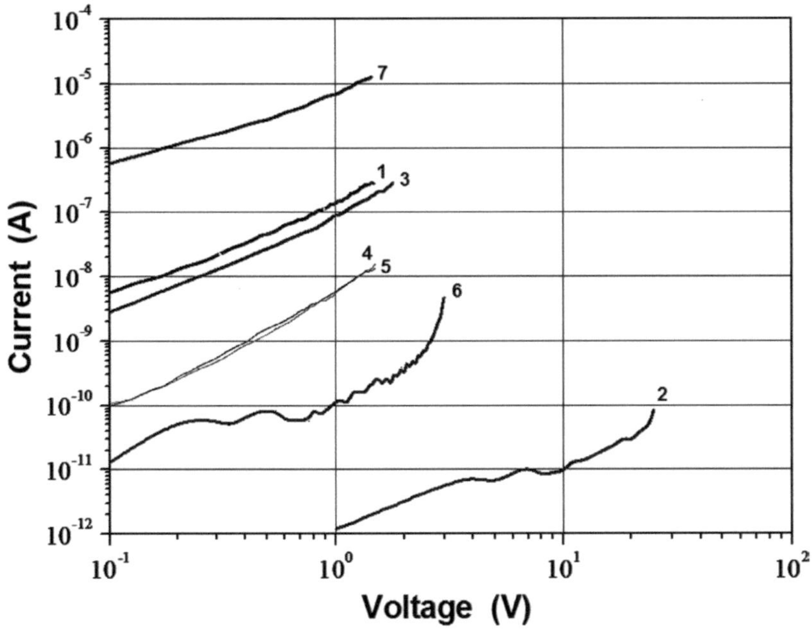

Figure 19. Current—voltage characteristics, in logarithmic representation, of the heterostructures based on arylenevinylene oligomer deposited on glass/ITO substrate by MAPLE using different no. of pulses (80,000 or 160,000): glass/ITO/Alq3/Al (curve 1—80,000 pulses, curve 2—160,000 pulses), glass/ITO/P13/Alq3/Al (curve 3—80,000 pulses, curve 4—160,000 pulses, curve 5—160,000 pulses, rev. bias), glass/ITO/P78/Alq3/Al (curve 6—80,000 pulses, curve 7—160,000 pulses).

Regarding the I-V characteristics of the heterostructures based on oligomers, it was remarked that their behavior is different. If, in the heterostructure with P78, a high current value was obtained for thicker layer (ITO/2P78/2Alq$_3$/Al), for the heterostructure with P13, the higher current value was recorded for thinner layers (ITO/1P13/1Alq$_3$/Al). No relevant changes are expected in the charge flow because there are no significant differences in the energetic barriers at IOT/P78 and ITO/P13 interfaces (**Figure 20**).

As was supposed by Gao [82], at the Al/Alq$_3$ interface, it can appear as a dipole layer, marked by a potential shift of −0.9 V, which determines a lowering of the Al cathode Fermi level to −5.2 eV, close to the Alq$_3$ HOMO level. The compounds have energy level positions that determine ohmic behavior or the appearance of the space charge limited currents (SCLC).

Analyzing at 1 V applied voltage, it is observed that in the heterostructure with Alq3 single layer formed with two layers, the current increases from 1.5×10^{-12} A for the ITO/2Alq$_3$/Al structure at 7×10^{-9} A and at 8×10^{-6} A for the ITO/2P13/2Alq$_3$/Al structure and for the ITO/2P78/2Alq$_3$/Al, respectively.

Thin films from oligomers and Alq$_3$ were deposited by MAPLE, and the films preserved the optical properties of the raw powders materials. In the heterostructure containing P78 and Alq$_3$ as stacked layers, the current value can be increased by using a thicker P78 layer.

http://dx.doi.org/10.5772/67659

4.2. Polymers based on arylene compounds

From the first generation of the organic photovoltaic cells realized with a single organic layer between anode and cathode [83], numerous attempts were made to improve the final cell parameters, either by developing new materials (as polymers with special properties) or using different cell architectures.

MAPLE method was used to obtain thin films from arylene-based polymers. The used polymers were poly[N-(2-ethylhexyl)2.7-carbazolyl vinylene]/AMC16 and poly[N-(2-ethylhexyl)2.7-carbazolyl 1.4-phenyleneethynylene]/AMC22 with the chemical structure presented in **Figure 21**. The same laser source mentioned above was used for the MAPLE deposition. The chloroform was used to obtain a target with 3 g/l concentration. We have deposited thin films on ITO, ITO covered with poly(3,4-ethylenedioxythiophene)-poly(styrenesulfonate) (PEDOT:PSS) and Si substrates in the following experimental conditions: 250 mJ/cm^2 laser fluence and 30,000 laser pulses [84].

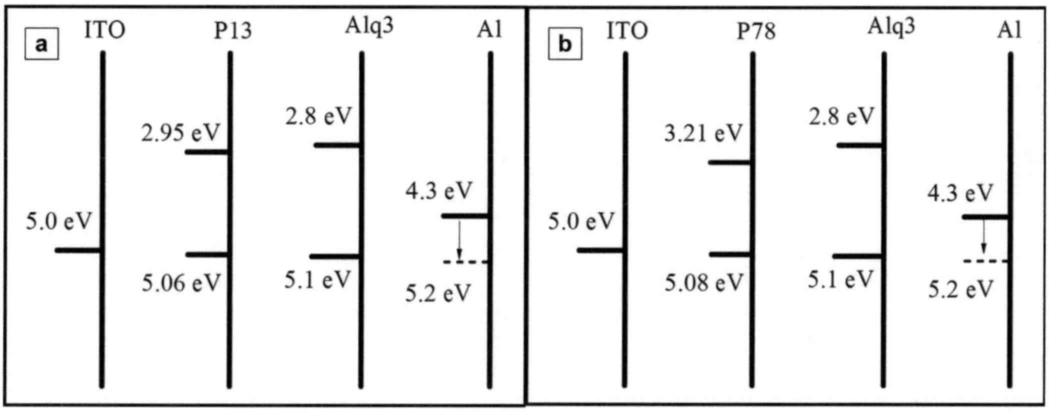

Figure 20. Band diagrams of the heterostructures based on arylenevinylene oligomer deposited on glass/ITO substrate by MAPLE: ITO/P13/Alq3/Al (a) and ITO/P78/Alq3/Al (b).

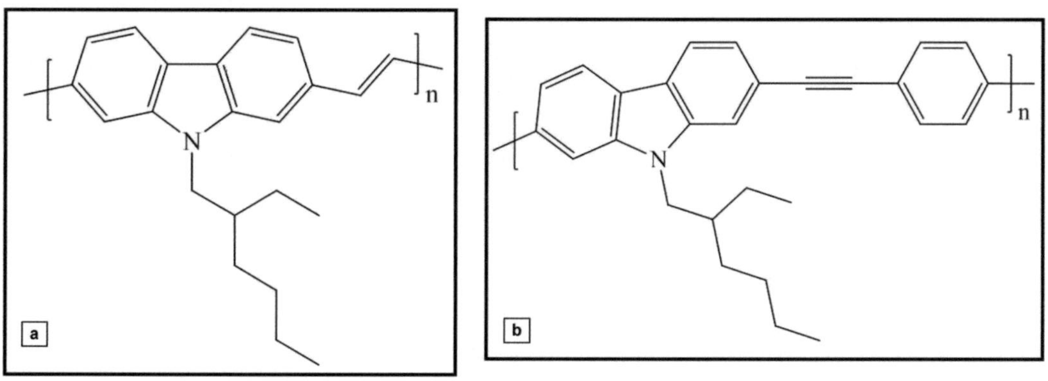

Figure 21. Chemical structures of the arylene polymers: AMC16 (a) and AMC22 (b).

The UV-VIS spectra (**Figure 22a**) and (**b**) of the organic film deposited by MAPLE were given in comparison with drop-cast realized films on glass substrate. A high transmittance is observed both for MAPLE films and for the drop-cast coatings (**Figure 22a**). The absence of the absorption maxima from 375 nm in the layers prepared by MAPLE (observed for the films deposited by drop-casting) is attributed to some modifications in the electronic structures of the polymeric films, determined by some differences in the arrangement of the molecules on the substrate surface. If for the preparing drop-cast film, the solvent is evaporated at room temperature, in MAPLE the solvent is thermally evaporated during the film deposition [85, 86]. Some cluster can appear, and this can affect the polymer backbone configuration [87, 88]. The polymers deposited by MAPLE on ITO and ITO/PEDOT:PSS (**Figure 22b**) exhibit a lower transmittance compared with the same layers prepared on glass.

The AFM images of the AMC 16 and AMC 22 films prepared by MAPLE on ITO and on ITO/PEDOT:PSS are given in **Figure 23**. The AMC22 polymer (with a reduced conjugated chain) forms layers with an increased roughness (RMS = 42 nm), when it is deposited on ITO/PEDOT:PSS substrate compared with the same layer deposited on ITO (RMS = 36 nm). In the case of AMC 16 with longer conjugated length, an increased roughness was obtained for the sample made on simple ITO substrate (RMS = 49.5 nm) compared with the sample realized on ITO/PEDOT:PSS substrate (RMS = 37 nm).

The I-V characteristics plotted in the dark and under illumination put into evidence the appearance of the photovoltaic effect in the polymeric films realized on ITO covered by an additional PEDOT:PSS layer (**Figure 24**). A higher current density (2.5×10^{-9} A/cm^2) was obtained in the dark at 0.5 V for the structure prepared with AMC 16 compared with that made with AMC 22 (~3×10^{-10} A/cm^2). The best photovoltaic parameters (open-circuit voltage-V_{OC}, short-circuit current-I_{SC}, and fill factor-FF) were shown by the cells based on AMC 16 ($V_{OC} = 0.303$, $I_{SC} = 12.7 \times 10^{-9}$ A, and FF = 29%) compared with the cell based on AMC22 ($V_{OC} = 0.073$, $I_{SC} = 8.2 \times 10^{-11}$ A, and

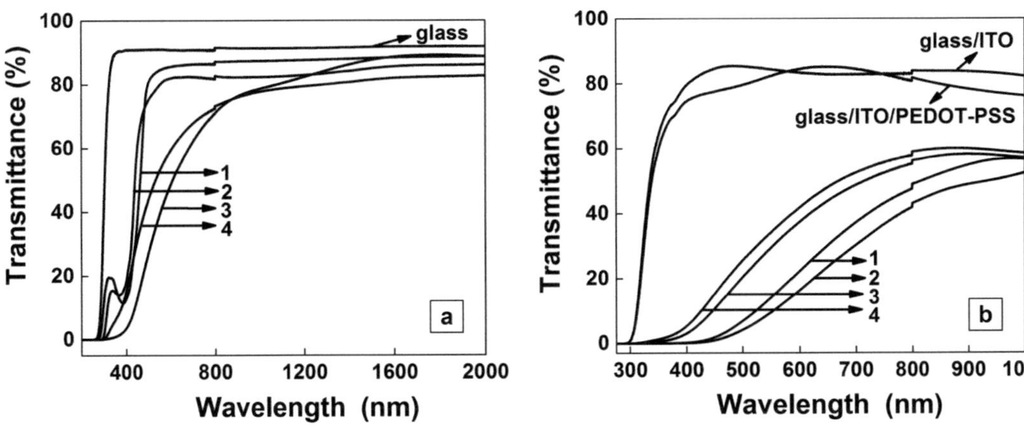

Figure 22. Transmission of the organic thin films based on arylene polymers (AMC16 and AMC22) deposited on glass (a), glass/ITO (b), and glass/ITO/PEDOT-PSS (b) substrates by drop-cast (only on glass) and MAPLE. On glass—AMC16 (curve 1—drop cast, curve 3) and AMC22 (curve 2—drop cast, curve 4), on glass/ITO—AMC16 (curve 1) and AMC22 (curve 3), on glass/ITO/PEDOT-PSS—AMC16 (curve 2) and AMC22 (curve 4).

FF = 25.9%). This sustains that a better collection of the charge appears in the cell structure realized with the polymer having a longer conjugation length. On the other hand, this is the sample with a lower roughness (RMS = 37 nm) of the active layer favoring the charge carrier transport.

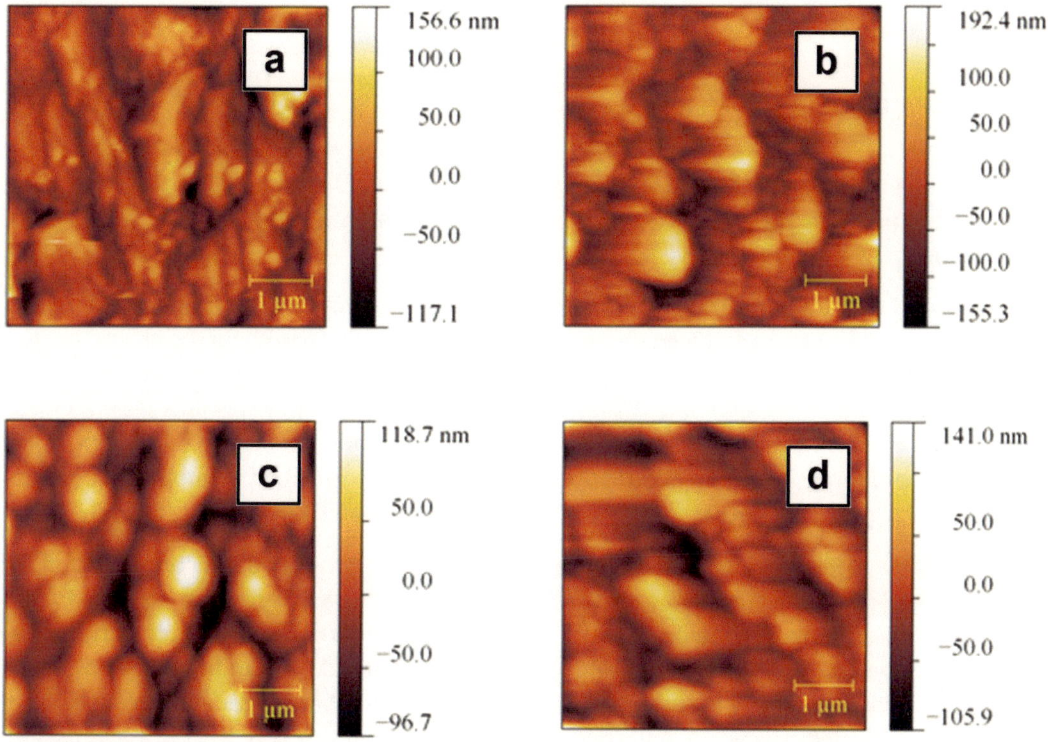

Figure 23. AFM images of the organic thin films deposited on glass/ITO (a, b) and glass/ITO/PEDOT-PSS (c, d) substrate by MAPLE: AMC16 (a, c) and AMC22 (b, d).

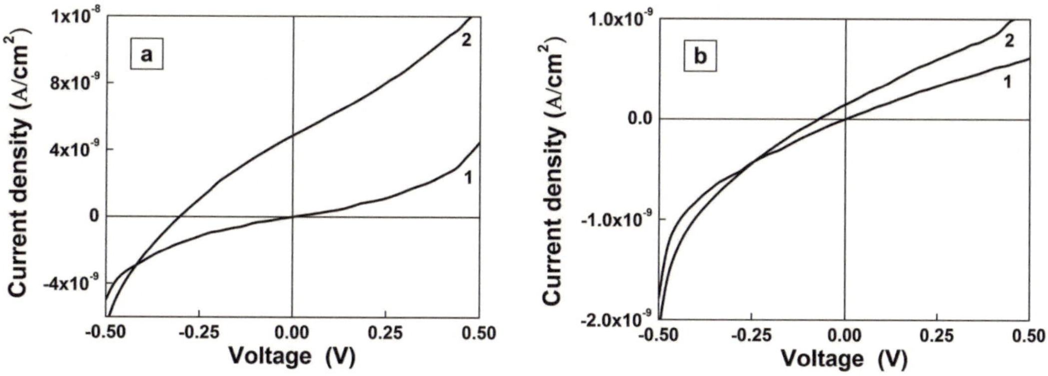

Figure 24. Current-voltage characteristics of the heterostructures based on arylene polymers deposited on glass/ITO/PEDOT-PSS substrate by MAPLE in dark (curves 1) and light (curves 2) conditions: glass/ITO/PEDOT-PSS/AMC16/Al (a) and glass/ITO/PEDOT-PSS/AMC22/Al (b).

Taking into account the presented results, it can be concluded that using MAPLE, polymeric thin films are deposited, characterized by good absorption in the blue-green region of the solar spectrum. The structures realized on ITO/PEDOT:PSS with AMC 16 and AMC 22 present photovoltaic effect, meaning that these materials can be taken into consideration for further applications in the OPV domain.

5. Conclusion

Summarizing, various thin films were deposited by PLD, CPLD, and MAPLE, the laser deposition technique being chosen in accordance with the material type (TCO or soft organic materials).

TCO films as ITO, AZO, and IZO are prepared by PLD on plastic substrate that presents a high transparency (~95%) and a reduced electrical resistivity (5×10^{-4} Ωcm), and the characteristics are very useful for integrating them in flexible electronics. An important parameter in the PLD experiments is the target-substrate distance. At a higher distance (8 cm), films free of cracks with a high transmittance and a reduced roughness were obtained. The exhibited electrical and optical properties are very good; all depositions were performed at room temperature without heating of the substrates.

$In_xZn_{1-x}O$ films were obtained by CPLD, using two targets with atomic In concentration In/(In + Zn), of 28 at.% and 56 at.% or 42 at.% and 70 at.%. The layers were analyzed from optical, electrical, and morphological point of view. We evidenced a high optical transmission, >95%. The lowest resistivity (8.6×10^{-4} Ωcm) was observed for an In content of about 44–49 at%. This technique is useful in the deposition of materials with different composition, each sample yielding practically a library of data, avoiding in this way the unnecessary loss of time. It is observed that the roughness of the samples decreases with the increase of the In content.

Layers based on arylenevinylene oligomers (P13, P78) and Alq3 were transferred by MAPLE without any material deterioration. Organic heterostructures with one or two organic layers were deposited. The optical properties of the start compounds were preserved. The prepared organic films present a good transmittance in the visible domain and the emission bands characteristic to the oligomers and to the Alq3. The globular morphology characteristic to MAPLE process was remarked. The I-V characteristics were symmetric, and the injector contact behavior was evidenced for the prepared heterostructures. For ITO/P78/Alq_3/Al heterostructures, the I-V plots evidenced dependence between the current value and the thickness of the organic layers.

Thin films from new arylene-based polymers were also processed by MAPLE. A good absorption was evidenced in the blue-green domain of the solar spectrum for the samples prepared with each polymer. The appearance of the photovoltaic effect was remarked for the heterostructures based on AMC16 film and AMC22 film deposited on ITO covered by a thin layer of PEDOT:PSS, which confirm that this buffer layer favors the charge carrier collection. The heterostructure based on the polymer with longer conjugation length present the higher dark current density. Due to their optical and electrical properties, such organic heterostructures can be interesting for OPV applications.

Acknowledgements

The authors thank to the group from ICMPP Iasi (led by Dr. M. Grigoras) for the synthesis of the oligomers and of the polymers. The work has been funded by the Romanian National Authority for Scientific Research, CNCS-UEFISCDI, projects TE 188/2014, PN-II-RU-TE-2014-4-1590 and the National Authority for Research and Innovation in the frame of Core Program - contract 4N/2016 and contract PN16-480102.

Author details

Marcela Socol[1]*, Gabriel Socol[2], Nicoleta Preda[1], Anca Stanculescu[1] and Florin Stanculescu[3]

*Address all correspondence to: marcela.socol@infim.ro

1 National Institute of Material Physics, Bucharest-Magurele, Romania

2 National Institute for Lasers, Plasma and Radiation Physics, Bucharest-Magurele, Romania

3 University of Bucharest, Faculty of Physics, Bucharest-Magurele, Romania

References

[1] Socol G., Socol M., Stefan N., Axente E., Popescu-Pelin G., Craciun D., et al. Pulsed laser deposition of transparent conductive oxide thin films on flexible substrate. Applied Surface Science. 2012; **260**:42–46. DOI: 10.1016/j.apsusc.2012.02.14.

[2] Stanculescu A., Socol M., Socol G., Mihailescu I. N., Stanculescu F., Girtan M. MAPLE prepared organic heterostructures for photovoltaic applications. Applied Physics A Materials Science & Processing. 2011; **104**(3):921–928. DOI: 10.1007/s00339-011-6440-y.

[3] Ristoscu C., Gyorgy E., Mihailescu I. N., Klini A., Zorba V., Fotakis C. Effects of pulse laser duration and ambient nitrogen pressure in PLD of AlN. Applied Physics A Materials Science & Processing. 2004; **79**:927–929. DOI: 10.1007/s00339-004-2857-x.

[4] Popescu-Pelin G., Sima F., Sima L. E., Mihailescu C. N., Luculescu C., Iordache I., et al. Hydroxyapatite thin films grown by pulsed laser deposition and matrix assisted pulsed laser evaporation: comparative study. Applied Surface Science. 2016. http://dx.doi.org/10.1016/j.apsusc.2016.10.043.

[5] Girtan M. Study of charge carriers' transport in organic solar cells by illumination area shifting. Solar Energy Materials and Solar Cells. 2017; **160**:430–434. DOI: 10.1016/j.solmat.2016.11.011.

[6] Kim S. I., Lee K. W., Sahu B. B., Han J. G. Flexible OLED fabrication with ITO thin film on polymer substrate. Journal of Appied Physics. 2015; **54**:090301–090304. http://dx.doi.org/10.7567/JJAP.54.090301.

[7] Zhang L., Li J.,Zhang X. W., Jiang X. Y., Zhang Z. L. High-performance ZnO thin film transistors with sputtering $SiO_2/Ta_2O_5/SiO_2$ multilayer gate dielectric. Thin Solid Films. 2010; **518**:6130–6133. DOI: 10.1016/j.tsf.2010.06.017.

[8] Granqvist C. G., Azens A., Heszler P., Kish L. B., Österlund L. Nanomaterials for benign indoor environments: electrochromics for "smart windows," sensors for air quality, and photo-catalysts for air cleaning. Solar Energy Materials and Solar Cells. 2007; **91**:355–365. DOI: 10.1016/j.solmat.2006.10.011.

[9] Dhara A., Alforda T. L. Optimization of $TiO_2/Cu/TiO_2$ multilayer as transparent composite electrode (TCE) deposited on flexible substrate at room temperature. ECS Solid State Letters. 2014; **3**:N33–N36. DOI: 10.1149/2.0061411ssl.

[10] Chen Z., Li W., Li R., Zhang Y., Xu G., Cheng H. Fabrication of highly transparent and conductive indium-tin oxide thin films with a high figure of merit via solution processing. Langmuir. 2013; **29**:13836–13842. DOI: 10.1021/la4033282.

[11] Stanculescu A., Socol M., Rasoga O., Mihailescu I. N., Socol G., Preda N., et al. Laser prepared organic heterostuctures on glass/AZO substrates. Applied Surface Science. 2014; **302**:169–176. http://dx.doi.org/10.1016/j.apsusc.2014.01.181.

[12] Leong W. L., Ren Y., Seng H. L., Huang Z., Chiam S. Y., Dodabalapur A. Efficient polymer solar cells enabled by low temperature processed ternary metal oxide as electron transport interlayer with large stoichiometry window. ACS Applied Materials &Interfaces. 2015;**7**:11099–11106. DOI: 10.1021/acsami.5b02215.

[13] Jun M. C., Park S. U., Koh J. H. Comparative studies of Al-doped ZnO and Ga-doped ZnO transparent conducting oxide thin films. Nanoscale Research Letters. 2012; **7**:639. DOI: 10.1186/1556-276X-7-639.

[14] Liu H., Wu Z., Hu J., Song Q., Wu B., Tam H. L., et al. Efficient and ultraviolet durable inverted organic solar cells based on an aluminium doped zinc oxide transparent cathode. Applied Physics Letters. 2013; **103**:043309. DOI: 10.1063/1.4816786.

[15] Bel-Hadj-Tahar R., Mohamed A. B. Sol–gel processed indium-doped zinc oxide thin films and their electrical and optical properties. New Journal of Glass and Ceramics. 2014; **4**:55–56. http://dx.doi.org/10.4236/njgc.2014.44008.

[16] Ellmer K., Mientus R. Carrier transport in polycrystalline transparent conductive oxides: a comparative study of zinc oxide and indium oxide. Thin Solid Films. 2008; **516**:4620–4627. DOI: 10.1016/j.tsf.2007.05.084.

[17] Zhang Y., Basel T. P., Gautam B. R., Yang X., Mascaro D. J., Liu F., et al. Spin-enhanced organic bulk heterojunction photovoltaic solar cells. Nature Communications. 2012; **3**:1043. DOI: 10.1038/ncomms2057.

[18] Bhagat S. A., Borghate S. V., Kalyani N. Thejo and Dhoble S. J. Novel Na^+ doped Alq_3 hybrid materials for organic light-emitting diode (OLED) devices and flat panel displays. Luminescence. 2013; **30**:251–256. DOI: 10.1002/bio.2721.

[19] Tiwari S., Singh A. K., Prakash R. Poly(3-hexylthiophene) (P3HT)/graphene nanocomposite material based organic field effect transistor with enhanced mobility. Journal of Nanoscience and Nanotechnology. 2014; **14**:2823–2828. DOI: https://doi.org/10.1166/jnn.2014.8570.

[20] Heliatek. Heliatek sets new organic photovoltaic world record efficiency of 13.2% [Internet]. 2016. Available from: http://www.heliatek.com/en/press/press-releases/details/heliatek-sets-new-organic-photovoltaic-world-record-efficiency-of-13-2.

[21] Xie T., Xie G., Su Y., Hongfei D., Ye Z., Jiang Y. Ammonia gas sensors based on poly(3-hexylthiophene)-molybdenum disulfide film transistors. Nanotechnology. 2016; **27**:065 502. DOI: 10.1088/0957-4484/27/6/065502.

[22] Ghorannevis Z., Akbarnejad E., Elahi Salar A., Ghoranneviss M. Application of RF magnetron sputtering for growth of AZO on glass substrate. Journal of Crystal Growth. 2016; **447**:62–66. http://dx.doi.org/10.1016/j.jcrysgro.2016.04.062.

[23] Meng L. J., Gao J., Silva R. A., Song S. Effect of the oxygen flow on the properties of ITO thin films deposited by ion beam assisted deposition (IBAD). Thin Solid Films. 2008; **516**:5454–5459. DOI: 10.1016/j.tsf.2007.07.071.

[24] Yadav S. C., Uplane M. D. Synthesis and properties of Boron doped ZnO thin films by spray CVD technique at low substrate temperature. International Journal of Engineering Science and Technology. 2012; **4**:4892–4898.

[25] Socol G., Craciun D., Mihailescu I. N., Stefan N., Besleaga C., Ion L., et al. High quality amorphous indium zinc oxide thin films synthesized by pulsed laser deposition. Thin Solid Films. 2011; **520**:1274–1277. DOI: 10.1016/j.tsf.2011.04.196.

[26] Pandey R., Yuldashev S., Nguyen H. D., Jeon H. C., Kang T. W. Fabrication of aluminium doped zinc oxide (AZO) transparent conductive oxide by ultrasonic spray pyrolysis. Current Applied Physics. 2012; **12**:S56–S58. DOI: 10.1016/j.cap.2012.05.027.

[27] Girtan M., Dabos-Seignon S., Stanculescu A. On morphological, structural and electrical properties of vacuum deposited pentacene thin films. Vacuum. 2009; **83**:1159–1163. DOI: 10.1016/j.vacuum.2009.03.001.

[28] Motaung D. E., Malgasa G. F., Arendse C. J. Comparative study: the effects of solvent on the morphology, optical and structural features of regioregular poly(3-hexylthiophene):fullerene thin films. Synthetic Metals. 2010; **160**:876–882. DOI: 10.1016/j.synthmet.2010.01.038.

[29] Acikbas Y., Erdogan M., Capan R., Yukruk F. Fabrication of Langmuir–Blodgett thin film for organic vapor detection using a novel *N,N'*-dicyclohexyl-3,4:9,10-perylenebis (dicarboximide). Sensors ans Actuator B-Chemical. 2014; **200**:61–68. http://dx.doi.org/10.1016/j.snb.2014.04.051.

[30] Eggenhuisen T. M., Galagan Y., Coenen E. W. C., Voorthuijzen W. P., Slaats M. W. L., Kommeren S. A., et al. Digital fabrication of organic solar cells by Inkjet printing using non-halogenated solvents. Solar Energy Materials and Solar Cells. 2015; 1**34**: 364–372. http://dx.doi.org/10.1016/j.solmat.2014.12.014.

[31] Caricato A. P., Cesaria M., Gigli G., Loiudice A., Luches A., Martino M., et al. Poly-(3-hexylthiophene)/6,6 -phenyl-C-61-butyric-acid-methyl-ester bilayer deposition by matrix-assisted pulsed laser evaporation for organic photovoltaic applications. Applied Physics Letters. 2012;**100**:073306. DOI: 10.1063/1.3685702.

[32] Norton D. P. Pulsed laser deposition of complex materials: progress toward applications. In: Eason R.. editor. Pulsed laser deposition of thin films: applications-led growth of functional materials. 1st ed. New York, United States: Wiley & Sons; 2006. pp. 3–32. DOI: 10.1002/9780470052129.ch1.

[33] Kim H. Transparent conducting oxide films. In: Easton R., editor. Pulsed laser deposition of thin films: applications-led growth of functional materials. 1st ed.. New York, United States: John Wiley & Sons Inc. 2006. pp. 240–255. DOI: 10.1002/9780470052129.ch11.

[34] Huc I., Lehn J. M. Virtual combinatorial libraries: dynamic generation of molecular and supramolecular diversity by self-assembly. Proceeding of the National Academy of Science USA. 1997; **94**:2106–2110.

[35] Plunkett M. J., Ellman J. A. Combinatorial chemistry and new drugs. Scientific American. 1997; **276**:68–73.

[36] Sakurai J., Hata S., Yamauchi R., Shimokohbe A. Search for novel amorphous alloys with high crystallization temperature by combinatorial arc plasma deposition. Applied Surface Science. 2007; **254**:738–742. DOI: 10.1016/j.apsusc.2007.05.092.

[37] Chikyow T., Ahmeta P., Nakajimab K., Koidac T., Takakurad M., Yoshimotod M., et al. A combinatorial approach in oxide/semiconductor interface research for future electronic devices. Applied Surface Science. 2002; **189**:284–291. http://dx.doi.org/10.1016/S0169-4332(01)01004-2.

[38] Martin T. P., Chan K., Gleason K. K. Combinatorial initiated chemical vapor deposition (iCVD) for polymer thin film discovery. Thin Solid Films. 2008; **516**:681–683. DOI: 10.1016/j.tsf.2007.06.113.

[39] Socol G., Socol M., Sima L., Petrescu S., Enculescu M., Sima F., et al. Combinatorial pulsed laser deposition of Ag-containing calcium phosphate coatings. Digest Journal of Nanomaterials and Biostructures. 2012; **7**:563–576.

[40] Axente E., Socol G., Beldjilali S. A., Mercadier L., Luculescu C. R., Trinca L. M., et al. Quantitative analysis of amorphous indium zinc oxide thin films, synthesized by combinatorial pulsed laser deposition. Applied Physics A-Materials Science & Processing. 2014; **117**:229–236. DOI: 10.1007/s00339-014-8427-y.

[41] Piqué A., McGill R. A., Chrisey D. B., Leonhardt D., Mslna T. E., Spargo B. J., et al. Growth of organic thin films by the matrix assisted pulsed laser evaporation (MAPLE) technique. Thin Solid Films. 1999; **355–356**:536–541.

[42] Epstein A. J. and Yang Y. Polymeric and organic electronic materials: from scientific curiosity to applications. Material Research Bulletin. 1997; **22**:13–14. DOI: https://doi.org/10.1557/S0883769400033571.

[43] Luches A.,Caricato A. P. Fundamentals and applications of MAPLE. In: Miotello A., Ossi P. M., editors. Laser-surface interactions for new materials production. 1st ed.. Berlin: Springer Berlin Heidelberg; 2010. pp.203–233. DOI: 10.1007/978-3-642-03307-0_9.

[44] Caricato A. P. Lasers in materials science. MAPLE and MALDI: theory and experiments. In: Castillejo M. O., Paolo L. Zhigilei, editors. Lasers in materials science. 1st ed. Switzerland: Springer International Publishing; 2014. pp. 295–323. DOI: 10.1007/978-3-319-02898-9_12.

[45] Caricato A. P., Luches A. Applications of the matrix-assisted pulsed laser evaporation method for the deposition of organic, biological and nanoparticle thin films: a review. Applied Physics A-Materials Science & Processing. 2011; **105**(3):565–582. DOI: 10.1007/s00339-011-6600-0.

[46] Bloisi F., Barra M., Cassinese A., Vicari L. R. M. Matrix-assisted pulsed laser thin film deposition by using Nd:YAG laser. Journal of Nanomaterials. 2012; **2012**:Article ID 395436. 9 p. DOI: 10.1155/2012/395436.

[47] Socol G., Mihailescu I. N., Albu A. M., Antohe S., Stanculescu F., Stanculescu A., et al. MAPLE prepared polymeric thin films for non-linear optics applications. Applied Surface Science. 2009; **255**:5611–5614. DOI: 10.1016/j.apsusc.2008.07.206.

[48] Stadler A. Transparent conducting oxides—an up-to-date overview. Materials. 2012; **5**:661–683. DOI: 10.3390/ma5040661.

[49] van Deelen J., Illiberi A., Hovestad A., Barbu I., Klerk L., Buskens P. Transparent conducting materials: overview and recent results. SPIE. 2012; **8470**:84700-1-8. DOI: 10.1117/12.929685.

[50] Chopra K. L., Major S., Pandya D. Transparent conductors—a status review. Thin Solid Films. 1983; 1**02**:1–46. http://dx.doi.org/10.1016/0040-6090(83)90256-0.

[51] Ellmer K. Past achievements and future challenges in the development of optically transparent electrodes. Nature Photonics. 2012; **6**:809–817. DOI: 10.1038/nphoton.2012.282.

[52] Löbmann P. Transparent conducting oxides. In: Schneller T., Waser R., Kosec M., Payne D., editors. Chemical solution deposition of functional oxide thin film. 1st ed. Vienna: Springer Vienna; 2013. pp. 655–672. DOI: 10.1007/978-3-211-99311-8_26.

[53] Sohn S., Han Y. S. Transparent conductive oxide (TCO) films for organic light emissive devices (OLEDs). In: Ko S. H., editors. Organic light emitting diode—material, process and devices. 1st ed.. Rijeka, Croatia: InTech; 2011. pp. 233–274. DOI: 10.5772/18545.

[54] Paine, D. C., Yeom, H. Y., Yaglioglu, B. Transparent conducting oxide materials and technology. In: Crawford G. P., editors. Flexible flat panel displays. 1st ed. Chichester, UK: John Wiley & Sons, Ltd; 2005. DOI: 10.1002/0470870508.ch5.

[55] Ellmer K., Klein A., Rech B., editors. Transparent conductive zinc oxide. Basics and applications in thin film solar cells. 1st ed. Berlin: Springer-Verlag Berlin Heidelberg; 2008. 446 p. DOI: 10.1007/978-3-540-73612-7.

[56] Hosono H., Paine D. C., Ginley D., editors. Handbook of transparent conductors. 1st ed. US: Springer US; 2011. 534 p. DOI: 10.1007/978-1-4419-1638-9.

[57] Badeker K. Concerning the electrical conductibility and the thermoelectric energy of several heavy metal bonds. Annals of Physics. 1907; **22**:749–766.

[58] Minami T. Transparent and conductive multicomponent oxide films prepared by magnetron sputtering. Journal of Vacuum Science and Technology. 1999; **17**:1765–1772. http://dx.doi.org/10.1116/1.581888.

[59] Yanagi H., Kawazoe H., Kudo A., Yasukawa M., and Hosono H. Chemical design and thin film preparation of p-type conductive transparent oxides. Journal of Electroceramics. 2000; **4**:407–414. DOI: 10.1023/A:1009959920435.

[60] Kawazoe H., Yasukawa M., Hyodo H., Kurita M., Yanagi H., Hosono H. P-type electrical conduction in transparent thin films of $CuAlO_2$. Nature. 1997; **389**:939–942. DOI: 10.1038/40087.

[61] Kudo A., Yanagi H., Hosono H., Kawazoe H. A p-type conductive oxide with wide band gap. Applied Physics Letters. 1998; **73**:220–222. DOI: 10.1063/1.121761.

[62] Nagrajan R., Draeseke A. D., Sleight A. W., Tate J. P-type conductivity in $CuCr_{1-x}Mg_xO_2$ films and powders. Journal of Applied Physics. 2001; **89**:8022–8025. DOI: 10.1063/1.1372636.

[63] Suzuki A., Mastsushita T., Aoki T., Yoneyama Y., Okuda M. Pulsed laser deposition of transparent conducting indium tin oxide films in magnetic field perpendicular to plume. Japanese Journal of Applied Physics. 2001; **40**:L401–L403. DOI: 10.1143/JJAP.40.L401.

[64] Kim, H., Piqué A., Horwitz J. S., Mattoussi H., Murata H., Kafafi Z. H., et al. Indium tin oxide thin films for organic light-emitting devices. Applied Physics Letters. 1999; **74**:3444. DOI: 10.1063/1.124122.

[65] Jee S. H., Kim S. H., Ko J. H., Yoon Y. S. Study on work function change of ITO modified by using a self-assembled monolayer for organic based devices. Journal of the Korean Physical Society. 2006; **49**(5). 2034–2039.

[66] Kim H., Gilmore C., Piqué A., Chrisey D. B. Electrical, optical, and structural properties of indium–tin–oxide thin films for organic light-emitting devices. Journal of Applied Physics. 1999; **86**(11):6451–6461. DOI: 10.1063/1.371708.

[67] Hu J., Gordon R. G. Textured aluminium doped zinc oxide thin films from atmospheric pressure chemical vapour deposition. Journal of Applied Physics. 1992; **71**:880–890. DOI: 10.1063/1.351309.

[68] Park J. W., Kim G., Lee S. H., Kim E. H., Lee G. H. The effect of film microstructures on cracking of transparent conductive oxide (TCO) coatings on polymer substrates. Surface and Coatings Technology. 2010; **205**:915–921. http://dx.doi.org/10.1016/j.surfcoat.2010.08.055.

[69] Chung W. S., Thompson M. O., Wickboldt P., Toet D., Carey P. G. Room temperature indium tin oxide by XeCl excimer laser annealing for flexible display. Thin Solid Films. 2004; **460**:291–294. http://dx.doi.org/10.1016/j.tsf.2004.01.050.

[70] Kim H., Horwitz J. S., Kushto G. P., Kafafi Z. H., Chrisey D. B. Indium tin oxide thin films grown on flexible plastic substrates by pulsed-laser deposition for organic light-emitting diodes. Applied Physics Letters. 2001; **79**:284–286. DOI: 10.1063/1.1383568.

[71] Mekhnache M., Drici A., Saad Hamideche L., Benzarouk H., Amara A., Cattin L, et al. Properties of ZnO thin films deposited on (glass, ITO and ZnO:Al) substrates. Superlattices and Microstructures. 2011;**49**(5):510–518. http://dx.doi.org/10.1016/j.spmi.2011.02.002

[72] Fernandez S., Naranjo F. B. Optimization of aluminum-doped zinc oxide films deposited at low temperature by radio-frequency sputtering on flexible substrates for solar cell applications. Solar Energy Materials and Solar Cells. 2010; **94**:157–163. DOI: 10.1016/j.solmat.2009.08.012.

[73] Socol G., Galca A. C., Luculescu C.R., Stanculescu A., Socol M., Stefan N., et al. Tailoring of optical, compositional and electrical properties of the $In_xZn_{1-x}O$ thin films obtained by combinatorial pulsed laser deposition. Digest Journal of Nanomaterials and Biostructures. 2011; **6**(1):107–115.

[74] Girtan M., Socol M., Pattier B., Sylla M., Stanculescu A. On the structural, morphological, optical and electrical properties of sol–gel deposited ZnOIn films. Thin Solid Films. 2010; **519**(2):573–577. DOI: 10.1016/j.tsf.2010.07.006.

[75] Minami T., Kakumu T., Takeda Y., Takata S. Highly transparent and conductive ZrO-In_2O_3 thin films prepared by dc magnetron sputtering. Thin Solid Films. 1996; **291**:1–5. DOI: 10.1016/S0040-6090(96)09094-3.

[76] Moriga T., Okamoto T., Hiruta K., Fujiwara A., Nakabayashi I., Tominaga K. J. Structures and physical properties of films deposited by simultaneous DC sputtering of ZnO and In_2O_3 or ITO targets. Journal of Solid State Chemistry. 2000; **155**(2):312–331. DOI: 10.1006/jssc.2000.8919.

[77] Socol M., Preda N., Vacareanu L., Grigoras M., Socol G., Mihailescu I.N., et al. Organic heterostructures based on arylenevinylene oligomers deposited by MAPLE. Applied Surface Science. 2014; **302**:216–222. DOI: 10.1016/j.apsusc.2013.12.091.

[78] Vacareanu L., Grigoras M. Electrochemical characterization of arylenevinylene oligomers containing triphenylamine and carbazole units. Journal of Applied Electrochemistry. 2014; **40**:1967–1975. DOI: 10.1007/s10800-010-0173-z.

[79] Sakurai Y., Hosoi Y., Ishii H., Ouchi Y., Salvan G., Kobitski A., et al. Study of the interaction of *tris*-(8-hydroxyquinoline) aluminum(Alq3) with potassium using vibrational spectroscopy: examination of possible isomerization upon K doping. Journal of Applied Physics. 2004;**96**:5534–5542. http://doi.org/10.1063/1.1776626.

[80] Stanculescu A., Vacareanu L., Grigoras M., Socol M., Socol G., Stanculescu F., et al Thin films of arylenevinylene oligomers prepared by MAPLE for applications in non-linear optics. Applied Surface Science. 2011;**257**:5298–5302. DOI: 10.1016/j.apsusc.2010.11.094.

[81] Stanculescu A., Stanculescu F., Tugulea L., Socol M. Optical properties of 3,4,9,10-perylenetetracarboxylic dianhydride and 8-hydroxyquinoline aluminium salt films prepared by vacuum deposition. Material Science Forum. 2006; **514–516**:956–960.

[82] Gao Y. Surface analytical studies of interfaces in organic semiconductor devices. Materials Science and Engineering: R: Reports. 2010; **68**(3):39–87. http://dx.doi.org/10.1016/j.mser.2010.01.001.

[83] Chamberlain G. A. Organic solar cells: a review. Solar Cells. 1983; **8**(1):47–83. http://dx.doi.org/10.1016/0379-6787(83)90039-X.

[84] Stanculescu F., Rasoga O., Catargiu A. M., Vacareanu L., Socol M., Breazu C., et al. MAPLE prepared heterostructures with arylene based polymer active layer for photovoltaic applications. Applied Surface Science. 2015; **33**:240–248. DOI: 10.1016/j.apsusc.2014.11.146.

[85] Piqué A. The matrix-assisted pulsed laser evaporation (MAPLE) process: origins and future direction. Applied Physics A-Materials Science & Processing. 2011; **105**(3):517–528. DOI: 10.1007/s00339-011-6594-7.

[86] Sellinger A. T., Leveugle E., Gogick K., Peman G., Zhigilei L. V., Fitz-Gerald J. M.. Journal of Physics: Conerence Series. 2007; **59**:314–317. DOI: 10.1088/1742-6596/59/1/066.

[87] Balakrishnan K., Datar A., Zhang W., Yang X., Naddo T., Huang J., et al. Nanofibril self-assembly of an arylene ethynylene macrocycle. Journal of the American Chemical Society. 2006; **128**(20):6576–6577. DOI: 10.1021/ja0618550.

[88] Dmitruk N. L., Borkovskaya O. Yu., Havrylenko T. S., Naumenko D. O., Petrik P., Meza-Laguna V., et al. Effect of chemical modification of thin C_{60} fullerene films on the fundamental absorption edge. Semiconductor Physics, Quantum Electronics and Optoelectronics. 2010; **13**:180–185.

Section 2

III-V Heterostructures

Chapter 4

Heteroepitaxy of III–V Zinc Blende Semiconductors on Nanopatterned Substrates

Thomas Riedl and Jörg K.N. Lindner

Additional information is available at the end of the chapter

http://dx.doi.org/10.5772/67572

Abstract

In the last decade, zinc blende structure III–V semiconductors have been increasingly utilized for the realization of high-performance optoelectronic applications because of their tunable bandgaps, high carrier mobility and the absence of piezoelectric fields. However, the integration of III–V devices on the Si platform commonly used for CMOS electronic circuits still poses a challenge, due to the large densities of mismatch-related defects in heteroepitaxial III–V layers grown on planar Si substrates. A promising method to obtain thin III–V layers of high crystalline quality is the growth on nanopatterned substrates. In this approach, defects can be effectively eliminated by elastic lattice relaxation in three dimensions or confined close to the substrate interface by using aspect-ratio trapping masks. As a result, an etch pit density as low as 3.3×10^5 cm^{-2} and a flat surface of submicron GaAs layers have been accomplished by growth onto a SiO_2 nanohole film patterned Si(001) substrate, where the threading defects are trapped at the SiO_2 mask sidewalls. An open issue that remains to be resolved is to gain a better understanding of the interplay between mask shape, growth conditions and formation of coalescence defects during mask overgrowth in order to achieve thin device quality III–V layers.

Keywords: heteroepitaxy, nanopatterning, crystal defects, defect trapping, compound semiconductors, strain relaxation

1. Introduction

III–V compound semiconductors are the materials of choice for making state-of-the-art power electronic and optoelectronic devices, which is due to the outstanding properties of this class of semiconductors. First, they have mostly direct bandgaps covering a broad range between 0.2 eV for InSb [1] and 5.3 eV for zinc blende AlN [2] (at room temperature, respectively).

Also, bandgaps and lattice parameters can be continuously tailored by forming ternary and quaternary alloys. This allows for continuous tuning of the wavelength of light-emitting or absorbing devices and of the lattice parameter in order to reduce lattice misfit to the substrate or between different layers in the device. By decreasing the size of the semiconductor close to the size of the electron wavelength quantum confinement effects come into play, providing an additional degree of freedom for tailoring the emission or absorption wavelength. Another benefit of the carrier confinement arises from the change in the density of states distribution leading to sub-band formation and increased exciton-binding energies in quantum wells, which enables the realization of high-performance LEDs, laser devices and solar cells. Moreover, GaAs, InAs and InSb have very high electron mobilities in the range between 9400 cm^2/Vs for GaAs [3] and 78,000 cm^2/Vs for InSb [4], which make them ideal for applications in high-speed power electronics. A particular advantage of zinc blende semiconductors as opposed to their wurtzite-structure counterparts is that piezoelectric fields occur to a much lesser extent in the former because of their high symmetry. In zinc blende layers grown along the [001] direction piezoelectric effects are completely absent [5].

In view of the compelling properties and the great potential for applications but also the high production cost of III–V semiconductor wafers there is a rising demand for integrating III–V structures into the less expensive and well established Si technology [6]. This is important e.g. for developing high-performance light sources on the Si platform. However, growth of high-quality III–V layers on planar Si substrates is hampered by the considerable mismatch of lattice parameters and thermal expansion coefficients as well as by the polar/non-polar surface incompatibility leading to the formation of defects, i.e. threading dislocations (TDs), stacking faults (SFs), twins and anti-phase boundaries (APBs). Threading defects, which cross the active layers of devices, are known to degrade their performance and lifetime by formation of electronic levels in the bandgap inducing non-radiative carrier recombination [7]. In addition, charge carriers are scattered at the defects, the diffusion of impurities is intensified, and defects reaching the layer surface increase its roughness.

In order to reduce the defect density in heteroepitaxial-mismatched layers on planar substrates different strategies have been pursued, like utilization of graded buffer layers [8], thermal cycle annealing and strained layer superlattices [9, 10]. Unfortunately, these approaches are time-consuming, yield limited defect density reductions or require thick buffer layers to be grown. For example, by application of complex thermal cycle annealing and strained layer superlattice processes in the case of GaAs growth on Si(001) an etch pit density lower than $\sim 1.2 \times 10^6$ cm^{-2} could not be achieved [9–11]. An alternative route of eliminating defects consists of using *nanopatterned* instead of planar substrates for the heteroepitaxial growth. In this way, III–V layers of remarkably high structural and morphological quality with respect to their low thickness have been achieved in the last ten years. Heteroepitaxial growth of c-GaN on nanopatterned 3C-SiC/Si(001) has also extensively been studied [12, 13] but in the following emphasis will be placed on arsenides, phosphides and anti-monides.

The book chapter is organized as follows. Section 2 reviews the different heteroepitaxial approaches for growing zinc blende structure III–V layers on nanopatterned mismatched substrates. Focus is put on approaches that aim at a reduction of defect densities by applying different defect elimination mechanisms. Also, the resulting structural, morphological and optical quality of the layers is assessed. After that, Section 3 highlights the relevance of such high-quality III–V layers for advanced applications.

2. Heteroepitaxial layer growth and quality on nanopatterned surfaces

In the following semiconductor, heteroepitaxy strategies are reviewed, which use nanopatterned substrates in order to minimize the density of misfit-related defects in III–V layers. **Figure 1** provides an overview. Most of these heteroepitaxy approaches rely on nanoscale selective growth (Section 2.1.1) on a mask-patterned surface, i.e. selective growth on nanoscale crystalline areas, which are laterally surrounded by an oxide or nitride mask layer (**Figure 1(a–c)**). This growth mode, which also enables selective area growth of nanowires and quantum dots (QDs), can be exploited for the epitaxial lateral overgrowth (Section 2.1.2) and aspect ratio trapping (ART) techniques (Section 2.1.3). Moreover, multiple QD layers on planar, buffer layer coated substrates can be exploited as dislocation filters (**Figure 1(d)**, Section 2.2). Yet another approach uses nanoporous substrate surfaces without any mask, where the local three-dimensional elastic lattice relaxation enhances the layer quality (**Figure 1(e)**, Section 2.3).

2.1. Growth on mask-patterned surfaces

The use of nanopatterned oxide or nitride masks offers a great potential for the heteroepitaxial growth of mismatched semiconductor layers. Low-defect III–V layers have been dem-

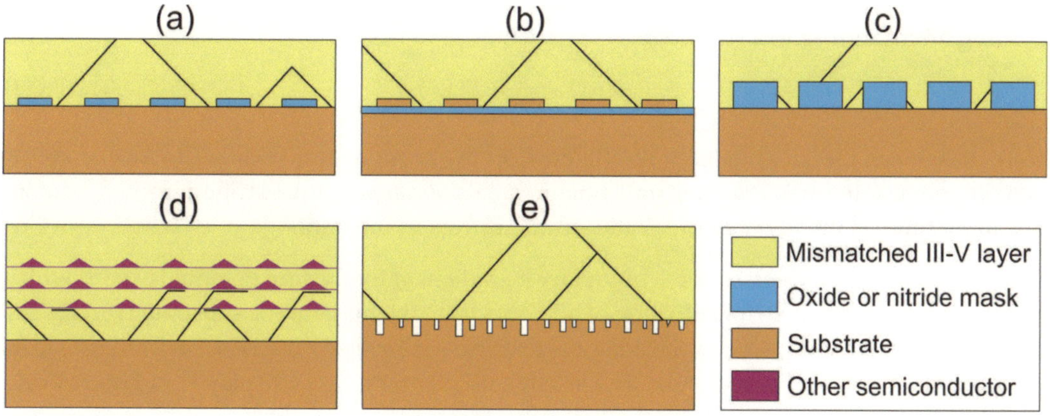

Figure 1. Overview of heteroepitaxial growth methods on nanopatterned substrates. The graphs show schematic cross-sections of the layer systems, respectively. (a–c) Three variants of mask-controlled growth with nucleation on nanoscale substrate areas: (a) thin mask film with nanoopenings, (b) continuous mask film with substrate nanoislands on top, (c) mask nanoopenings with aspect ratio >1 (aspect ratio trapping technique), (d) growth using multiple QD layers as dislocation filters and (e) growth on a mask-free nanoporous substrate. In (a–e) the inclined black lines indicate the propagation of threading crystal defects.

onstrated on substrates covered with (i) a mask layer having nanoscale line or round shaped openings, or with (ii) semiconductor islands or mesas on top of a continuous mask layer. In both variants the nanoscale semiconductor areas on the surface serve as nucleation sites for the III–V growth. Due to lateral growth the overgrown islands eventually coalesce to form continuous III–V layers, bridging the mask areas in between them.

2.1.1. *Nanoscale selective growth*

Nanoscale selective growth (NSG) describes selective homoepitaxial or heteroepitaxial growth on the unmasked areas of a partially masked substrate surface. It represents a key mechanism in the growth of mismatched layers on mask patterned substrates, first proposed by Luryi and Suhir [14]. In the following two subsections, the conditions required for NSG are analysed theoretically and corresponding experimental observations are described [15].

2.1.1.1. *Theoretical description*

On an ideal surface, i.e. in absence of nucleation centres such as impurities, defects or surface steps, selective growth can be achieved (i) by surface out-diffusion of the adatoms from the masked to the unmasked areas and/or (ii) by adatom desorption before the cluster formation starts. The surface out-diffusion mechanism works if the lateral dimensions of the masked areas are chosen smaller than the surface migration length of the adatoms, i.e. the average diffusion length before desorption or agglomeration with other atoms. This allows the adatoms to diffuse to the unmasked areas and to suppress crystal nucleation on the masked areas resulting in selective growth in the mask openings [15]. In the case (ii) the surface diffusion length of the adatoms without coming across other adatoms has to be less than a characteristic length describing the average distance between adjacent adatoms [16]. NSG can be described by an equation analogous to that for step-flow growth with a critical cluster size of 1, where arriving atoms diffuse to a step edge:

$$\frac{\partial n}{\partial t} = F - \frac{n}{\tau_{\mathrm{des}}} + D\,\nabla^2 n \qquad (1)$$

Here, n denotes the areal density of adatoms on the mask at time t, F is the flux of atoms arriving on a unit mask area per unit time t. τ_{des} designates the desorption-limited adatom residence time on the surface, and D the surface diffusion constant.

In the vicinity of the substrate-mask boundary a gradient of the surface potential occurs, which provides the driving force for diffusion of adatoms from the mask to the substrate [16, 17]. The potential gradient also prevents the adatoms to diffuse in reverse direction, i.e. from the GaAs surface onto the mask. For a circular SiO_2 mask with diameter L_M as displayed in **Figure 2** and the boundary condition $n(L_M/2, t) = 0$, the steady state is given for [15]:

$$\frac{D}{F} = \frac{\pi L_M^4}{4N} \sum_{m=1}^{\infty} \left[z_m^2 + \left(\frac{L_M}{2L_d} \right)^2 \right]^{-1} \sim L_M^{\alpha} \qquad (2)$$

For selective growth the total number of Ga adatoms on the SiO_2 mask N is small and can be set to 1 without affecting the results for α. z_m is the mth zero of the zero-order Bessel function, L_d is the diffusion length of a Ga atom on a planar SiO_2 surface and α is a

http://dx.doi.org/10.5772/67572

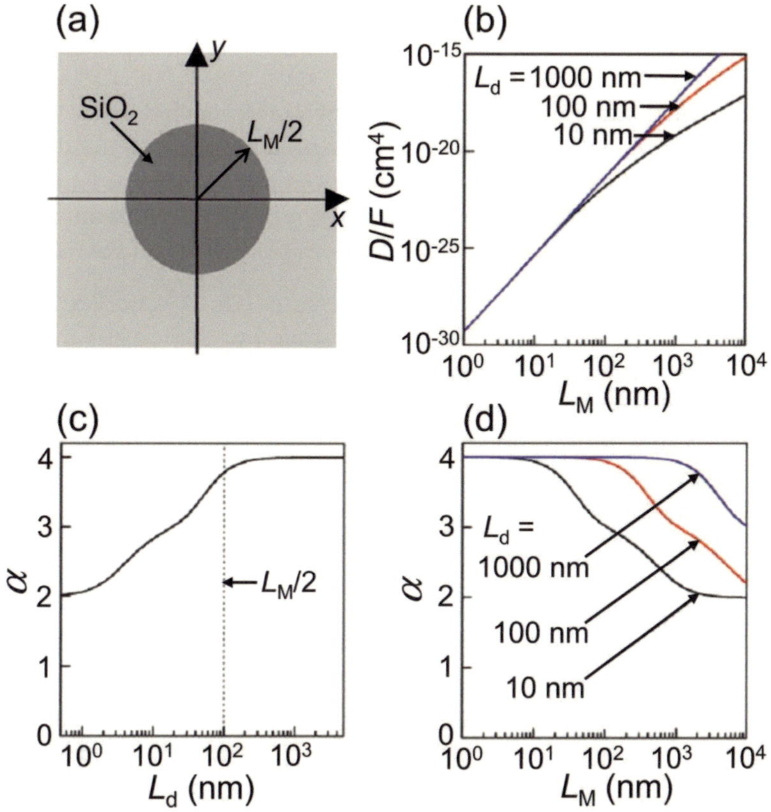

Figure 2. (a) Schematic of a SiO_2 disc of diameter L_M deposited on a GaAs surface, (b) calculated ratio D/F versus L_M for different L_d according to Eq. (2). (c, d) Diagrams showing the exponent α as a function of (c) L_d and (d) L_M, obtained from (b), respectively. In (c) $L_M = 200$ nm ($\alpha = 3.8$), where the dashed line marks $L_d = L_M/2$. (Reproduced from [15], with the permission of AIP Publishing.)

dimensionless exponent [15]. **Figure 2(b)** plots D/F as a function of L_M for different L_d, and **Figure 2(c)** and **(d)** display the resulting exponent α versus L_d and L_M, respectively. With increasing L_M and decreasing L_d α decreases from 4 to 2, owing to desorption. In the limit $\alpha = 4$ ($L_M \ll L_d$) suppression of growth on the mask is achieved only by surface out-diffusion, whereas for $\alpha = 2$ ($L_M \gg L_d$) selective growth is accomplished solely by desorption. In the latter case the critical flux below which selective growth occurs obeys an exponential scaling law:

$$F_c \sim \frac{\nu_0^{\,2}}{C} \exp\left[-\frac{2E_{des} - E_{diff}}{kT}\right], \tag{3}$$

where E_{des} denotes the activation energy for adatom desorption, E_{diff} is the activation energy for adatom surface diffusion, T is the absolute temperature, k is the Boltzmann constant and ν_0 is a desorption rate constant. Commonly, $2E_{des} > E_{diff}$, meaning that the critical flux below which nucleation on the SiO_2 mask can be suppressed increases with rising temperature.

2.1.1.2. *Experimental observations*

Experimentally, two main types of nanopatterned substrates have been utilized for oxide mask-based NSG/nanoheteroepitaxial growth: (i) substrates covered with a mask film having round or line shaped openings and (ii) substrates capped with a continuous mask layer with nanoscale semiconductor islands or horizontal wires on top. On the one hand, the fabrication of type (i) substrates is more facile. On the other hand, type (ii) offers the advantage of compliant nanoscale substrate islands, which can accommodate a portion of the misfit strain.

Approach (i) has been extensively studied by Lee et al. [16, 17] and Lee and Brueck [15]. They used SiO_2 films deposited on GaAs(001) substrate by electron beam evaporation. For the patterning of the SiO_2 films interference lithography followed by dry etching [15] was deployed, resulting in arrays of approximately circular mask openings with pattern periods between 260 and 350 nm. By varying the dry etching time different opening sizes were realized. For the longer etching times the openings coalesced leaving isolated SiO_2 islands on the GaAs substrate. Scanning electron microscopy (SEM) images of the surface after molecular beam epitaxy (MBE) growth at 570°C using a Ga flux of $F = 0.33\times10^{14}$ atoms/(cm^{2}s) show that the growth becomes increasingly selective [15], when reducing the mask dimension L_M from infinity to 70 nm (**Figure 3(a–f)**). For these conditions the critical mask dimension $L_{M,c}$ below which growth on the mask is completely inhibited amounts to ~120 nm. The experimental $L_{M,c}$ results for different growth temperatures and Ga fluxes confirm that the temperature has the strongest influence on $L_{M,c}$. **Figure 3(g)** depicts a plot of $\ln(L_{M,c}\cdot F^{1/\alpha})$ over $1/T$ for $\alpha = 3.8$ and $\alpha = 6$, where $\alpha = 3.8$ is the exponent, which the calculations (Eq. (2), **Figure 2**) predict for the experimental mask dimensions, and $\alpha \approx 6$ has been found by means of growth kinetic MBE simulations [18] for a critical cluster size of 1 by using different boundary conditions than in the study of Lee et al. [15]. From the slope the activation energy E_{diff} for surface diffusion of a Ga adatom on a SiO_2 surface was determined in the range between ~4.9 eV for $\alpha = 3.8$ and ~6.8 eV for $\alpha = 6$, which significantly exceeds the activation energy for surface diffusion of a Ga adatom on a GaAs surface of ~1 eV [19]. The short diffusion lengths of Ga adatoms on SiO_2 related to the large E_{diff} comply with the small critical mask dimensions observed in the experiments.

Regarding the morphology of the GaAs islands selectively grown in circular mask openings it has been found that the height as well as the shape of the islands varies with the diameter of the openings [17]. With decreasing opening diameter the island height increases, because of the larger mask area and related increased number of Ga atoms diffusing from the mask to the GaAs surface (**Figure 4(a)**). Side-view SEM images show that the cross-sections of the GaAs islands change from a round profile for small opening diameters to trapezoidal shapes for larger diameters (**Figure 4(b)**). In all cases, the total GaAs volume deposited in the mask openings is considerably smaller than the estimated volume of a continuous GaAs layer on a mask-free GaAs surface using the same deposition time, indicating that not all Ga adatoms impinging on the SiO_2 areas migrate to the mask openings. According to the estimated volume of GaAs islands deposited at 630°C on a continuous SiO_2 film a sticking coefficient of Ga atoms on SiO_2 of ~0.13 was estimated [17]. A more recent study found a value of ~0.007 at 632°C and confirmed the expected exponential dependence on the inverse absolute temperature [20].

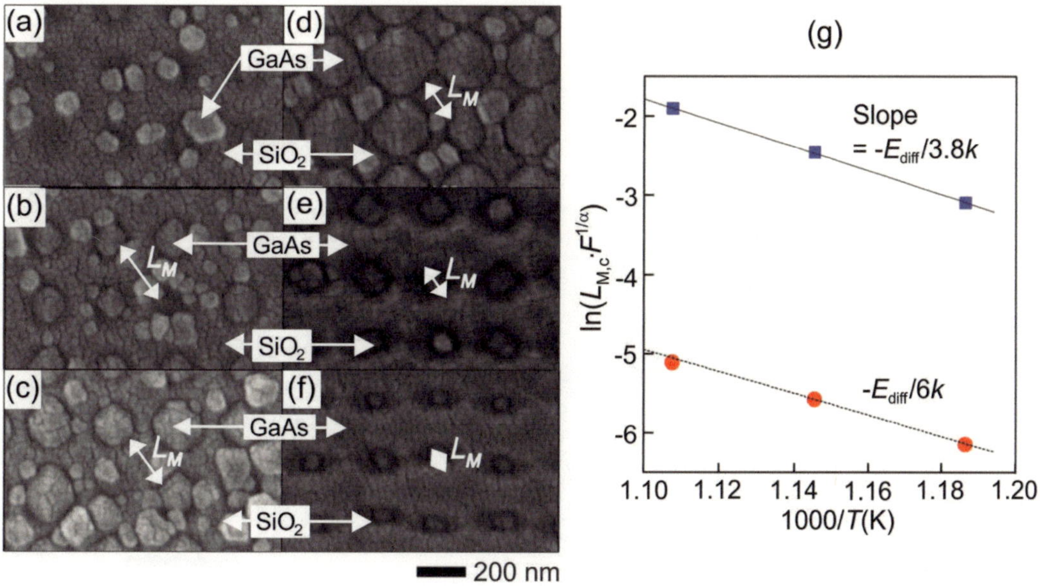

Figure 3. Top-view SEM images showing GaAs grown by MBE at 570°C on GaAs(001) covered with SiO_2 masks having different lateral dimensions L_M. (a) Continuous SiO_2 film, i.e. infinite L_M, (b) $L_M \approx 270$ nm, (c) $L_M \approx 220$ nm, (d) $L_M \approx 180$ nm, (e) $L_M \approx 120$ nm, and (f) $L_M \approx 70$ nm. The periodic features in (b) through (d) are SiO_2 mask openings, while those in (e) and (f) are SiO_2 islands. (g) A plot of $\ln(L_{M,c} \cdot F^{1/\alpha})$ over $1/T$ for NSG of GaAs on SiO_2-nanopatterned GaAs(001) and two different exponent values $\alpha = 3.8$ and $\alpha = 6$. The values of the critical mask dimension $L_{M,c}$, the flux F and the growth temperature T are deduced from MBE growth experiments conducted at three different temperatures. Straight lines are Arrhenius fits. (Reproduced from [15], with the permission of AIP Publishing.)

In the second approach, nanoscale Si islands on a continuous SiO_2 film mask serve as nucleation sites for the heteroepitaxial growth. Such patterned surfaces can be obtained by patterning of silicon-on-insulator (SOI) wafers, i.e. by removing parts of the Si top layer. Zubia et al. studied the metalorganic vapour phase epitaxy (MOVPE) growth of GaAs on square arrays of SOI islands with diameters between 100 and 280 nm, and a pitch of 500 nm (**Figure 5(a)**) [21, 22].

During the initial growth stage crystalline GaAs nuclei form on the Si islands, where the number and size of the nuclei strongly depend on the temperature-controlled surface diffusion (**Figure 5(b)**) [21]. For low temperatures multiple GaAs crystals were observed on each Si island, while only a fraction of the Si islands was covered with GaAs for high temperatures. At an optimum temperature of 605°C each Si island has one GaAs crystal on top. The size, shape and structural quality of these GaAs nuclei play an important role for the subsequent lateral growth (performed at optimum temperature). For an initial growth temperature below 605°C various types of defects such as twins appear in the overgrown GaAs, while for 605°C (**Figure 6**) only stacking faults on the {111} planes have been found by transmission electron microscopy (TEM) diffraction [21]. In contrast, threading dislocations represent the dominant defect in GaAs growth on planar Si. The spacings of the Moirée fringes visible in TEM images as well as the diffraction spot spacings consistently indicate a residual lattice strain of 0.4–0.7% in the GaAs [21], i.e. given the GaAs-Si lattice misfit of ~4% the nanoheteroepitaxial GaAs is predominantly relaxed. However, it has not been clarified whether the relaxation originates from misfit dislocations (MDs) at the Si/GaAs heterointerface, from elastic relaxation or from both.

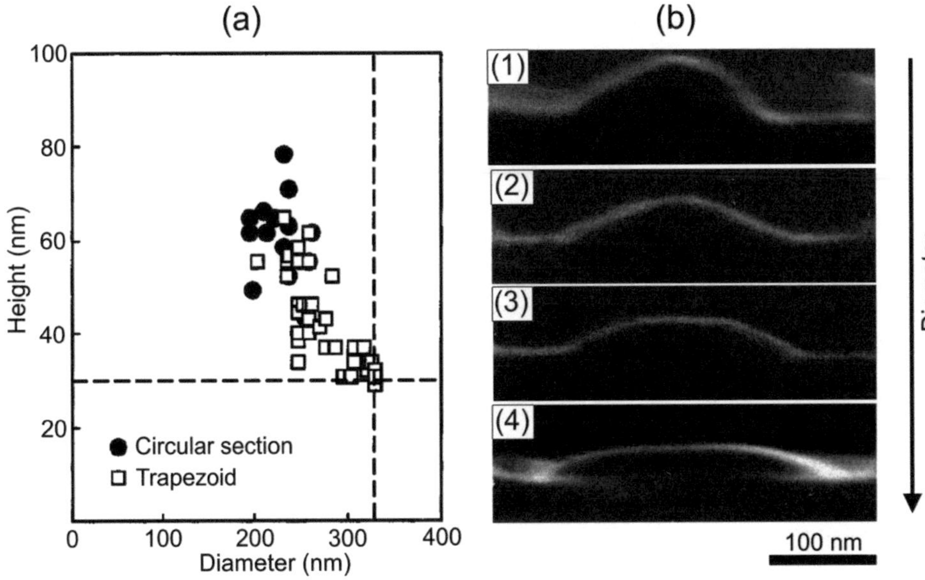

Figure 4. (a) Diagram plotting the height of GaAs islands grown by MBE selectively in SiO_2 mask openings as a function of opening diameter. The vertical and horizontal dotted lines represent the pattern period and the thickness of a GaAs layer on a mask-free GaAs substrate using the same growth parameters, respectively. (b) Side-view SEM images showing the diameter-dependent transition from round to trapezoidal cross-section profiles of the GaAs islands selectively grown in the mask openings. (Reproduced from [17], with the permission of AIP Publishing.)

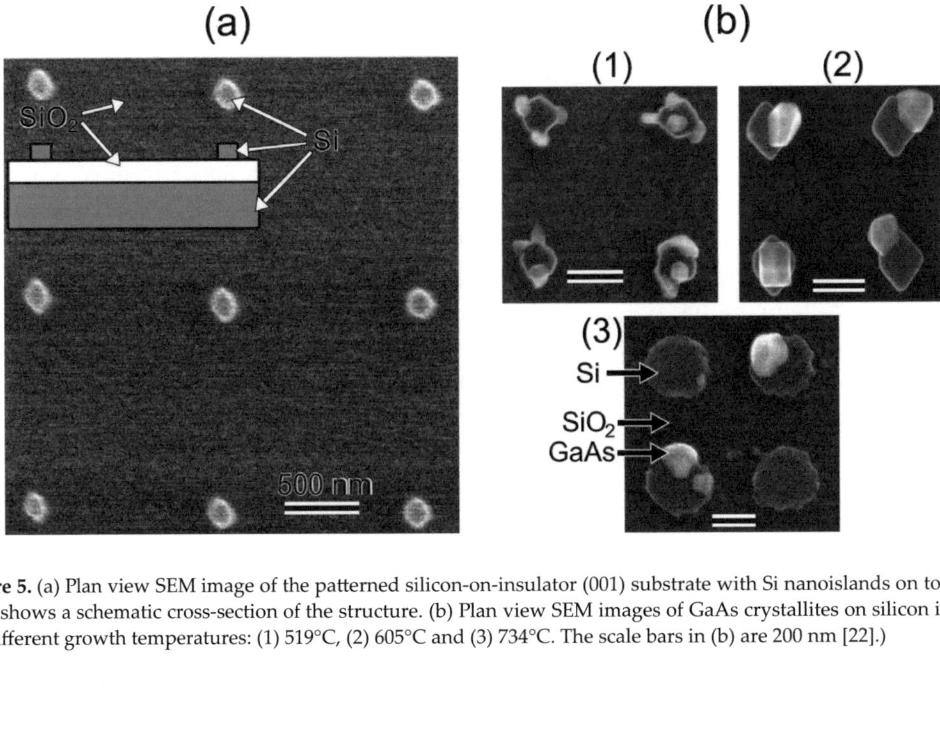

Figure 5. (a) Plan view SEM image of the patterned silicon-on-insulator (001) substrate with Si nanoislands on top. The inset shows a schematic cross-section of the structure. (b) Plan view SEM images of GaAs crystallites on silicon islands for different growth temperatures: (1) 519°C, (2) 605°C and (3) 734°C. The scale bars in (b) are 200 nm [22].)

Apart from nanoscale semiconductor islands also lateral nanowires on a continuous mask film can be used for selective three-dimensional growth of III–V semiconductors. On Si(001) substrate Chin et al. fabricated $Si_{0.35}Ge_{0.65}$-on-SiO_2 structures by deposition of a $Si_{0.85}Ge_{0.15}$ layer onto a SOI substrate and subsequent two-step Ge condensation process [23]. After photolithography and dry etching lateral $Si_{0.35}Ge_{0.65}$ nanowires on SiO_2 were obtained, which served for the study of selective migration enhanced epitaxy (MEE) growth of GaAs onto the nanowires (**Figure 7**). Owing to the pulsed Ga and As fluxes and intermittent annealing phases in MEE, selective growth is achieved even in the case of masks patterned on the micrometre scale. Cross-sectional TEM images reveal that GaAs grows pseudomorphically and defect-free in the form of facetted planar nanowires on top of ~75 nm wide $Si_{0.35}Ge_{0.65}$ nanowires (**Figure 7(c)**). Defects that relieve the strains due to the ~1.5% lattice misfit are absent because both $Si_{0.35}Ge_{0.65}$ and GaAs lattice regions relax elastically in the lateral and vertical directions.

2.1.2. *Lateral overgrowth of oxide mask and layer coalescence*

By means of continued deposition on a mask nanopatterned substrate the selectively formed heteroepitaxial crystals grow and finally coalesce to form a continuous layer. In 1991 Ismail et al. were probably the first who—unintentionally—performed heteroepitaxial overgrowth of a nanometre thin holey SiO_x film [24]. GaAs was deposited by two-step MOVPE growth onto a sawtooth-patterned vicinal Si(001) substrate defined by {111} surface facets. The two-step growth was followed by either an ex situ rapid thermal annealing at 900°C or a thermal cycle growth. In TEM analyses, it was found that a ~1 nm thin SiO_x film covered the Si/GaAs interface. The GaAs layer exhibited an exceptionally low etch pit density of <10^4 cm^{-2} at 0.8 μm distance from the heterointerface, and a dislocation density evaluated from plan-view TEM of

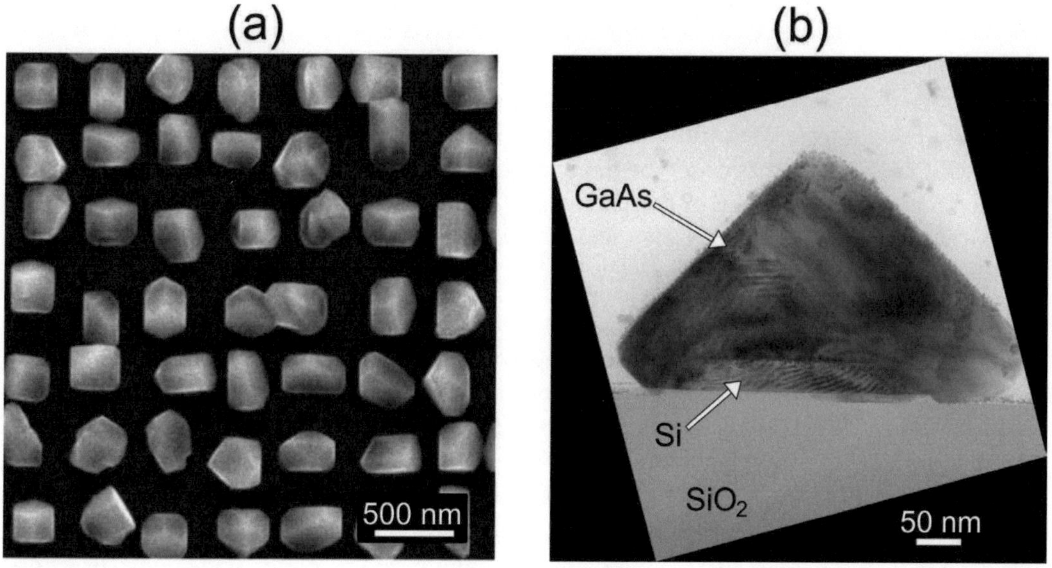

Figure 6. Plan view SEM (a) and cross-sectional TEM image (b) of 100 nm GaAs MOVPE growth on nanopatterned silicon-on-insulator (001) substrate. The GaAs nucleation was performed at 605°C [22].)

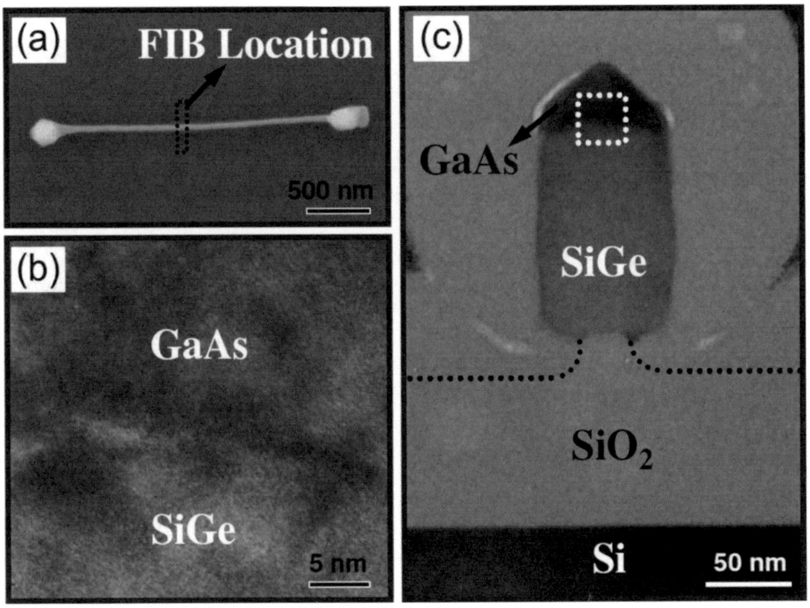

Figure 7. (a) Top-view SEM image depicting a $Si_{0.35}Ge_{0.65}$ nanowire with a GaAs layer grown on top, (b, c) cross-sectional TEM images confirming pseudomorphic, defect-free GaAs on the $Si_{0.35}Ge_{0.65}$. Image (b) covers the region defined by the white dotted rectangle in (c). The TEM specimen was prepared by focussed ion beam at the position marked by an arrow in (a). (Reproduced from [23], with the permission of AIP Publishing.)

$<10^5$ cm^{-2} for 1 μm layer thickness. Based on the limited TEM data, two different mechanisms were considered to account for the occurrence of such a low-defect density: firstly, the orientation of the Si substrate could be transferred to the GaAs layer via Si-GaAs contacts in small pinholes in the interfacial oxide, where MDs formed. Because of the small area fraction of the pinholes, the overall MD density was strongly reduced. The also observed twins and SFs are likely to originate from the roughness of the oxide layer [25]. Secondly, the obtained GaAs layer could have resulted from a type of graphoepitaxy, where the GaAs 'perceives' only the symmetry and the inclination of the {111} facets and not the exact positions of the Si atoms. As the SiO_x film separated the GaAs and Si crystals, relaxed dislocation-free GaAs formed directly above the SiO_x during the final annealing step.

Following a theoretical study on the concept of nanoheteroepitaxy by Zubia and Hersee [26], Hersee et al. reported that the faceting of the forming heteroepitaxial NSG crystals plays a decisive role for the island coalescence and thus for the surface morphology of the coalesced layer [27]. Like in the earlier work of Zubia et al. [21] heterogeneous faceting was observed, which leads to locally different growth rates. To analyse explicitly the layer coalescence behaviour, GaAs and $Al_{0.4}Ga_{0.6}As$ marker layers were grown by MOVPE onto a GaAs(001) substrate nanopatterned with a holey SiO_2 film [27]. After coalescence different growth habits of the layer surface were observed along two orthogonal cleaving directions by using cross-section SEM [27] (**Figure 8**): in one direction a rapid surface planarization occurred, while the growth remained conformal to the substrate geometry in the orthogonal direction. This can provoke again surface roughening in the former direction, which slowly decays with increasing layer thickness. However, still significant large-scale variations of the layer thickness in the order of 100 nm were present even for layer thicknesses in excess of 1 μm.

More recently, the growth of coalesced heteroepitaxial III–V layers on highly mismatched substrates patterned with nanoporous or nanohole oxide masks was investigated in the cases of GaSb on GaAs(001) (misfit ~7.8%) and GaSb on Si(001) (misfit ~12%).

Jha et al. deployed a SiO_2 nanohole mask (hole diameter ~20 nm) fabricated by blockcopolymer lithography and reactive ion etching for MOVPE heteroepitaxy of GaSb on GaAs(001) and reported an improved quality of the GaSb layers [28]. For a layer thickness of 200 nm the width of the (004) X-ray rocking curve decreased to ~40% of that of GaSb grown on non-patterned GaAs. Moreover, the GaSb layer on the nanopatterned GaAs exhibited a strongly reduced surface roughness as measured by atomic force microscopy (AFM), and the absence of threading dislocations as indicated by cross-sectional TEM. However, twins on the {111} planes were observed, which seem to nucleate on the SiO_2 mask.

Nanoheteroepitaxial growth with reduced surface roughness and defect density was demonstrated by Nakamura et al. for even larger misfits, i.e. for the GaSb/Si system [29]. They used ultrathin SiO_2 films (thickness ~0.3 nm) with openings of a high areal density of 10^9 to 10^{12} cm^{-2} as masks on the Si(001) substrate. The growth was performed in three stages by means of MBE (**Figure 9(a)**): In the first stage elastically strain-relaxed GaSb nanodots formed in the SiO_2 openings, followed by nanodot coalescence in the second stage. During the third stage, which was conducted at a higher temperature than stages 1 and 2, the final layer with a flattened surface was obtained. For optimum growth temperatures a root mean square surface roughness of ~7 nm was achieved for a total layer thickness of ~90 nm. Cross-sectional HRTEM images indicated a reduced threading dislocation density of ~10^8–10^9 cm^{-2}

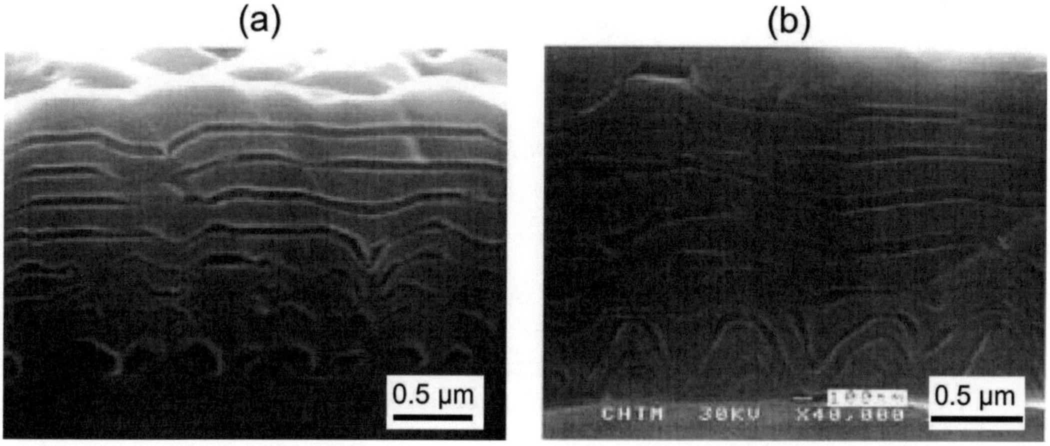

Figure 8. Cross-section SEM images of a coalesced GaAs layer grown on oxide nanopatterned GaAs(001) substrate along two orthogonal cleaving directions (a, b). The darker layers are intentionally grown $Al_{0.4}Ga_{0.6}$ As layers serving as thickness markers. (Copyright 2002 IEEE. Reprinted, with permission, from [27].)

as compared to 10^9 cm^{-2} for a ~1 μm thick GaSb layer grown by using an AlSb initiation layer (**Figure 9(b)**). According to Fourier transformation analysis of HRTEM images the GaSb lattice was almost completely relaxed (**Figure 9(c)** and **(d)**). As, except for the openings, GaSb and Si were separated by the ultrathin SiO_2 layer, the misfit was not accommodated by MDs. Rather, the GaSb lattice planes above the SiO_2 exhibited close-to-equilibrium spacings, i.e. ~12% wider spacings than the corresponding Si planes beneath the oxide (**Figure 9(e)**), which can be attributed to the elastic relaxation in the nanodots during the growth. The direct Si/GaSb contacts in the openings, but also the very thin oxide layer enable a transfer of the epitaxial relationship.

Heteroepitaxial growth and continuous layer formation has also been studied by using grating-shape nanopatterned oxide masks. He et al. [30] applied a three-stage MOVPE growth scheme to obtain a smooth GaAs layer surface on a Si(001) substrate: the first stage served for selective deposition of GaAs in the trenches, while in the second and third stages coalescence and planarization of the GaAs layer were achieved. Repetition of the experiment for different trench orientations showed that this has a strong impact on the coalescence and consequently on the surface roughness of the layer. Minimum surface roughness was observed for a trench orientation parallel to [410], i.e. ~31° inclined to the [110] direction, due to maximum and uniform lateral growth rate. **Figure 10** compares the GaAs layer morphology for different trench orientations on the basis of AFM and cross-sectional SEM images.

Another approach compassing nanoscale selective growth and filtering of defects relies on MBE growth through the openings of a silica colloidal monolayer [31]. First, a GaAs buffer layer is grown on a planar Si(001) substrate, partially relieving the misfit by formation of extended defects. The subsequently deposited silica sphere monolayer enables selective growth of GaAs in the monolayer openings on the GaAs buffer surface (**Figure 11**). Due to the conical shape of the openings, propagating defects are largely stopped at the silica surfaces.

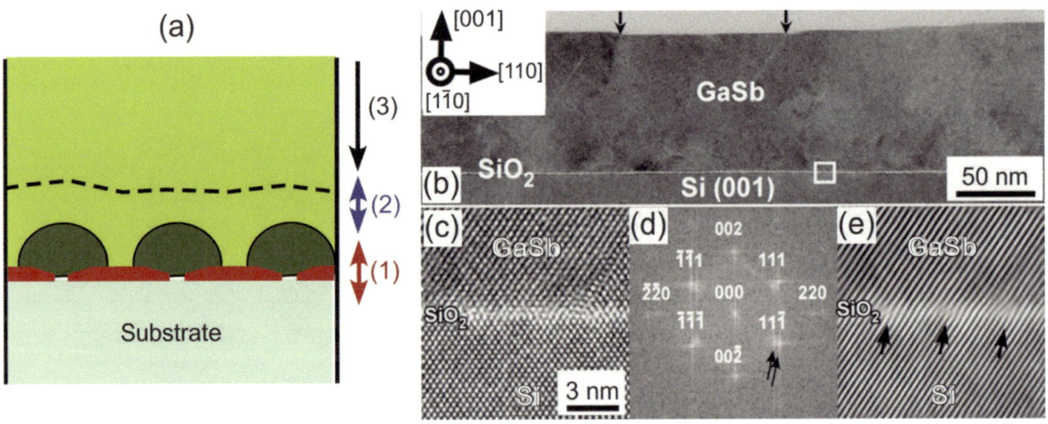

Figure 9. (a) Schematic of the three-step MBE growth procedure, (b) cross-sectional HRTEM image of a GaSb layer on Si(001) substrate obtained by the procedure of (a), (c) Detail of (b) showing the marked square area in (b), (d) Fourier transform of (c), (e) inverse Fourier transform using the (11$\bar{1}$) and ($\bar{1}\bar{1}$1) lattice planes.

When the growing GaAs surface has passed half the height of the spheres elastic relaxation in the lateral directions contributes to reduce the misfit strains.

2.1.3. Aspect ratio-trapping techniques

Although three-dimensional elastic relaxation of the lattice relieves the misfit strains during lateral overgrowth of thin masks, defects at the heterointerface may still form threading segments, which extend far into the layer. In order to stop the propagation of such defects ART techniques have been developed, where the growth areas are situated at the bottom of cylindrical or trench shaped mask openings whose height exceeds their lateral width. Therefore, defects that propagate in the layer on planes inclined to the substrate surface, such as {111} slip planes in the case of a {001} substrate, get trapped at the intersections of the slip planes with the mask walls and are hence hindered to further spread into the layer. Essentially, the lattice undergoes a plastic relaxation where the defects are confined to a region close to the heterointerface and mask sidewall regions. Owing to the efficient plastic relaxation and considerable mask height the layer is nearly completely relaxed when reaching the mask top so that elastic relaxation during mask overgrowth can be neglected.

2.1.3.1. Growth on surfaces patterned with a trench mask

Among the ART techniques those deploying SiO_2 or SiN_x masks with trench shaped openings along [110] or $[1\bar{1}0]$ have been most extensively studied. The trench masks can be fabricated (i) by oxidation of the Si substrate surface, lithography and reactive ion etching, or alternatively (ii) by using processes based on the established shallow trench isolation technology [32]. In particular, selective-area MOVPE growth of InP and GaAs on trench mask patterned Si(001) substrate has attracted great interest [25, 33–43]. Although the detailed defect reduction mechanism depends on the trench dimensions, on the shape of the trench bottom as well as on the heteroepitaxial growth sequence and conditions, the common characteristic

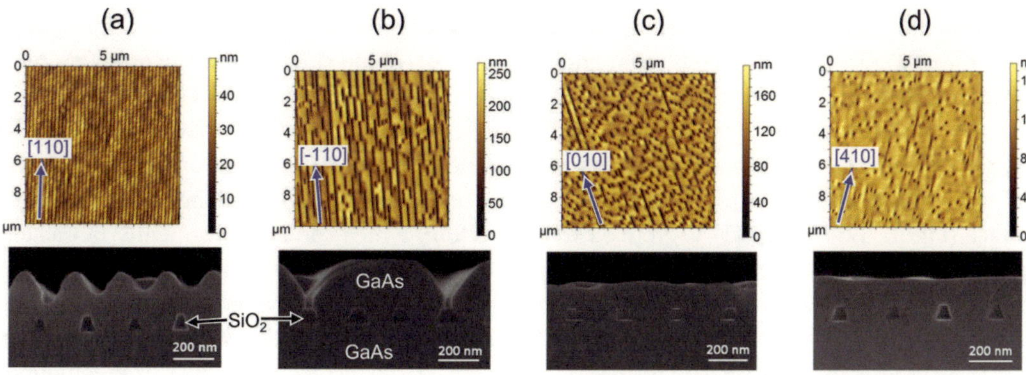

Figure 10. AFM images (top row) and cross-sectional SEM images (bottom row) of GaAs layers after the second MOVPE growth stage on SiO_2 stripe patterned Si(001), with stripe orientations: (a) [110], (b) $[\bar{1}10]$, (c) [010] and (d) [410]. (Reproduced from [30], with the permission of AIP Publishing.)

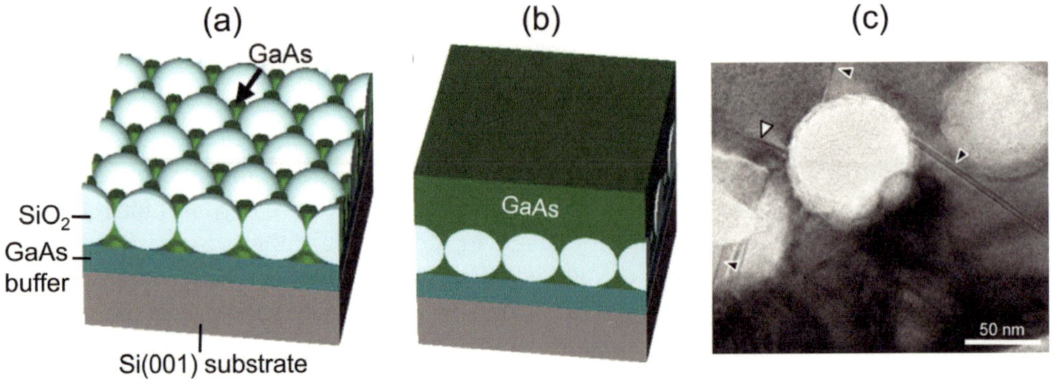

Figure 11. (a, b) Schematics illustrating the growth of a GaAs layer onto a SiO_2 nanosphere layer covered Si(001)/GaAs buffer template, (c) cross-section TEM image corresponding to (b). The black arrowheads and the white arrowhead point to {111} stacking faults and to a defect off from {111}, respectively. (Reprinted (adapted) with permission from [31]. Copyright 2011 American Chemical Society.)

of trench-based ART techniques consists in the stopping of propagating misfit defects, i.e. dislocations, stacking faults and twins, at the trench sidewalls. This applies to defects on {111} planes parallel to the trench direction. However, threading defects running on {111} planes inclined to the trenches are not trapped and can cross the layer (**Figure 12**). XRD ω-2θ scans of the (004) reflection demonstrated that the selectively grown InP and GaAs regions are almost fully relaxed [36, 40]. Obviously, the aspect ratio of the trenches (height divided by the width) determines the fraction of threading defects that are trapped by the sidewalls. For aspect ratios larger than 2 misfit defects are confined to the lower part of the trenches, whereas the upper part has a high crystal quality with an etch pit density close to zero [43].

One of the key factors for the formation and trapping efficiency of defects is the shape of the trench bottom. Orzali et al. found that in the case of a recess in the Si substrate defined by a Si(001) bottom surface and Si{111} facets in the vicinity of the sidewalls MOVPE GaAs nucleated preferentially on the {111} facets, thus forming a central void after growth front coalescence [25]. SFs and TDs originating from the trench bottom annihilated at the oxide sidewalls, while only few twin lamellae crossed the layer and reached the surface. Merckling et al., who used a Ge buffer layer in a 'round' Si recess defined by {111}, {112} and (001) facets, evidenced a MD network at the Si/Ge interface (**Figure 12(a)**). Threading defects on the trench parallel {111} planes are stopped by the SiO_2 sidewalls (**Figure 12(a)**), whereas some defects on the inclined {111} planes reach the layer surface (**Figure 12(d)**). A {111} V grooved trench bottom, which can be conveniently fabricated by anisotropic wet etching of the Si substrate, permits a further reduction of the defect density in the III–V layer, for several reasons: firstly, the {111} surface, especially when covered with an As monolayer, energetically facilitates the nucleation of III–V layers [25]. Secondly, single steps on Si{111} surfaces do not induce APB formation [44]. Thirdly, the incorporation and arrangement of MDs in the {111} heterointerface is alleviated, because dislocations can glide on the {111} planes. In the case of MOVPE GaAs deposited in such trenches propagating SFs and TDs are confined to the lower part of the trench, leaving the upper part defect-free with very few twins [25].

http://dx.doi.org/10.5772/67572

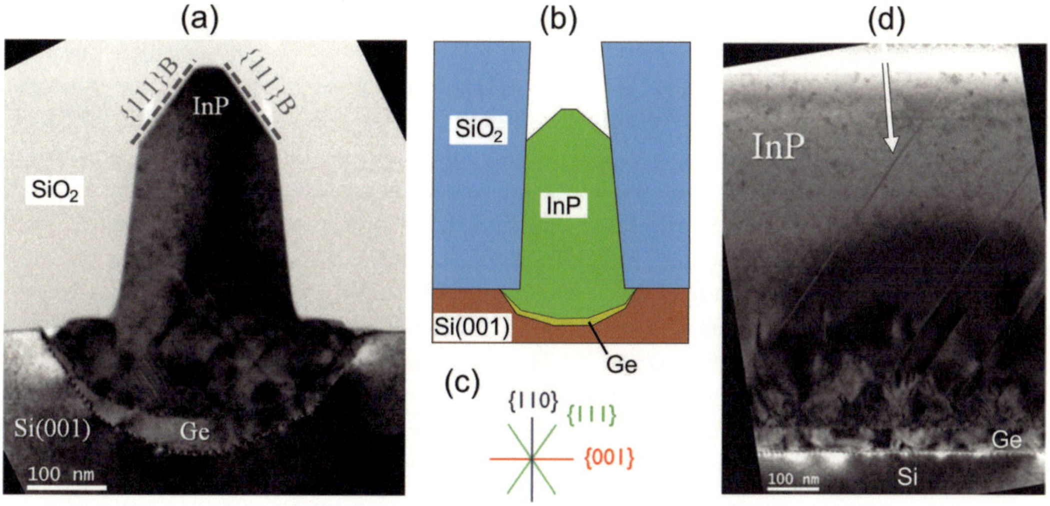

Figure 12. (a, d) Cross-section TEM images of InP/Ge/Si(001) heterostructures (Ge: buffer layer) in 150 nm wide in SiO_2 trenches with viewing direction (a) parallel and (d) perpendicular to the trenches, (b) schematic of the structure, dashed lines indicate threading defects, (c) orientation of lattice planes in (a–c). In (a) the vertical surfaces of the SiO_2 mask are invisible. In (d) the white arrow marks a twin lamella crossing the entire layer. (Reproduced from [36], with the permission of AIP Publishing.)

As shown by XRD pole figure analysis for the InP/Si(001) system the defects are distributed anisotropically on the {111} planes, i.e. more defects populate the $(\bar{1}11)$ and $(1\bar{1}1)$ planes parallel to the trench bottom faces (for trenches along [110]) than the inclined $(11\bar{1})$ and (111) planes [37]. This finding agrees with TEM analyses, which prove the presence of a dense nanotwin network within 10 nm distance from the trench-parallel Si/InP{111} interfaces, effecting the lattice relaxation. On the other inclined {111} planes stacking faults and twins propagate, which extend from the Si/InP interface towards the layer top surface, where they could limit the performance of devices. Regarding the determination of defect densities of heteroepitaxial nanostructures the width of XRD rocking curves provides qualitative data useful for the comparison of samples having the same layer thickness, as the X-ray penetration depth is in the μm range. For the quantification of defect densities plan-view TEM of larger areas represents the most reliable technique, which yields a density of layer-crossing twin lamellae in the 10^8–10^9 cm^{-2} range for GaAs fins grown in {111} V grooved trenches on Si(001) [25].

Furthermore, it has been found that the surface morphology of InP layers deposited selectively in trenches critically depends on the MOVPE growth conditions, especially on temperature and pressure [36, 38]. These parameters determine the precursor/adatom diffusivity and the heat transfer from the substrate to its surroundings, thus the growth rates of different facets. Before starting layer growth, a high temperature annealing step in the H_2 carrier gas is helpful to remove surface oxide layers and to promote the formation of double steps in the case of a Si(001) surface, which prevents the development of APBs [45]. Concerning the main growth step similar, but slightly different layer morphologies have been reported by different groups, depending on the structure of the patterned samples and the growth setup. By using phosphine as group-V precursor, a SiN_x trench mask with rounded rectangular recess in the Si(001) substrate and GaP and InP low-temperature buffer layers Lee et al. obtained wedge-shaped InP ridges bounded by {111} facets for a high growth temperature of 650°C (**Figure 13(a)**), whereas an atomically smooth InP(001) top surface resulted for 550°C (**Figure 13(b)**) [38]. In the latter case, GaP islands appeared on the Si sidewalls and in the course of continued InP deposition the lateral growth led to coalescence of opposing sidewall deposits before the recess was fully filled (**Figure 13(c)**). Consequently, a void together with a (001) growth surface was formed. Merckling et al. in contrast deployed a SiO_2 trench mask with shallow curved recess in the Si(001) substrate, a Ge buffer layer with As termination for improved InP wetting, and used a two-stage InP deposition and tert-butylarsine and tert-butylphosphine as group-V precursors [36]. These authors reported highest uniformity of the ridge-shaped crystals bounded by {111} planes for a growth temperature of 550°C. At higher temperatures large three-dimensional islands protruding from the trenches were observed while at lower temperatures non-uniform faceting occurred due to a large density of defects.

Clear evidence for the correlation between crystal defects and surface morphology has been established Orzali et al. for GaAs grown by MOVPE on SiO_2 trench-patterned Si(001) [25]. Analysis of high-resolution STEM images reveals that single twins intersecting the surface give rise to steps, whereas twins that intersect beneath the surface produce surface pits (**Figure 14**).

In order to keep the defect density low, two distinct types of buffer layers are frequently used in combination with the trenches. On the one hand, buffer layers of a different material with a lattice parameter between those of the substrate and of the layer deposited on the trench bottom reduce the effective misfit. For the InP/Si system (i) Ge with a Ge/InP misfit of ~4% [33, 36] (**Figure 12**) and (ii) GaP avoiding the polarity mismatch at the InP/GaP interface [38] (**Figure 13**) have been used as buffer materials. On the other hand, buffer layers of the layer material grown at low temperature help to achieve a relatively smooth layer surface, since at low temperature the reduced Ga adatom diffusivity promotes the formation of a large density of small nucleation islands, which rapidly coalesce. Low temperature buffer layers contain a lot of defects leading to a relief of misfit strains and to an improved crystal quality in the following homoepitaxial layer grown at higher temperatures.

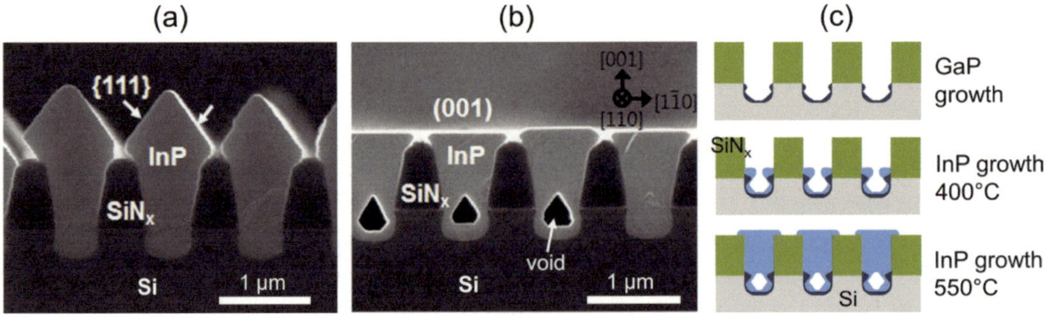

Figure 13. Cross-sectional SEM images of InP layer on SiN_x trench patterned Si(001) for a growth temperature of (a) 650°C and (b) 550°C. (c) Schematic growth evolution for (b). (Reprinted from [38], Copyright 2015, with permission from Elsevier.)

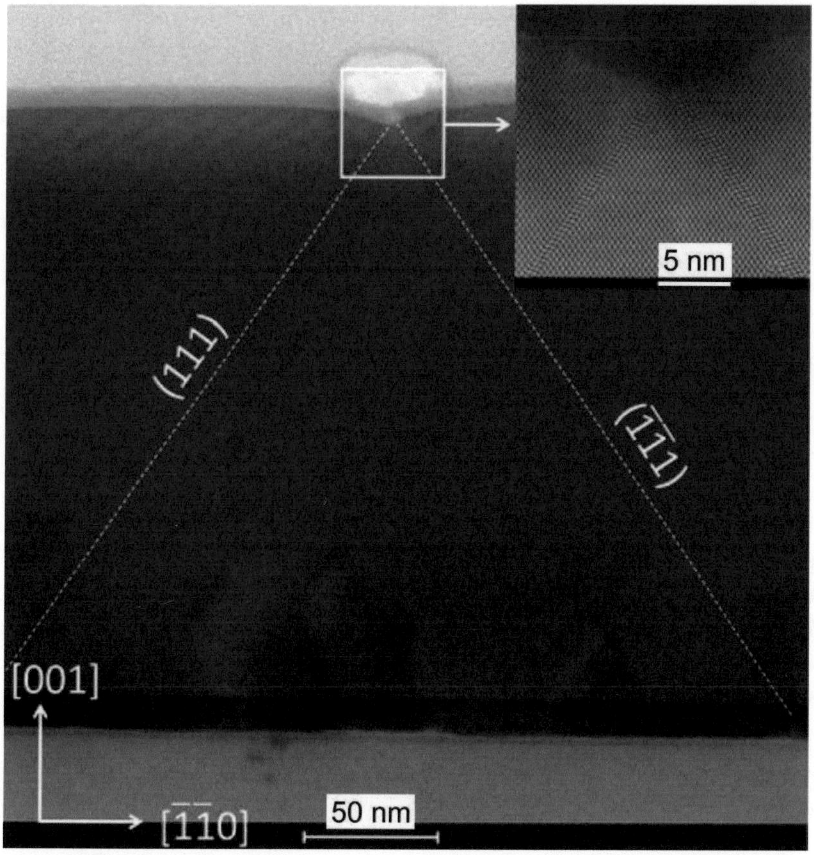

Figure 14. Bright-field STEM image of a GaAs layer in a SiO_2 trench on Si(001) depicting a shallow surface pit with twin lamellae intersecting underneath. The viewing direction is perpendicular to the trench direction. Inset: High magnification high-angle annular dark field STEM image of the pit. (Reproduced from [25], with the permission of AIP Publishing.)

After prolonged growth on trench mask patterned substrates the III–V layer grows above the mask and coalesces to form a continuous layer, where the coalescence often induces the development of further defects. In the case of GaAs MOVPE growth over a SiO_2 trench mask Li et al. observed {111} twins and stacking faults initiating at the SiO_2 top surface [46]. By using optimized MOVPE growth parameters for the coalescence stage, i.e. a slightly lower temperature and a higher V/III ratio as compared to the growth in the trenches the density of these defects was reduced. In this way, Li et al. obtained a high-quality GaAs layer on patterned Si(001), however with considerable large-scale thickness variations.

Very recently, it has been demonstrated that GaAs nanoridge arrays grown by the ART technique on exactly oriented Si(001) substrate can be used after removal of the SiO_2 mask as a GaAs-on-Si compliant substrate for a coalescence growth step in order to obtain APB-free, low-defect GaAs layers with flat (001) surfaces [47]. In the SiO_2 mask trenches V-grooved {111} bottom faces were generated by KOH etching (**Figure 15**). The SiO_2 walls were underetched, and beneath the SiO_2 walls a curved Si surface formed, which undertakes the task of defect trapping instead of the SiO_2 walls. Remarkably, during the coalescence growth step no defects originated from the top of the Si ridges in between neighbouring V grooves, as seen in TEM images (**Figure 15**). Furthermore, the TEM analysis shows that the misfit was accommodated by a large density of SFs within the first few nanometres from the Si/GaAs interface. This method has the advantages that no new defects appear in the coalesced layer in the course of oxide overgrowth and that the resulting layer surface has a low root-mean-square roughness of 1.9 nm measured in a 5×5 μm² scan area for a layer thickness of only ~300 nm.

2.1.3.2. Growth on surfaces patterned with a cylinder hole mask

In contrast to trench masks cylindrical mask openings allow to trap misfit defects running on all lattice planes that are inclined against the substrate normal. Despite this benefit, ART with cylinder hole masks has been investigated only by few research groups [48–50]. Hsu et al. [48] reported the MOVPE growth of smooth continuous 900 nm thick GaAs layers of low defect density on Si(001) substrate patterned with a round nanohole SiO_2 mask having a hole diameter of 55 nm and aspect ratio of 4.7. Like in the case of ART trenches, threading misfit defects were confined to the lower part of the holes, giving monocrystalline, relaxed GaAs in the upper part and a very low etch pit density of 3.3×10^5 cm^{-2} of the coalesced continuous GaAs layer. Application of a two-stage growth scheme (nucleation layer at low temperature, main growth at higher temperature) allowed to avoid thermal tensile strains in GaAs arising from the thermal expansion mismatch between Si and GaAs, as observed in GaAs layers on planar Si substrates by a redshift of the photoluminescence bandgap emission energy. However, the surface morphology of the coalesced GaAs layer and its defect structure above the SiO_2 holes has not been addressed in the work of Hsu et al. [48].

Chang et al., who studied the selective MBE growth of GaAs on SiO_2 nanohole patterned Si(001) found that the top surface of the GaAs nanostubs was not planar but bounded by inclined facets [50]. These authors stated that SFs starting near the Si/GaAs interface originated from disorders in the stacking sequence during merging of GaAs nuclei, owing to

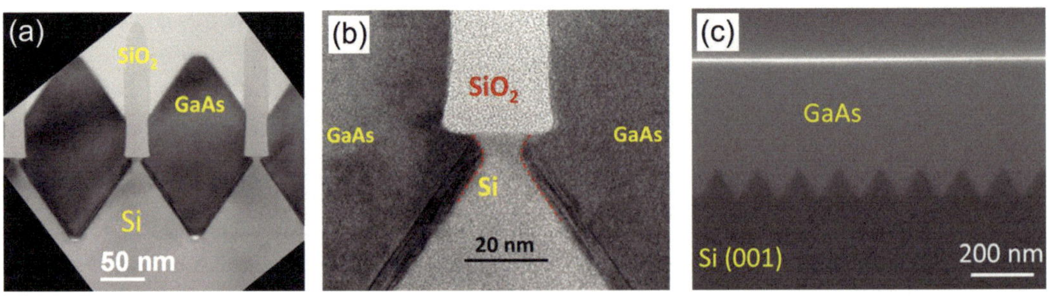

Figure 15. (a, b) Cross-section TEM images of GaAs nanoridges grown by the ART technique on Si(001). The zoomed-in TEM image (b) reveals SF trapping by the curved Si undercuts. (c) Cross-section SEM image of a coalesced GaAs layer obtained from the nanoridges shown in (a, b) by overgrowth after SiO_2 mask removal. (Reproduced from [47], with the permission of AIP Publishing.)

different Si surface planes generated by the reactive ion etching of the holes. In contrast, SFs originating at the SiO_2 sidewalls formed due to local stresses in the GaAs induced by SiO_2 surface roughness.

Although Si(111) does not constitute the main substrate in complementary metal oxide semiconductor technology, a method for the growth of small GaAs layer volumes on Si(111) having structural and morphological characteristics highly attractive for applications has been developed: Chu et al. [49] deployed a two-stage MBE growth to obtain monocrystalline GaAs with atomically smooth surface and low defect density in the openings of a SiO_2 hole mask on Si(111). The growth mechanism and thus the properties of the resulting layer can be understood as follows (**Figure 16(a)**): During the first low-temperature growth stage numerous small GaAs islands form by homogeneous nucleation, which then coalesce. During the temperature ramping from the lower to the higher temperature used in the second growth stage the nucleation layer regrows epitaxially and, as a consequence of the thermal expansion coefficient mismatch between Si and GaAs, MDs are introduced at the GaAs/Si interface. As revealed by cross-section TEM analyses, these dislocations are confined to within ~3 nm from the GaAs/Si interface (**Figure 16(c)**). In the second stage growth proceeded in a layer-by-layer fashion, which yielded monocrystalline GaAs of high quality and low roughness (**Figure 16(b)**).

2.2. Defect filtering by multiple quantum dot layers

Another approach to enhance the quality of III–V semiconductor layers grown on mismatched substrates, exploits multiple layers of QDs, which act as a filter for threading defects [51]. QDs form by self-organized Stranski-Krastanov growth of a semiconductor material which shows (i) good wettability and (ii) considerable lattice misfit to the host crystal on which it is deposited. Typically, the QD layers are not grown directly on the substrate, but on a low-temperature buffer of the III–V layer material to be deposited (**Figure 17(a)**). In this way, the QD layers hinder the threading defects coming from the plastically relaxed buffer in extending to the subsequently grown III–V layer. Moreover, misfit strain in the vicinity of the QD layer originates only from the misfit between the QD and the III–V layer material, and not additionally from misfit between different lower and upper layer materials. The defect blocking effect has been explained by a deflection of the defect propagation induced by the strain fields of the QDs [52, 53]. In these strain fields dislocations experience shear Peach-Koehler forces, which lead to a sideward deflection so that the dislocation is redirected towards the material edges or subjected to annihilation before reaching the surface (**Figure 17(a)**). Calculations based on continuum elasticity predict increasing effectiveness of a single QD layer to bend TDs with increasing QD base area, but only a weak increase with QD height [52]. Besides the QD base area, the dislocation bending effectiveness is enhanced by using a larger number of QD layers. However, in the case of too large QDs or too large number of QD layers strain accumulates and gives rise to MD formation. Therefore, an optimum layer quality is expected for a QD size and number of QD layers just below the critical values for dislocation formation. For the Si/GaAs system, Yang et al. reported that a defect filter layer consisting of ten InAs QD layers with a QD width of 20–30 nm separated by 50 nm spacer layers on 2 μm GaAs buffer layer

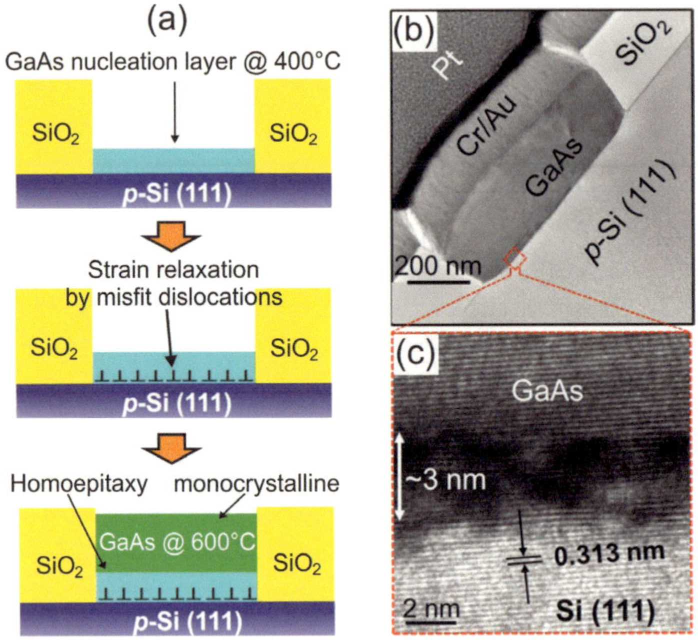

Figure 16. (a) Schematic of the two-stage MBE growth on SiO_2 hole mask patterned Si(111), (b) cross-section TEM image of the Si/GaAs heterostructure fabricated by the procedure of (a), (c) HRTEM image from the marked area in (b). (Reprinted (adapted) with permission from [49]. Copyright 2014 American Chemical Society.)

was most effective [52]. Their TEM analysis demonstrated that 60° dislocations were bent to the side (i.e. the dislocation labelled 'B' in **Figure 17(b)**), and moreover, that pure edge dislocations were stopped at the interface between the GaAs spacer layer and the QD (the dislocation labelled 'C' in **Figure 17(b)**). The detailed mechanism of this blocking of edge dislocations is not well understood. Apart from dislocation filtering, QD layers also contribute to reduce the density of point defects in the layer, because the QD strain field imposes a driving force for point defects with compressive (dilatative) strain to migrate to elastically dilated (compressed) lattice regions. Further reduction of the defect density, measured as the etch pit density on the GaAs layer surface, was enabled by a combined application of QD dislocation filter layers and a buffer layer grown in three stages, starting with low temperature, followed by intermediate and eventually high-temperature growth [54]. However, the etch pit density of 9×10^5 cm^{-2} resulting for a GaAs top layer thickness of 1 μm [54] still exceeds the value achieved with ART growth of GaAs using a SiO_2 cylinder hole mask [48] (Section 2.1.3.2).

2.3. Growth on mask-free nanopatterned surfaces

Although the growth on mask-free nanopatterned surfaces has been frequently used for site controlled fabrication of semiconductor QDs or nanoislands [55], it is also beneficial for improving the quality of continuous III–V layers [56–59]. Here, 'nanopatterned' means that the substrate surface exhibits ordered or random topographical features with

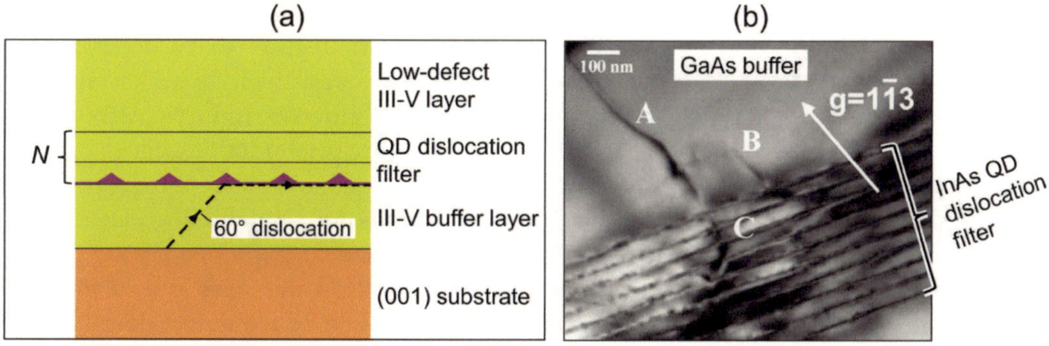

Figure 17. (a) Schematic cross-section of a heterostructure, in which multiple QD layers are deployed as dislocation filter, (b) cross-section TEM image using (1$\bar{1}$3) diffracting planes, showing the bending or blocking effect of a ten-fold InAs QD layer stack on threading dislocations (A, B, C) in GaAs grown on Si(001) substrate. ((b) (Copyright 2007 IEEE. Reprinted, with permission, from [52].)

dimensions well below 1 μm. In general, the insufficient blocking of threading defect propagation in the growth on mask-free nanopatterned surfaces leads to a lower increase of the layer quality as compared to the heteroepitaxy approaches based on: (i) selective growth on mask-patterned substrates (Section 2.1) and (ii) on growth using QD dislocation filter layers (Section 2.2). Nevertheless, growth on mask-free nanopatterned surfaces is interesting for reducing defect densities in III–V layers because of the simplicity and low cost of the processes needed for the substrate nanopatterning. Essentially, the active defect elimination mechanisms originating from the reduced growth area size and the geometry of the substrate surface rely on three-dimensional elastic lattice relaxation of the substrate and/or the layer as well as on defect trapping at steep or concave topographic features.

So far, morphologically different surfaces fabricated by different methods have been deployed in studies of the quality of III–V layers grown on mask-free nanopatterned surfaces. One possibility to create nanoroughened surfaces is *in-situ* annealing of Si(001) in H_2/O_2 ambient in the MOVPE reactor, which effects the formation of volatile Si compounds such as SiO and thus a nanoscale roughening (pitting) of the Si surface. By using spectroscopic polarimetry Liu et al. [56] showed that polar GaP deposited by MOVPE on non-polar nanoroughened vicinal Si(001) (misfit ~0.4%) formed continuous layers, and attributed the continuous layer growth to a large number of nucleation sites owing to the higher reactivity between nanoroughened Si and the TMGa precursor, whose mobility was seen to decrease.

Nohavica et al. [57] performed liquid phase epitaxy growth of InAs on electrochemically etched micro- to nanoporous InP substrates (misfit ~3.2%). Due to the enhanced elastic relaxation of misfit strains an improved InAs layer morphology and a 20–35% reduction of the dislocation density of etched surfaces were obtained as compared to an InAs layer grown on a planar InP substrate. Similarly, the same group reported MOVPE growth and characterization of $In_xGa_{1-x}As$ layers ($0.2 \le x \le 1$) on electrochemically etched nanoporous GaAs(001) substrates [58]. Such nanoporous GaAs surfaces showed low roughness areas between the pore openings and are therefore suitable for nucleation of a layer on these small smooth areas. SEM images revealed a relatively smooth surface of a deposited 0.4 µm thick $In_{0.2}Ga_{0.8}As$ layer (**Figure 18(b)**). However, the mosaic-shaped pattern visible in top-view SEM images indicated the presence of grains (**Figure 18(a)**). InAs layers grown on nanoporous GaAs exhibited enhanced low-temperature photoluminescence emission as compared to an InAs layer of same thickness grown on planar substrate (**Figure 18(c)**). As crystal defects boost non-radiative electronic transitions, the photoluminescence intensity represents an indicator of the crystal quality. The observed improved crystal quality can be assigned to elastic relaxation of misfit strains both in the layer and in the substrate during lateral pore overgrowth leading to a reduction of the defect density. Due to the high porosity of the GaAs substrate of ~50% compliancy of the substrate lattice is expected.

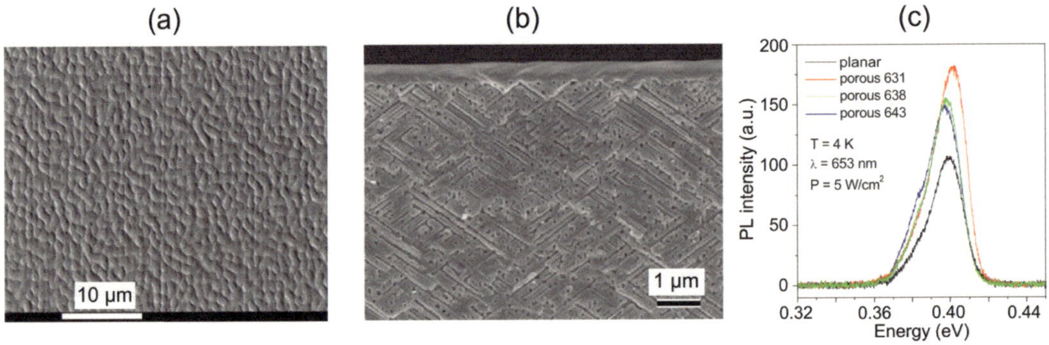

Figure 18. Plan view (a) and cross-section (b) SEM images of a 0.4 µm thick $In_{0.2}Ga_{0.8}As$ layer on nanoporous GaAs(001) substrate (misfit ~1.4%), (c) photoluminescence spectra of 1.2 µm thick InAs layers grown on planar and on porous GaAs(001) substrates with different pore layer thicknesses (610–631, 635–638, 641–643 µm). (Reprinted from [58], Copyright 2013, with permission from Elsevier.)

3. Relevance of nanoheteroepitaxial layers for advanced optoelectronic applications

It has been recognized long ago that excellent crystalline quality is essential for a high performance of semiconductor power and optoelectronic devices. In particular, minority carrier devices such as insulated-gate bipolar transistors, as well as light-emitting and absorbing devices require dislocation densities below 10^5 cm^{-2} [24, 60], which has not been achieved by traditional methods of defect reduction in planar III–V heteroepitaxy on the common Si platform. However, in the last 10 years, substantial progress has been made in the heteroepitaxial growth of low-defect, mismatched semiconductors by using nanopatterned substrates, which allowed to accomplish a good crystalline quality of III–V layers close to that needed for long-lifetime devices. Recently, these nanoheteroepitaxy methods have been exploited for the realization of several III–V devices integrated on a Si platform. Examples include laser diodes [59, 61], GaAs solar cells [49], and GaAs/$In_xGa_{1-x}As$ tunnel diodes for digital circuits applied e.g. in inverters or random-access memory cells [62, 63]. Further applications have been proposed such as In-rich $In_xGa_{1-x}As$ channels for n-type metal oxide semiconductor field effect transistors with enhanced electrical characteristics [64] or GaAs waveguides for lasers [42]. Exemplarily, we will briefly describe the fabrication and performance of two such unique devices, one using oxide-separated, disc-shaped III–V volumes as the active region of a solar cell, and one exploiting the V-groove defect reduction technique as the basis for a QD micro-disc laser structure.

3.1. GaAs nanodisc *p-i-n* solar cell on Si(111) nanopatterned with a SiO_2 hole mask

A prototype GaAs nanodisc *p-i-n* solar cell was fabricated on p-doped Si(111) substrate by using two-stage MBE growth for selective deposition of an intrinsic GaAs buffer layer (nucleation layer: 25 nm, main layer: 150 nm) in the circular openings of a SiO_2 hole mask, as described in Section 2.1.3.2 [50]. On the (111) surface of the buffer layer 50 nm of strongly n-doped (n+) GaAs was grown. To complete the device, indium tin oxide (ITO) and indium contacts were deposited on the top GaAs and bottom Si, respectively (**Figure 19(a)**). In the dark, the current density-voltage characteristic of the solar cell showed a good rectification behaviour with a current ratio > 10^2 for ±1 V bias (**Figure 19(b)**). Under illumination of one sun AM 1.5G a short circuit current of 18.4 mA/cm^2, an open circuit voltage of 0.18 V, an ideality factor n = 1.6 and a fill factor of 28% were obtained. The resulting energy conversion efficiency of 0.9% is similar to the values reported for nanostructured solar cells, but much lower than those for conventional Si solar cells. The poor conversion efficiency and fill factor can be explained by surface states at the GaAs/SiO_2 interface and carrier-trapping dislocations at the Si/GaAs interface. It is expected that these characteristic values can be enhanced by suitable SiO_2 surface passivation and improved contacts.

3.2. QD microdisc laser using nanoheteroepitaxial buffer on Si(001)

GaAs buffer layers on V-grooved exact-oriented Si(001) templates have been used to fabricate InAs/$In_xGa_{1-x}As$ dot-in-a-well microdisc laser structures emitting at ~1.2 μm wavelength with almost similar characteristics as devices fabricated on native GaAs substrate [59]. For the growth of the low-defect GaAs buffer layer a two-step deposition was used starting with selective growth in SiO_2 trenches followed by SiO_2 removal and the mask-free GaAs coalescence growth step by MOVPE, as described in Section 2.1.3.1. The active region comprised five InAs QD layers embedded in $In_xGa_{1-x}As$ quantum wells separated by GaAs spacer layers, respectively, grown by MBE (**Figure 20(b)**). **Figure 20(a)** depicts a cross-section TEM image of the as-deposited layer structure. In order to obtain the micro-discs (**Figure 20(c)**) colloidal lithography and reactive ion etching were deployed.

Optical emission spectra were measured with a micro-photoluminescence setup in surface-normal pump/collection configuration under continuous wave excitation at 10 K. A low average threshold power of 50 μW has been achieved for 1 μm disc diameter, as compared to 33 μW for

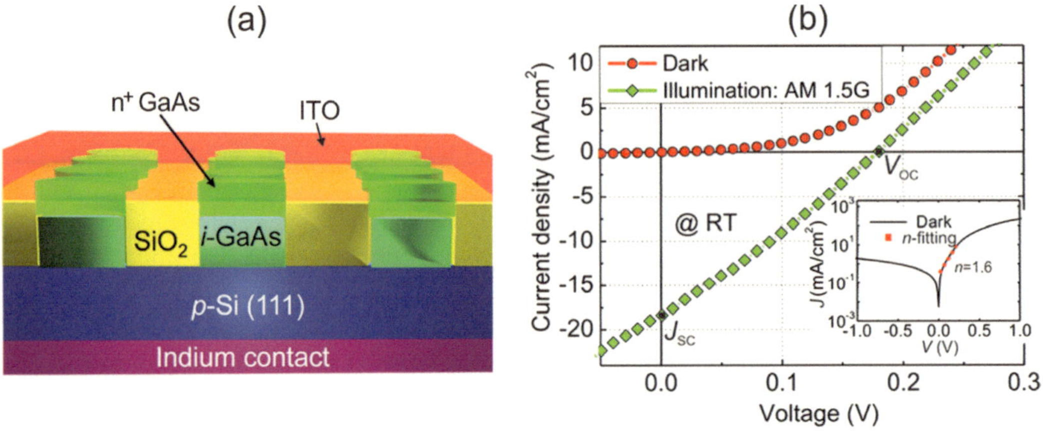

Figure 19. (a) Schematic cross-section of the p-i-n solar cell grown on Si(111), (b) current density-voltage characteristics of the device in the dark and under illumination of one sun AM 1.5G. The inset shows a semilogarithmic current density-voltage curve for the non-illuminated state. (Reprinted (adapted) with permission from [49]. Copyright 2014 American Chemical Society.)

lasers on GaAs substrate (**Figure 20(d, e)**). With increasing disc diameter the threshold power rose due to smaller mode separation, occurrence of multimode lasing and absorption in the microdisc centre. The larger lasing thresholds on the Si substrate can be explained by the presence of crystal defects (i) inside the active region, reducing the modal gain and (ii) outside the active region promoting carrier recombination, thus lowering the injection efficiency of the carriers.

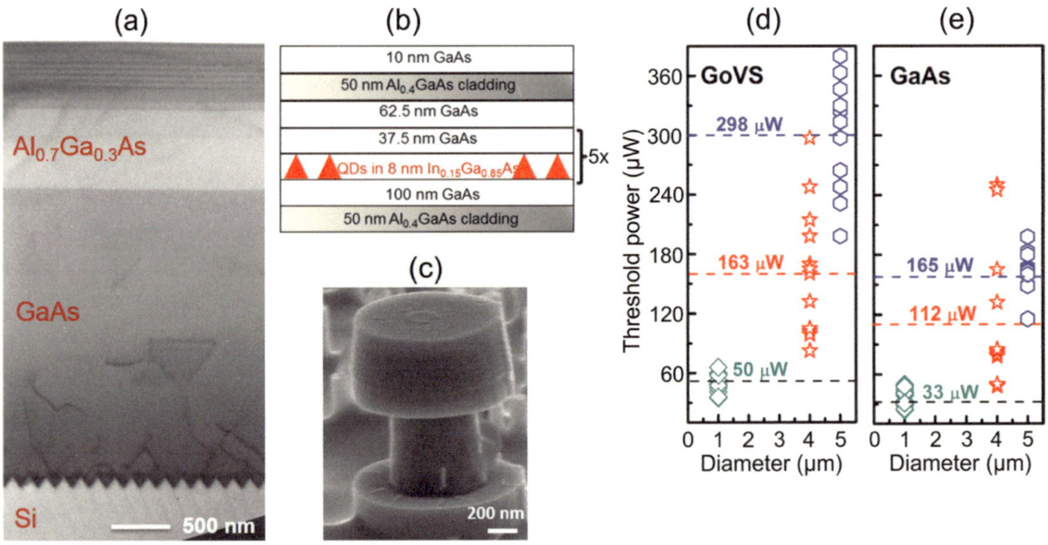

Figure 20. (a) Cross-section TEM image of the as deposited laser structure on V-grooved Si(001), (b) Schematic cross-section of layers in the disc region, (c) oblique-view SEM image of the microdisc, (d, e) threshold power for different disc diameters on GaAs-on-V-groove-Si (GoVS) template (d) and on GaAs substrate (e). The dashed lines denote the average thresholds. (Reproduced from [59], with the permission of AIP Publishing.)

4. Summary

In summary, heteroepitaxial approaches deploying nanopatterned substrates for the growth of mismatched III–V zinc blende semiconductor layers of low defect density are reviewed. As an important growth mode occurring in some of these approaches, nanoscale selective growth is explained with focus on its scaling with adatom surface migration and desorption. Efficient trapping and elimination of misfit-related defects has been demonstrated by ART techniques using nanopatterned oxide or nitride masks as well as by utilization of multiple QD defect filter layers. The active mechanisms of defect elimination in these growth methods are described. Besides, heteroepitaxial growth on mask-free nanoporous substrates also reduces the defect density, albeit in a less pronounced fashion. Finally, selected applications of low-defect III–V layers integrated on the common Si platform are presented. Remarkably, laser structures on the Si substrate have been realized exhibiting characteristics not far from those of structures grown on native GaAs substrates.

Acknowledgements

The authors acknowledge financial support of the German Science Foundation (DFG) under project numbers Ri2655/1-1 and Li449/16-1.

Author details

Thomas Riedl* and Jörg K.N. Lindner

*Address all correspondence to: thomas.riedl@uni-paderborn.de

Department of Physics, University of Paderborn, Paderborn, Germany

References

[1] Vurgaftman I, Meyer JR, Ram-Mohan LR. Band parameters for III–V compound semiconductors and their alloys. Journal of Applied Physics. 2001;**89**:5815–5875. DOI: 10.1063/1.1368156

[2] Adachi S. Properties of Group-IV, III-V and II-VI Semiconductors. Chichester: J. Wiley & Sons; 2005. 387 p. DOI: 10.1002/0470090340

[3] Stillman GE, Wolfe CM, Dimmock JO. Hall coefficient factor for polar mode scattering in n-type GaAs. Journal of Physics and Chemistry of Solids. 1970;**31**:1199–1204. DOI: 10.1016/0022-3697(70)90122-8

[4] Rode DL. Electron transport in InSb, InAs and InP. Physical Review B. 1971;**3**:3287–3299. DOI: 10.1103/PhysRevB.3.3287

[5] Duggen L, Willatzen M, Lassen B. Crystal orientation effects on the piezoelectric field of strained zinc-blende quantum-well structures. Physical Review B. 2008;**78**:205323. DOI: 10.1103/PhysRevB.78.205323

[6] Lourdudoss S. Heteroepitaxy and selective area heteroepitaxy for silicon photonics. Current Opinion in Solid State and Materials Science. 2012;**16**:91–99. DOI: 10.1016/j.cossms.2012.01.003

[7] Mahajan S. Defects in semiconductors and their effects on devices. Acta Materialia. 2000;**48**:137–149. DOI: 10.1016/S1359-6454(99)00292-X

[8] Groenert ME, Leitz CW, Pitera AJ, Yang V, Lee H, Ram RJ, Fitzgerald EA. Monolithic integration of room-temperature cw GaAs/AlGaAs lasers on Si substrates via relaxed graded GeSi buffer layers. Journal of Applied Physics. 2003;**93**:362–367. DOI: 10.1063/1.1525865

[9] Hayafuji N, Miyashita M, Nishimura T, Kadoiwa K, Kumabe H, Murotani T. Effect of employing positions of thermal cyclic annealing and strained-layer superlattice on defect reduction in GaAs-on-Si. Japanese Journal of Applied Physics. 1990;**29**:2371–2375. DOI: 10.1143/JJAP.29.2371

[10] Yamaguchi M. Dislocation density reduction in heteroepitaxial III-V compound films on Si substrates for optical devices. Journal of Materials Research. 1991;**6**:376–384. DOI: 10.1557/JMR.1991.0376

[11] Takano Y, Hisaka M, Fujii N, Suzuki K, Kuwahara K, Fuke S. Reduction of threading dislocations by InGaAs interlayer in GaAs layers grown on Si substrates. Applied Physics Letters. 1998;**73**:2917–2919. DOI: 10.1063/1.122629

[12] Kemper RM, Hiller L, Stauden T, Pezoldt J, Duschik K, Niendorf T, Maier HJ, Meertens D, Tillmann K, As DJ, Lindner JKN. Growth of cubic GaN on 3C–SiC/Si (001) nanostructures. Journal of CrystalGrowth. 2013;**378**:291–294. DOI: 10.1016/j.jcrysgro.2012.10.011

[13] Kemper RM, As DJ, Lindner JKN. Cubic GaN on nanopatterned 3C-SiC/Si (001) substrates. In: Li H, Wu J, Wang ZM, editors. Silicon-based Nanomaterials. Springer Series in Materials Science 187; 2013. Chapter 15, pp. 381–405. DOI: 10.1007/978-1-4614-8169-0_15

[14] Luryi S, Suhir E. New approach to the high quality epitaxial growth of lattice mismatched materials. Applied Physics Letters. 1986;**49**:140–142. DOI: 10.1063/1.97204

[15] Lee SC, Brueck SRJ. Scaling of the surface migration length in nanoscale patterned growth. Applied Physics Letters. 2009;**94**:153110. DOI: 10.1063/1.3117366

[16] Lee SC, Dawson LR, Brueck SRJ, Stintz A. Heteroepitaxial selective growth of $In_xGa_{1-x}As$ on SiO_2-patterned GaAs(001) by molecular beam epitaxy. Journal of Applied Physics. 2004;**96**:4856–4865. DOI: 10.1063/1.1786677

[17] Lee SC, Malloy KJ, Brueck SRJ. Nanoscale selective growth of GaAs by molecular beam epitaxy. Journal of Applied Physics. 2001;**90**:4163–4168. DOI: 10.1063/1.1401805

[18] Schroeder M, Wolf DE. Magic islands and submonolayer scaling in molecular beam epitaxy. Physical Review Letters. 1995;**74**:2062–2065. DOI: 10.1103/PhysRevLett.74.2062

[19] Neave JH, Dobson PJ, Joyce BA, Zhang J. Reflection high-energy electron diffraction oscillations from vicinal surfaces—A new approach to surface diffusion measurements. Applied Physics Letters. 1985;**47**:100–102. DOI: 10.1063/1.96281

[20] Spirkoska D, Colombo C, Heiss M, Abstreiter G, Fontcuberta i Morral A. The use of molecular beam epitaxy for the synthesis of high purity III-V nanowires. Journal of Physics: Condensed Matter. 2008;**20**:454225. DOI: 10.1088/0953-8984/20/45/454225

[21] Zubia D, Zhang S, Bommena R, Sun X, Brueck SRJ, Hersee SD. Initial nanoheteroepitaxial growth of GaAs on Si(100) by OMVPE. Journal of Electronic Materials. 2001;**30**:812–816. DOI: 10.1007/s11664-001-0062-7

[22] Zubia D. Personal communication; 2017

[23] Chin HC, Gong X, Ng TK, Loke WK, Wong CP, Shen Z, Wicaksono S, Yoon SF, Yeo YC. Nanoheteroepitaxy of gallium arsenide on strain-compliant silicon-germanium nanowires. Journal of Applied Physics. 2010;**108**:024312. DOI: 10.1063/1.3465327

[24] Ismail K, Legoues F, Karam NH, Carter J, Smith HI. High-quality GaAs on sawtooth-patterned Si substrates. Applied Physics Letters. 1991;**59**:2418–2420. DOI: 10.1063/1.106034

[25] Orzali T, Vert A, O'Brien B, Herman JL, Vivekanand S, Hill RJW, Karim Z, Rao SSP. GaAs on Si epitaxy by aspect ratio trapping: analysis and reduction of defects propagating along the trench direction. Journal of Applied Physics. 2015;**118**:105307. DOI: 10.1063/1.4930594

[26] Zubia D, Hersee SD. Nanoheteroepitaxy: The application of nanostructuring and substrate compliance to the heteroepitaxy of mismatched semiconductor materials. Journal of Applied Physics. 1999;**85**:6492–6496. DOI: 10.1063/1.370153

[27] Hersee SD, Zubia D, Sun X, Bommena R, Fairchild M, Zhang S, Burckel D, Frauenglass A, Brueck SRJ. Nanoheteroepitaxy for the integration of highly mismatched semiconductor materials. IEEE Journal of Quantum Electronics. 2002;**38**:1017–1028. DOI: 10.1109/JQE.2002.800987

[28] Jha S, Liu CC, Kuan TS, Babcock TS, Nealey PF, Park JH, Mawst LJ, Kuech TF. Defect reduction in epitaxial GaSb grown on nanopatterned GaAs substrates using full wafer block copolymer lithography. Applied Physics Letters. 2009;**95**:062104. DOI: 10.1063/1.3204013

[29] Nakamura Y, Miwa T, Ichikawa M. Nanocontact heteroepitaxy of thin GaSb and AlGaSb films on Si substrates using ultrahigh-density nanodot seeds. Nanotechnology. 2011;**22**:265301. DOI: 10.1088/0957-4484/22/26/265301

[30] He Y, Wang J, Hu H, Wang Q, Huang Y, Ren X. Coalescence of GaAs on (001) Si nanotrenches based on three-stage epitaxial lateral overgrowth. Applied Physics Letters. 2015;**106**:202105. DOI: 10.1063/1.4921621

[31] Lee SC, Dawson LR, Huang SH, Brueck SRJ. Lithography-free nanoscale patterned growth of GaAs on Si(001) with sub-100-nm silica nanoparticles by molecular beam epitaxy. Crystal Growth and Design. 2011;**11**:3673–3676. DOI: 10.1021/cg101363q

[32] May GS, Sze SM. Fundamentals of Semiconductor Fabrication. New York: J. Wiley & Sons; 2007. 320 p. ISBN: 978-0-471-23279-7

[33] Wang G, Leys MR, Loo R, Richard O, Bender H, Waldron N, Brammertz G, Dekoster J, Wang W, Seefeldt M, Caymax M, Heyns MM. Selective area growth of high quality InP on Si (001) substrates. Applied Physics Letters. 2010;**97**:121913. DOI: 10.1063/1.3491554

[34] Wang G, Leys MR, Nguyen ND, Loo R, Brammertz G, Richard O, Bender H, Dekoster J, Meuris M, Heyns MM, Caymax M. Growth of high quality InP layers in STI trenches on miscut Si (0 0 1) substrates. Journal of Crystal Growth. 2011;**315**:32–36. DOI: 10.1016/j.jcrysgro.2010.07.039

[35] Wang G, Leys M, Loo R, Richard O, Bender H, Brammertz G, Waldron N, Wang WE, Dekoster J, Caymax M, Seefeldt M, Heyns M. Selective area growth of InP and defect elimination on Si (001) substrates. Journal of the Electrochemical Society. 2011;**158**:H645–H650. DOI: 10.1149/1.3571248

[36] Merckling C, Waldron N, Jiang S, Guo W, Richard O, Douhard B, Moussa A, Vanhaeren D, Bender H, Collaert N, Heyns M, Thean A, Caymax M, Vanderhorst W. Selective area growth of InP in shallow trench isolation on large scale Si(001) wafer using defect confinement technique. Journal of Applied Physics. 2013;**114**:033708. DOI: 10.1063/1.4815959

[37] Merckling C, Waldron N, Jiang S, Guo W, Barla K, Heyns M, Collaert N, Thean A, Vandervorst W. Selective-area metal organic vapour-phase epitaxy of III-V on Si: What about defect density? In: Harame D, Caymax M, Heyns M, Masini G, Miyazaki S, Niu G, Reznicek A, Saraswat K, Tillack B, Vincent B, Yeo YC, Ogura A, Murota J, editors. SiGe, Ge, and Related Compounds 6: Materials, Processing, and Devices. Book Series: ECS Transactions Vol. 64, pp. 513–521. Pennington: The Electrochemical Society; 2014. DOI: 10.1149/06406.0513ecst

[38] Lee SM, Cho YJ, Park JB, Shin KW, Hwang E, Lee S, Lee MJ, Cho SH, Shin DS, Park J, Yoon E. Effects of growth temperature on surface morphology of InP grown on patterned Si(001) substrates. Journal of Crystal Growth. 2015;**416**:113–117. DOI: 10.1016/j.jcrysgro.2015.01.027

[39] Li JZ, Bai J, Park JS, Adekore B, Fox K, Carroll M, Lochtefeld A, Shellenbarger Z. Defect reduction of GaAs epitaxy on Si (001) using selective aspect ratio trapping. Applied Physics Letters. 2007;**91**:021114. DOI: 10.1063/1.2756165

[40] Guo W, Date L, Pena V, Bao X, Merckling C, Waldron N, Collaert N, Caymax M, Sanchez E, Vancoille E, Barla K, Thean A, Eyben P, Vandervorst W. Selective metal-organic chemical vapour deposition growth of high quality GaAs on Si(001). Applied Physics Letters. 2014;**105**:062101. DOI: 10.1063/1.4892468

[41] David S, Roque J, Rochat N, Bernier N, Piot L, Alcotte R, Cerba T, Martin M, Moeyaert J, Bogumilowizc Y, Arnaud S, Bertin F, Bassani F, Baron T. Spatially correlated structural and optical characterization of a single InGaAs quantum well fin selectively grown on Si by microscopy and cathodoluminescence techniques. APL Materials. 2016;**4**:056102. DOI: 10.1063/1.4949761

[42] Kunert B, Guo W, Mols Y, Tian B, Wang Z, Shi Y, Van Thourhout D, Pantouvaki M, Van Campenhout J, Langer R, Barla K. III/V nano ridge structures for optical applications on patterned 300 mm silicon substrate. Applied Physics Letters. 2016;**109**:091101. DOI: 10.1063/1.4961936

[43] Hsu CW, Chen YF, Su YK. Heteroepitaxy for GaAs on nanopatterned Si (001). IEEE Photonics Technology Letters. 2012;**24**:1009–1011. DOI: 10.1109/LPT.2012.2192726

[44] Paladugu M, Merckling C, Loo R, Richard O, Bender H, Dekoster J, Vandervorst W, Caymax M, Heyns M. Site selective integration of III-V materials on Si for nanoscale logic and photonic devices. Crystal Growth and Design. 2012;**12**:4696–4702. DOI:10.1021/cg300779v

[45] Ting SM, Fitzgerald EA. Metal-organic chemical vapour deposition of single domain GaAs on Ge/Ge_xSi_{1-x}/Si and Ge substrates. Journal of Applied Physics. 2000;**87**:2618–2628. DOI: 10.1063/1.372227

[46] Li JZ, Bai J, Major C, Carroll M, Lochtefeld A, Shellenbarger Z. Defect reduction of GaAs/Si epitaxy by aspect ratio trapping. Journal of Applied Physics. 2008;**103**:106102. DOI: 10.1063/1.2924410

[47] Li Q, Ng KW, Lau KM. Growing antiphase-domain-free GaAs thin films out of highly ordered planar nanowire arrays on exact (001) silicon. Applied Physics Letters. 2015;**106**:072105. DOI: 10.1063/1.4913432

[48] Hsu CW, Chen YF, Su YK. Nanoepitaxy of GaAs on a Si(001) substrate using a round-hole nanopatterned SiO_2 mask. Nanotechnology. 2012;**23**:495306. DOI: 10.1088/0957-4484/23/49/495306

[49] Chu CP, Arafin S, Nie T, Yao K, Kou X, He L, Wang CY, Chen SY, Chen LJ, Qasim SM, BenSaleh MS, Wang KL. Nanoscale growth of GaAs on patterned Si(111) substrates by molecular beam epitaxy. Crystal Growth and Design. 2014;**14**:593–598. DOI: 10.1021/cg401423d

[50] Chang YJ, Simmonds PJ, Beekley B, Goorsky MS, Woo JCS. Selective-area growth of heavily n-doped GaAs nanostubs on Si(001) by molecular beam epitaxy. Applied Physics Letters. 2016;**108**:163106. DOI: 10.1063/1.4947436

[51] Huang D, Reshchikov MA, Yun F, King T, Baski AA, Morkoc H. Defect reduction with quantum dots in GaN grown on sapphire substrates by molecular beam epitaxy. Applied Physics Letters. 2002;**80**:216–218. DOI: 10.1063/1.1432445

[52] Yang J, Bhattacharya P, Mi Z. High-performance $In_{0.5}Ga_{0.5}As$/GaAs quantum-dot lasers on silicon with multiple-layer quantum-dot dislocation filters. Transactions on Electron Devices. 2007;**54**:2849–2855. DOI: 10.1109/TED.2007.906928

[53] Bhattacharya P, Mi Z, Yang J, Basu D, Saha D. Quantum dot lasers: from promise to high-performance devices. Journal of Crystal Growth. 2009;**311**:1625–1631. DOI: 10.1016/j.jcrysgro.2008.09.035

[54] Hu H, Wang J, He Y, Liu K, Liu Y, Wang Q, Duan X, Huang Y, Ren X. Modified dislocation filter method: toward growth of GaAs on Si by metal organic chemical vapor deposition. Applied Physics A. 2016;**122**:588. DOI: 10.1007/s00339-016-0116-6

[55] Usman M, Reithmaier JP, Benyoucef M. Site-controlled growth of GaAs nanoislands on pre-patterned silicon substrates. Physica Status Solidi A. 2015;**212**:443–448. DOI: 10.1002/pssa.201431459

[56] Liu X, Kim IK, Aspnes DE. Investigation of heteroepitaxy on nanoscopically roughened (001)Si by real-time spectroscopic polarimetry. Physica Status Solidi C. 2008;**5**:1312–1315. DOI: 10.1002/pssc.200777896

[57] Nohavica D, Grym J, Hulicius E, Pangrac J, Gladkov P, Jarchovsky Z. Epitaxial overgrowth of InP and GaAs micropores, microcavities and microlamellas by InAs and InGaAs. In: Proceedings of the 2nd NANOCON International Conference; 12–14 October 2010; Olomouc/Czech Republic. pp. 28–33. ISBN: 978-80-87294-19-2

[58] Grym J, Nohavica D, Gladkov P, Hulicius E, Pangrac J, Piksova K. Epitaxial growth on porous GaAs substrates. Comptes Rendus Chimie. 2013;**16**:59–64. DOI: 10.1016/j.crci.2012.06.012

[59] Wan Y, Li Q, Liu AY, Chow WW, Gossard AC, Bowers JE, Hu EL, Lau KM. Subwavelength InAs quantum dot micro-disk lasers epitaxially grown on exact Si (001) substrates. Applied Physics Letters. 2016;**108**:221101. DOI: 10.1063/1.4952600

[60] Kroemer H. MBE growth of GaAs on Si: Problems and progress. MRS Symposium Proceedings. 1986;**67**:3–14. DOI: 10.1557/PROC-67-3

[61] Wang Z, Tian B, Pantouvaki M, Guo W, Absil P, Campenhout JV, Merckling C, Van Thourhout D. Room-temperature InP distributed feedback laser array directly grown on silicon. Nature Photonics. 2015;**9**:837–843. DOI: 10.1038/NPHOTON.2015.199

[62] Li Q, Han Y, Lu X, Lau KM. GaAs-InGaAs-GaAs fin-array tunnel diodes on (001) Si substrates with room-temperature peak-to-valley current ratio of 5.4. IEEE Electron Device Letters. 2016;**37**:24–27. DOI: 10.1109/LED.2015.2499603

[63] Han Y, Li Q, Lau KM. Monolithic integration of tunnel diode-based inverters on exact (001) Si substrates. IEEE Electron Device Letters. 2016;**37**:717–720. DOI: 10.1109/LED.2016.2552219

[64] Cipro R, Baron T, Martin M, Moeyaert J, David S, Gorbenko V, Bassani F, Y. Bogumilowicz, Barnes JP, Rochat N, Loup V, Vizioz C, Allouti N, Chauvin N, Bao XY, Ye Z, Pin JB, Sanchez E. Low defect InGaAs quantum well selectively grown by metal organic chemical vapor deposition on Si(100) 300 mm wafers for next generation non planar devices. Applied Physics Letters. 2014;**104**:262103. DOI: 10.1063/1.4886404

Chapter 5

Surface Modification of III-V Compounds Substrates for Processing Technology

Rodica V. Ghita, Constantin Logofatu,
Constantin-Catalin Negrila, Lucian Trupina and
Costel Cotirlan-Simioniuc

Additional information is available at the end of the chapter

http://dx.doi.org/10.5772/67916

Abstract

Semiconductor materials became a part of nowadays life due to useful applications caused by characteristic properties as variable conductivity and sensitivity to light or heat. Electrical properties of a semiconductor can be modified by doping or by the application of electric fields or light; and from this view, devices made from semiconductors can be used for amplification or energy conversion. The compound semiconductor materials from III-V class experienced a qualitative leap from promising potential to nowadays technologic environment. The III-V semiconductor compounds are the material bases for electronic and optoelectronic devices such as high-electron-mobility transistors (HEMT), bipolar heterostructure transistors, IR light-emitting diodes, heterostructure lasers, Gunn diodes, Schottky devices, photodetectors, and heterostructure solar cells for terrestrial and spatial operating conditions. Among III-V semiconductor compounds, gallium arsenide (GaAs) and gallium antimonide (GaSb) are of special interest as a substrate material due to the lattice parameter match to solid solutions (ternary and quaternary) whose band gaps cover a wide spectral range from 0.8 to 4.3 μm in the case of GaSb. The solid/solid interfaces could play a key part in the development of microelectronic device technology. In most of the cases, the initial surface of III-V compounds exposed to laboratory conditions is covered usually with native oxide layers. Various techniques for performing the surface cleaning process are used, e.g., controlled chemical etching, *in situ* ion sputtering, coupled with controlled annealing in vacuum and often these classic techniques are combined in order to prepare an eligible semiconductor surface to be exposed to a technological device chain. The evolution of surface native oxides in different cleaning procedures and the characteristics of as-prepared semiconductor surface were investigated by modern surface investigation techniques, i.e., X-ray photoelectron spectroscopy (XPS), atomic force microscopy (AFM), Rutherford backscattering spectrometry (RBS) combined with electrical characterization. Surface preparation of semiconductors in particular for III-V compounds is a necessary requirement in device technology due to the existence of surface impurities and the presence of native oxides. The impurities can affect the adherence of ohmic and Schottky contacts and due to thermal decomposition of native oxides (e.g., GaSb) it also affect the interface metal/semiconductor. The practical experience reveals that the simple preparation of a surface is a nonrealistic expectation, i.e., surface preparation is a result of combined treatments, namely chemical etching and thermal treatment, ion beam sputtering and thermal reconstruction procedure.

Keywords: III-V compounds, surface oxides, surface etching, XPS, AFM, RBS, metallic nanolayers

---

1. Introduction

Semiconductor materials became a part of nowadays life due to useful applications caused by characteristic properties as variable conductivity and sensitivity to light or heat. Electrical properties of a semiconductor can be modified by doping or by the application of electric fields or light; and from this view, devices made from semiconductors can be used for amplification, switching, and energy conversion. A pure semiconductor is not very useful as it is neither a very good insulator nor a very good conductor. The conductivity of semiconductors may easily be modified by introducing impurities into their crystal lattice, implying a controlled doping process. The materials suitable as dopants depend on the semiconductor nature (e.g., for silicon the common are group III *p*-type dopants, acceptor impurities and group V-*n*-type dopants, donor impurities). Electrical conductivity is a consequence of the presence of electrons in states that are partly delocalized, however in order to transport electrons a state must be partially filled. The energies of these quantum states are critical due to the fact that a state is partially filled only if its energy is near the Fermi level (Fermi-Dirac statistics). In a semiconductor, the number of partially filled states is increased by doping or by gating move either the conduction or valence band closer to the Fermi level. Electron-hole pair generation can appear when a semiconductor is exposed to ionizing radiation, and as a consequence an electron is excited out of its energy level and consequently leaving a hole. The number of electron-hole pairs in a steady state at a given temperature is established by quantum statistics, where the precise quantum mechanism of generation and recombination are governed by conservation laws of energy and momentum. In electronics, the applications are based in the most part on current flow through material in different *p-n* junction geometries. Heterojunctions occur when two different doped semiconducting materials are jointed in an intimate contact. Operating different heterojunction devices is based on the current flow between *p*-type and *n*-type regions caused by an applied electric field. In general, a difference in electric potential would cause the appearance of a nonequilibrium situation that can introduce electrons and holes to the system, which interact through ambipolar diffusion. The change of thermal equilibrium as a result of a temperature difference or to exciting photons caused the process of generation and recombination of carriers (i.e., change of electrons and holes number). In certain semiconductors, exciting electrons relax by emitting light and this process is the fundament of light-emitting diode operation. Certain semiconductors have large thermoelectric power factors, therefore they can be used in thermoelectric generators (e.g., thermoelectric coolers).

http://dx.doi.org/10.5772/67916

The compound semiconductor materials from III-V class experienced a qualitative leap from promising potential use in semiconductor devices as they were viewed at the mid of twentieth century to extensively fabrication in optoelectronics industry at the beginning of twenty-first century. As it was stated Ref. [1], III-V semiconductor compounds are the material bases for electronic and optoelectronic devices as high-electron-mobility transistors (HEMT), bipolar heterostructure transistors, IR light-emitting diodes, heterostructure lasers, Gunn diodes, Schottky devices, photodetectors, heterostructure solar cells for terrestrial and spatial operating conditions, microwaves devices, electro-optical modulators, and frequency-mixing components. In principal, the main characteristics of these devices are dependent on the physical properties of III-V materials and their solid solutions, mostly ternary and quaternary alloys. It is to be remarked the presence of quantum effects in quantum heterostructure devices design that extended the operating and reliability features. In order to exploit this flexibility of III-V materials it is necessary to use a band parameter data base as the starting point to analyze, compute, and simulate the operation of a device in its way to a reliable technology. The III-V compounds of interest are those between Al, Ga (group III) and P, As and Sb (group V), i.e., AlP, AlAs, AlSb, GaP, GaAs, GaSb, InP, InAs, InSb and their ternary and quaternary solid solutions, e.g., $Al_xIn_{1-x}P$, $Al_xGa_{1-x}As$, $Al_xIn_{1-x}As$, $Al_xGa_{1-x}Sb$, $Al_xIn_{1-x}Sb$, InGaAsP, AlGaPAs, GaInAsSb, AlGaAsSb. Isoperiodic epitaxial heterostructures composed of quaternary solid solutions as GaInAsP/InP, GaInAsSb/GaSb, or AlGaAsSb/GaSb are the basis for optoelectronic devices of a given spectral range extending over a wide spectrum and their study determined a progress in semiconducting band engineering. Some important characteristics of classic III-V compounds are the lattice structure of zincblende type, the same space group $T^2{}_d$-F-$43m$, lattice parameter a_0 at 300 K in the range (5.451 Å, for GaP to 6.47 Å, for InSb) and energy gap at 300 K in the range (0.17 eV, for InSb to 2.45 eV, for AlP with different direct and indirect features. Also, an important class of wurtzite structure is AlN and GaN based compounds that have been used for short wavelength light emitters. It is worth to mention that incorporation of N and In into GaAs lattice results in lowering the band gap energy of GaAs, whereas minimizing the strain as In tends to increase and N tends to reduce the lattice constant of GaAs. The lattice matching InGaAsN layers to GaAs conduced to first (In)GaAsN lasers presented in 1996 [2].

Among III-V semiconductor compounds, gallium arsenide (GaAs) and gallium antimonide (GaSb) are of a special interest as a substrate material due to the lattice parameter match to solid solutions (ternary and quaternary) whose band gaps cover a wide spectral range from 0.8 to 4.3 μm in the case of GaSb [3]. Regarding devices perspectives, GaSb have shown applications in laser diodes with low threshold voltage [4], photodetectors with high quantum efficiency [5], high frequency devices [6] or to high efficiency thermophotovoltaic (TPV) cells [7]. In this perspective, GaSb is a III-V semiconductor compound with zincblende crystal structure and has an energy gap of 0.726 eV and is worth to mention that the structure GaAs/GaSb had set a record for solar cell efficiency of 35% opening a new era for photovoltaics applications. We can say that the GaSb photosensitive structures offers the possibility of an almost total conversion of sun energy from visible spectrum to heat transform in electricity by TPV effect. Another property to mention is related to GaSb lattice limited electron mobility that is greater than GaAs, which make it available in fabrication microwave devices. The property of GaSb band structure related to the spin-orbit splitting of the valence band, that is almost equal to the gap conduced to high hole ionization coefficients leading to an improvement of signal-to-noise ratio at $\lambda > 1.3$ μm in AlGaSb/GaSb avalanche photodetectors [5]. GaSb-based devices are candidates in infrared (IR) imaging sensors, fire detection, and monitoring environmental pollution. We remark that the absorption wavelength of different industrial toxic gases are in the IR range where GaSb devices are recommended due to band gap structure, the near IR regime is also suitable for biological and medical applications. Some interesting aspects of GaSb are related to the fact that undoped semiconductor is always of *p*-type irrespective of growth procedure, and sulfur-doped GaSb is the only III-V compound with high concentration of donor-related deep traps (*DX* centers) at atmospheric pressure [3]. At this status, the GaSb compound study is of interest due to effort for understanding material preparation and processing in order to develop a competitive technology for optoelectronic devices, in our case the competitive Schottky devices for terrestrial applications. As it was stated, the solid/solid interfaces could play a key part in the development of microelectronic device technology. In most cases, the initial surface of III-V compounds exposed to laboratory conditions is covered usually with native oxide layers. In this view, the III-V surface oxides are unstable from a chemical point of view. Various techniques for performing the surface cleaning process are used, e.g., controlled chemical etching, *in situ* ion sputtering, cycle of controlled annealing in vacuum and often these classic techniques are combined in order to prepare an eligible semiconductor surface to be exposed to a technological device chain.

Generally speaking obtaining good ohmic contacts on semiconductor material surfaces is dependent on the ability of the cleaning procedure to remove the native oxides and in the case of III-V compound semiconductors to keep the surface stoichiometry. Deposition of metallic layers on clean semiconductor surface in order to obtain good ohmic or Schottky contacts is an important goal for device and structure preparation [8]. The evolution of surface native oxides in different cleaning procedures and the characteristics of as-prepared semiconductor surface were investigated by modern surface investigation techniques, i.e., X-ray photoelectron spectroscopy (XPS), atomic force microscopy (AFM), Rutherford backscattering spectrometry (RBS) combined with electrical characterization.

2. Preparation and investigation techniques

The III-V semiconductor compounds as solid state materials are covered by a surface that acts as an interface to the surrounding environment. Generally speaking, the surface modification is the action upon the surface of a material by bringing physical or chemical characteristics that are slightly different from the ones initially found on the surface. Another view of surface modification is represented by surface functionalization that introduces chemical functional groups to a surface. In this way, materials with functional groups on their surface can be designed from substrates with standard bulk material properties, e.g., in semiconductor industry. Commonly speaking, the surface modification is made to the semiconductor materials for altering characteristics as roughness, surface energy, or reactivity. From this point of view, surface engineering involves the alteration for the surface phase properties in order to reduce degradation over time. Surface engineering techniques that are commonly used in electronics and optoelectronics implied in this work, the controlled action on semiconductor surface by chemical and physical etching.

Etching procedure is used in microelectronics to chemically remove layers from the surface of a wafer during fabrication. Etching is a peculiar and important process where generally the wafer undergoes many etching steps before it is complete. From a historical point of view, chemical etching or so-called wet etching, plays an important role in both preparation and the utilization of thin films. In this perspective, for thin film deposition, e.g., metallic layers, the substrate has to be first suitable prepared, i.e., the removal of a work damaged surface layers or by creating a relief structure of specific geometry. In both cases, the method used is either a chemical polish etching or is a structural etching. One important application of chemical etching is in the structural characterization of materials, especially in detection of lattice defects in semiconductors, in the distribution study of localized impurities, in the determination of layer structures or *p-n* junctions, and last but not least in the composition determination. Regarding the chemical etching [9], the simplest way involves only the dissolution of the material in a solvent without changing the chemical nature of the dissolved species. Most part of etching processes involves one or more chemical reactions. A true etching reaction is characterized by the formation of an etchant product that must be soluble in the etchant medium. Different types or reaction involved are of oxidation-reduction type, electrochemical type, complexation, and gas phase etching. A redox etching process implies conversion of the material being etched to a soluble higher oxidation state:

$$M \rightarrow M^{n+} + ne^{-}. \tag{1}$$

It is worth to mention that complex formation is frequently involved in etching processes, often in conjunction with a redox act, where the formation of a complex ion or molecule is accompanied by solubilization in the etchant medium. Gas phase etching may involve vaporization of the material that has been etched in a vacuum or in inert atmosphere. In another situation, gas phase etching is involving the reaction of gaseous etchants with the surface to produce volatile products. Etching reactions are characterized by a process involving several sequential steps [10]. The dissolution kinetics is depending on the nature of the rate-limiting step of the process in the following view. If the rate of this step is a function of the chemical reactivity of the species involved, the process is said to be activation limited. On the other hand, if the rate is determined by the velocity at which a fresh reactant can be supplied to the surface, the process is said to be diffusion limited. It is observed that an increase in etching temperature is a cause for change in the etching kinetics [11]. Other factors that can influence the etching process are the presence of catalytic species in the etchant, solution agitation, or localized solution heating. It is important to mention that adsorption and desorption processes can affect the etching kinetics, i.e., adsorption of a reactant from the etchant solution onto the substrate may produce surface complexes that can facilitate the etching process. The kinematic aspect of etching is related to the tendency of various crystallographic planes to etch at different rates. Various orientations of single-crystal semiconductors may etch differently in a given etchant and substrates with nonuniform roughness exhibit large differences in etch rates. It is remarkable that the simplest etching technique is liquid chemical immersion or dip etching. In this process, mechanical agitation is usually desirable, because it improves the uniformity and control of chemical etching by enhancing the spread of etching solution at the solid surface. Mechanical agitation generally avoids local overheating in case of exothermic reactions a fact that conduced to a control of etching rate. The addition of a surface-active agent to etch solution can prevent bubble accumulation (e.g., formation of H_2 bubbles that can inhibit uniform etching).

The subject of contamination and cleaning of surfaces before and after etching is closely related with practical etch process. Surface contamination as related to etching has two aspects: (a) initially presence of contaminants and their removal prior to etching, and (b) residual contaminants arising from etching treatments and their removal as a post-etch step. Contamination prior to etching consists of particulate materials, organic residues, or inorganic surface films different from the semiconductor to be etched. Particulate removal is generally accomplished by ultrasonic treatment in cleaning solution, use of compressed gas jets, application of liquid sprays or jets, or by mechanical means as brushing. Organic residues are removable down to monolayer levels by dissolution in suitable organic solvents (e.g., trichloroethylene, acetone or propranolol) or by vapor refluxing in organic solvents as azeotropic solvent mixtures. Complete removal is generally possible only by plasma etching, glow discharge sputtering, or chemical reaction leading to dissolution. Inorganic surface films (an example may be oxides) had to be attacked by specific chemical reagents designed to produce soluble reaction products that can be flushed away. While the treatments of initial contaminants are generally simple, the effective removal of residual trace contaminants resulting from etching process is more difficult to accomplish. It is remarkable that the surface contaminants are very critical in technologies where high-temperature processing may cause residual surface impurities to penetrate into the substrate and to be the origin of undesirable effects as electrical instability of semiconductor devices. Surface cleaning by glow discharge sputtering techniques is effective on semiconductor surface. Most organic surface contaminants are removable by chemical sputtering in O_2 and sputter etching in Ar^+ removes residual oxide layers on metals. However, surface recontamination due to backscattering or ion migration can occur during RF sputtering treatments, unless carefully optimized processing conditions are employed. Ultrahigh vacuum heating after sputter cleaning is effective for desorbing gases that may become incorporated into substrates surface during operation [12]. Plasma etching is to be noted as one of the most effective methods for surface decontamination with application to semiconductor device processing. It is remarked that for surface cleaning two aspects are important, i.e., terminal treatment and storage of the cleaned material. The final rising in wet cleaning and etching process is usually in water. In this view, cleaning procedure is used only in deionized and distilled water in order to avoid contamination. High-purity electronic-grade isopropyl alcohol is a good alternative final rinse after DIW (deionized water). Storage of cleaned materials should be minimized, but if the storage is necessary, should be used chemically cleanedclosed glass containers (e.g., Petri dishes) kept in a contamination-free clean-room ambient. In etching processes, selectivity is an important factor in experimental applications. In III-V compound semiconductors etching reactions are complicated due to the effect of crystallographic orientations. Chemical etching of GaAs, GaP, or GaSb proceeds by oxidation-reduction complex reactions analogous in the main part with the mechanism for Si and Ge etching. The most common etchant in this system is based on H_2SO_4-H_2O_2-H_2O in

http://dx.doi.org/10.5772/67916

different volume ratios as a function of orientation-dependent etching characteristics. In this work will be analyzed also chemical etchants in citric acid base or sulfur base organic and inorganic etchants for GaAs or GaSb.

Physical action on the surface, the so-called "dry etching," refers to the removal of material typically a masked pattern of semiconductor by exposing it to the bombardment of ions, e.g., plasma of reactive gases that dislodge portions of the material from the exposed surface. Dry etching is a process with characteristics of anisotropy and is particularly useful for semiconductors which are chemically resistant.

In modern VLSI (very large scale integration) processes, plasma etchers, i.e., charged ions or neutral atoms and radicals, can operate in several modes by adjusting plasma parameters. Plasma etching operates at pressure in the range (0.1÷5) torr by producing free radical, neutrally charged, that react at the substrate surface in a process with an isotropic character. In plasma, there are present different species as electrons, ions, radicals, and neutral particles that are constantly interacting with each other. In plasma etching interaction can be divided into two types, namely, generation of chemical species and interaction with the surrounding surfaces. The operation mode of plasma system will change if the operating pressure changes and it is also dependent on the reaction chamber geometry. The main factors that influence the plasma process are electron source, pressure, gas species, and vacuum. The success of a plasma etching process is dependent on appropriate etch chemistry that will form volatile products with the exposed semiconductor to be etched. The surface temperature also affects the reaction of products. The plasma etching process is also affected by different properties such as volatility, adsorption, or chemical affinity. In semiconductors the processing plasma etching can be used to create different patterns on a fabrication chain playing a major role in microelectronic production.

Ion milling or sputter etching uses lower pressures in the range of 10^{-4} torr and it acts by the bombardment of the substrate with energetic ions of noble gases, more often Ar^+, which knock atoms from the substrate by momentum transferring. This process performed by ions which act approximately from one direction has a highly anisotropic character but it tends to present a poor selectivity characteristic.

Reactive-ion etching (RIE) operates under pressure conditions intermediate between sputter and plasma etching, i.e., in the pressure range 10^{-3}–10^{-1} torr. The upgrade technique is deep reactive-ion etching (DRIE) that modifies RIE technique to produce deep, narrow features. From this point of view, DRIE is a highly anisotropic etch process to create deep penetration, steep-sided holes and trenches in substrates with high aspect accuracy. This technique was developed initially for microelectromechanical systems (MEMS) which require these features. There are two main technologies for high-rate DRIE, namely, cryogenic and Bosch, and both of them can fabricate truly vertical walls, but in practice, the walls are slightly tapered, e.g., 88° ("reentrant") or 92° ("retrograde").

It is worth to mention that the dry etching technology is expensive compared to wet etching procedure, but regarding the resolution in thin film structures is to be consider as possible solution.

Surface cleaning by physical means, in our case, sputtering technique, can also be effective in the case of removal of oxide layers on III-V semiconductors by Ar^+ ion beam sputtering. Sputtering is a process whereby particles are ejected from a target by bombardment of gas ions. The incident ions set off collision cascades in target and the number of atoms ejected from target (sputter yield) is depending on ion incident angle, ion energy, ion mass, target atom mass, and surface biding energy of target atoms. Ion sputtering has a defined threshold energy depending on biding energy of surface atom with a value in the range 10–100 eV. The process of atom removing from by sputtering using inert gas is an ion etching process. For semiconductors (peculiar for III-V compounds), ion sputtering is used to etch the target and sputter etching is chosen in cases with a high degree of etching anisotropy and where selectivity is not a concern. For Ar^+ ion beam, sputtering of carbon and oxygen presence from GaAs, GaSb, or GaP surface is needed an accelerating voltage of 0.5 kV for few minutes. The sputtering process is an *in situ* cleaning procedure where the usual Ar^+ accelerating voltage is in the range 1–3–5 kV and ion beam current I in the range 25–50–100 μA. The rate used for Ar^+ ion beam voltage is usually 1 kV and the sputtering time used for cleaning is in the range 1–5 minutes. The main advantages of sputtering are related to the fact that it is a rapid and *in situ* cleaning process (the process has a high rate cleaning degree). The important disadvantages are related to the fact that at the semiconductor surface may appear stoichiometric modifications and Ar^+ ion can damage the semiconductor by penetrating the lattice. For III-V compounds in Ar^+ ion sputtering process the Ga atoms are preferentially present on the surface (e.g., GaAs, GaSb, GaP), and it is necessary for a thermal reconstruction treatment at temperatures in the range 200–400°C for few hours. In other words, one major drawback of sputtering as an etching cleaning technique for semiconductors is the wafer damage.

In the case of III-V semiconductor compounds technology, have been used combined *ex situ* techniques as chemical etching and thermal treatments cycles and *in situ* cleaning as sputtering and reconstruction treatment in order to prepare a semiconductor surface for device technology requirements.

The characteristics of a native semiconductor surface and of a prepared surface for different technological device requirements may be investigated by surface characterizing techniques, i.e., XPS, AFM, SE, or RBS.

X-ray photoelectron spectroscopy (XPS) is a sensitive quantitative spectroscopic technique that detects all elements and can be used to analyze the surface chemistry of a material in its as-received state or of a surface exposed to different chemical or physical treatments. The presence of native oxides (e.g., GaAs, GaSb) with a structure and a well-defined composition may be analyzed by angle-resolved XPS (ARXPS), for example in a XPS SPECS system. It is worth to say that a typical XPS spectrum is a plot of the number of electron detected (y-axis) versus binding energy of detected electrons (x-axis). Each element is characterized by a set of XPS peaks at a specific binding energy values that directly identify each element that is situated inside or on the surface of the as-analyzed material. The characteristic spectral peaks correspond to the electron configuration of the electrons within the atoms, e.g., 1s, 2p, 2p, 3s, 3d, etc. The number of detected electrons in each characteristic peaks is related to the amount of element within the XPS sampling volume. In order to obtain atomic percentage values, each raw XPS signal must be corrected by dividing its signal intensity by a relative sensitivity factor (RSF) and normalized over all the elements detected (except hydrogen that is not detected). In order to count the number of electrons during acquisition of spectrum with a minimum of error, the XPS detectors have to operate under ultra-high vacuum (UHV) conditions (p ~ 10^{-9} torr) because the electron counting detectors in XPS instruments are typically 1 m away from the material irradiated with X-rays and this is the condition why this long path length for detection requires such low pressures. In this work, the XPS recorded spectra were obtained in the as-mentioned SPECS Spectrometer based on a Phoibos analyser with monochromatic X-rays emitted by an anticathode of Al (1486.7 eV). The hemispherical analyzer operated in the constant energy mode with a pass energy of 5 eV, giving an energy resolution of 0.4 eV, which was established as FWHM (full width half maximum) of the Ag *3d* 5/2 peak. The analysis chamber was maintained as specified at ultrahigh vacuum conditions (~10^{-9} torr). The recorded spectra were processed using Spectral Data Processor v 2.3 (SDP) software, and the good fit for experimental data is ensured by a specific ratio between a Lorentzian and a Gaussian shape in the deconvolution of XPS peaks. The XPS measurements were carried out at different TOA (take-off angles), where TOA is defined as well as its description in the ASTM E 673-98 related to standard terminology regarding surface analysis. In this perspective TOA represents the angle between the beam of emitted photoelectrons and the sample surface. As regarding surface sensitivity, we remark that XPS detects only those electrons that have actually escape from the sample into vacuum of the instrument, and reach the detector. The photoelectron that escapes from the sample into the vacuum actually travels through the sample. These photo-emitted electrons may undergo inelastic collisions, recombination, excitation of the sample, recapture, or trapping in various excited states within material, where all these processes can reduce the number of escaping photoelectrons. These effects are present as an exponential attenuation function as the depth increases, the result is that the signals detected from analyzed element at the surface is much stronger than the signals detected from the same element deep under sample surface. The consequence is that the XPS signal has a characteristic of an exponentially surface-weighted signal and may be used to estimate different elements depths in layered materials. The main characteristic of XPS analysis as a surface-sensitive investigation technique was intensively used for the quantitative analysis of as-received semiconductor surface, namely the structure of native oxides developed at the surface of GaAs, GaSb, GaP single crystals. The effect of different surface preparation procedures was also put into evidence by XPS and ARXPS analyses.

Atomic force microscopy (AFM) is a powerful technique for investigating the surface morphology of materials at nanoscale level. AFM can image the sample surface topography with extremely high—up to atomic—resolution which is comparable with the scanning electronic microscopy (SEM). The advantage of AFM compared to SEM is that the measurement is made in three dimensions. The vertical resolution can be on the order of Angstroms while the lateral resolution depends on the tip radius. The working principle of AFM is based on force measurement between a small tip and a sample surface. The tip placed at the end of a silicon cantilever can have a nominal radius of curvature at the tip apex of 10–40 nm. High-resolution images can be obtained with cantilevers having an extremely sharp diamond-like carbon tip with typical curvature radius of 1 nm [13]. The image of the sample surface is obtained by monitoring the bending of the cantilever as it scans along topographic features. The cantilever displacement is measured by detecting the deflection of a laser beam which is reflected from the back side of the cantilever and collected in a photodetector. The output of the photodetector is used to control the cantilever deflection by adjusting the height of the probe via a piezoelectric actuator.

Depending on tip-sample separation during scanning three different modes of operation exists: noncontact, intermittent contact, and contact. The noncontact mode operates at distances between 0.1 and 10 nm where attractive Van der Waals force occurs between tip and sample surface. In this mode the tip oscillates above the surface during scanning and a feedback loop is used to track changes in the amplitude due to tip-sample interaction. The advantage of this operating mode is the low force applied on the sample, which can be important when soft surfaces are investigated, but usually images with lower resolution are obtained. At distances of 0.5–2 nm between the tip and surface sample the AFM works in intermittent contact. In this mode the tip is oscillating at its resonance frequency and just barely touching the surface while is at the bottom of its oscillation. Thus, compared with noncontact and contact mode, higher resolution and greatly reduced lateral force are obtained, respectively. As the tip moves closer to sample surface, at distances smaller than 0.5 nm, the tip-sample interaction force becomes predominantly repulsive due to the repulsion generated by the overlap between the electron clouds of the tip and surface atoms. This is known as the contact mode, which is the best mode for scanning hard and flat surfaces. In this study the surface morphology and roughness of the samples were investigated by atomic force microscopy using an MFP 3D SA, Asylum Research microscope operated in the intermittent contact mode under ambient atmosphere. The tip used had a radius of curvature of 10 nm, spring constant k = 3.4 N/m and resonance frequency about 107 kHz.

Rutherford backscattering spectrometry (RBS) is an analytical technique used in materials science. The elemental composition of thin metals layers were investigated by RBS using a $^{4}He^{++}$ beam extracted from the Alphatross ion source of the 3 MV Tandetron accelerator of IFIN-HH. The beam energy was 3 MeV. The alpha particles were detected with a passivated ion implanted silicon detector. The detector had an energy resolution of 16 keV and was placed at 165° with respect to the beam direction. The solid angle was 1.641 mSr. In order to avoid channeling effects the sample was tilted at 7° with respect to the beam direction. The elemental concentrations and the layer thickness were determined using code RUMP [14] by fitting the RBS experimental spectrum of the sample. The RBS analysis was corroborated with XPS data in obtaining the depth profile of AuGeNi/n-GaSb, the metal layers being vacuum deposited on a clean n-GaSb surface prepared by chemical etching. The depth profile shows that Au, Ge, and Ni are uniformly distributed in matrix and an Au-Ga alloy that assures a good ohmic contact was detected.

3. Initial stage of the III-V semiconductor surface-presence of native oxides

A major problem to be overcome in the technology of III-V compounds is the poor quality of oxide/semiconductor interfaces [15]. For devices of MIS (metal-insulator-semiconductor), the presence of a stable interface containing a low density of electronic defect states around band gap is essential for device operating characteristics. The main source of defects is known to be the oxidation of III-V semiconductor surface and in particular at GaSb surface the oxidation is an exothermic reaction. Generally speaking the III-V surface oxides are chemically unstable in air. It is stated that for GaSb [16], although the native oxides can be stabilized against chemical attack, this stabilization procedure requires high temperatures over 600°C, a temperature above the allowed temperature range for some device structure technology. In GaSb oxidation, the preferential reaction of gallium with oxygen and gallium segregation at the surface is attributed to the affinity of gallium to oxygen: Ga_2O_3 (−238 Kcal/mol) compared to antimony atoms Sb_2O_5 (−198.2 Kcal/mol) [17]. Also, the spontaneous reaction of Sb_2O_3 with GaSb ($\Delta G = -12$ Kcal/mol) is attributed to the preferential reaction of gallium with oxygen in the amorphous layer [3, 18]. The exposure of GaSb to air conduced to enhance the Ga fraction at the surface [19]. In Ref. [15] are presented data concerning experimental efforts to reduce the interface defect densities (D_{it}) to a comparable level as it is today's SiO_2/Si interfaces (e.g., $D_{it} \sim 10^{13}$ cm^2/eV, as grown) and after hydrogen passivation, this is reduced to the mid-10^{10} range. For III-V semiconductor/oxide interfaces the lowest reported D_{it} values are in the range 10^{11}–10^{12} cm^2/eV, and that is why the challenge in III-V compounds technology, peculiar for GaSb is to find a suitable oxide that allowed interface passivation by an appropriate passivating agent.

Most of characteristics of GaSb compound is determined by the arrangement of oxygen atoms on semiconductor surface. In GaSb Sb is octahedral coordinated by oxygen, a situation also existing in its oxygen compounds. Reaction kinetic between GaSb surface and oxygen is expressed by oxidation that takes place in two steps [3]:

$$2GaSb + 3O_2 \rightarrow Ga_2O_3 + Sb_2O_3 \quad (2)$$

$$2GaSb + Sb_2O_3 \rightarrow Ga_2O_3 + 4Sb \quad (3)$$

Semiconductor surface structure, besides semiconductor bulk, has an important influence on the electrical properties of semiconductor devices. In this view, the study of native oxides and passivating oxides prepared in different GaSb surface oxidation procedures is necessary for the reproducibility of surface properties.

Figure 1 shows the Ga *3d* spectrum from the native oxides of n-GaSb(100) Te doped substrates. The lines corresponding to 18.98 and 19.50 eV are related to metallic Ga in GaSb, and the lines corresponding to 20.27 and 20.99 eV are related to Ga in Ga_2O_3.

The quantitative analysis indicates a percentage of 71.5% Ga in GaSb and 28.5% Ga in Ga oxide, a nonstoichiometric ratio in native oxide.

Figure 2 shows the Sb *4d* lines of GaSb with native oxides having the following peaks namely 32.00 and 33.27 eV for Sb in GaSb and the peaks 34.56 and 35.74 eV for Sb in Sb oxide, with an estimated composition of 71.1% Sb in GaSb and 28.8% Sb in oxide. The FWHM for Sb *4d* lines is extended by 0.6–0.7 eV in oxide than Sb in GaSb for our experiment.

The morphological aspect of native oxide film as a conglomerate of Ga and Sb oxides defining a surface roughness of 1.854 nm is presented in **Figure 3** in an AFM image.

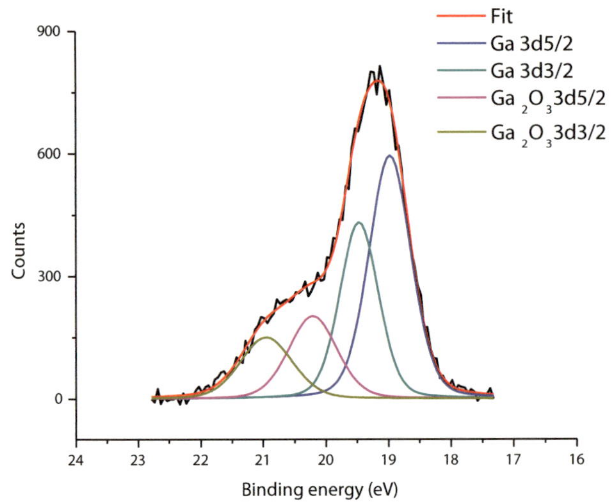

Figure 1. XPS spectrum for Ga *3d* lines on n-GaSb.

Among III-V semiconductor compounds, GaAs is probably the most studied due to its importance in modern microelectronics and optoelectronics applications. The GaAs(100) surfaces have a high surface energy [20], and the result is high reactivity and chemical unstableness. Due to the reactivity of native oxides, it is difficult to reach the passivation of GaAs surfaces. The native oxide characterization in XPS measurements led to the determination of different chemical species. The obtained data [21] indicate that the native oxide layer is complex containing amounts of carbon and oxygen. The discussions regarding surface composition are related in this work with the As $2p^{3/2}$ and Ga $2p^{3/2}$ signals in native oxides and subsequent changes related to different etching processes. As presented [22] the C *1s* peak shape has predominant contribution from C-C bond on the surface with binding energy (BE = 285 eV) that

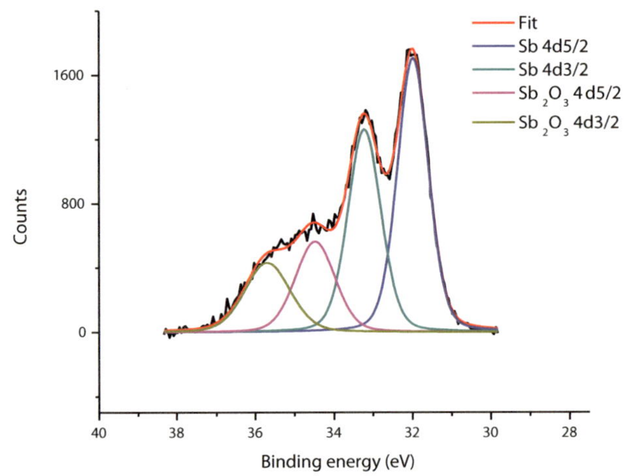

Figure 2. XPS spectrum for Sb *4d* line on n-GaSb.

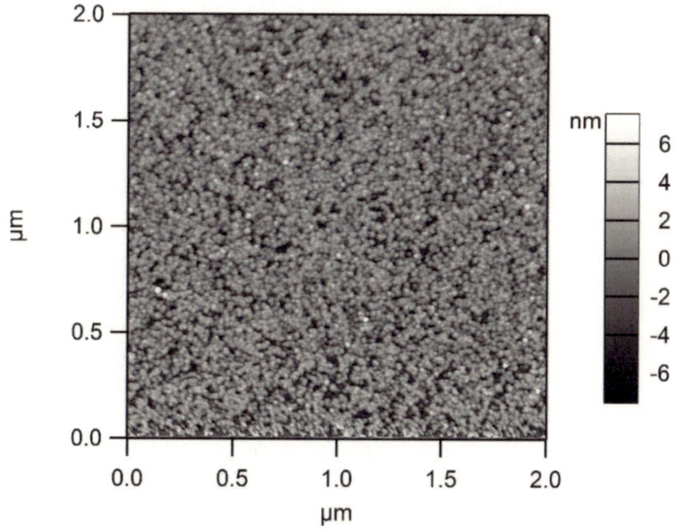

Figure 3. AFM image of GaSb covered with native oxides.

was used for BE calibration scale. The O *1s* is relatively broad and its main part arises from the chemisorbed oxygen at BE = 532 eV. The Ga and As *2p* spectra from n-GaAs(100) Te doped substrates, that contains the effect of naturally oxidized semiconductor surfaces are taken at TOA = 90°.

Figure 4 presents the As *2p* spectra from native oxides on GaAs where the line related to BE = 1323.36 eV arises from As in GaAs and the line corresponding to BE = 1326.06 eV is related to As in As_2O_3. The quantitative analysis indicates a percentage of 21.4% As in GaAs and a percentage of 78.6% that belongs to a majority of As signal from oxide (As_2O_3).

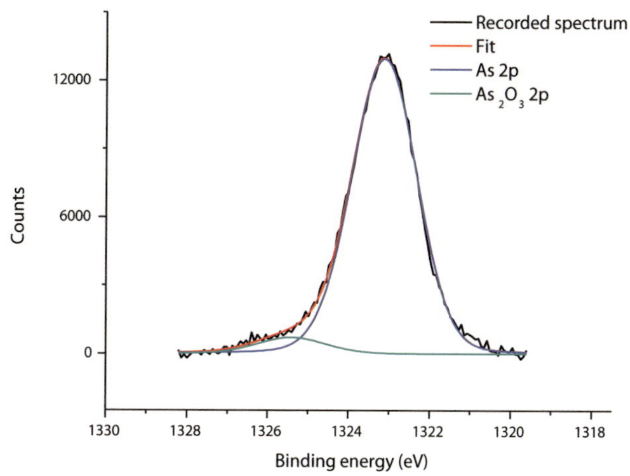

Figure 4. XPS spectra of As *2p* in GaAs and in As_2O_3.

Figure 5 presents the Ga *2p* spectra from native oxides on GaAs where the line related to BE = 1116.80 eV it arises from Ga in GaAs and the line corresponding to BE = 1118.37 eV is related to Ga in Ga_2O_3. The quantitative analysis indicates a percentage of 11.7% Ga in GaAs and a percentage of 88.3% that belongs to Ga in Ga_2O_3.

As can be observed in **Figure 6** the as-received surface of n-GaAs is covered by a conglomerate of Ga_2O_3 and As_2O_3, determining on the surface a roughness characteristic of RMS ~ 0.21 nm.

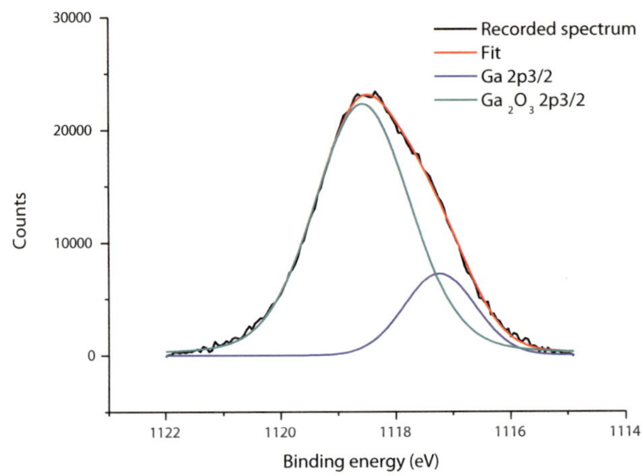

Figure 5. XPS spectra of Ga *2p* in GaAs and in Ga_2O_3.

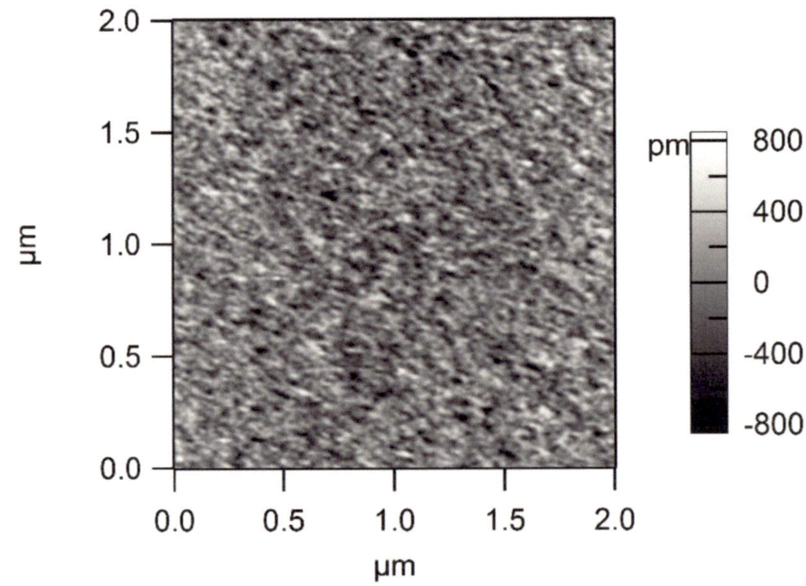

Figure 6. AFM image of naturally oxidized n-GaAs(100) surface.

http://dx.doi.org/10.5772/67916

Then n-GaP(111) native oxides structure composition were studied by XPS analysis on Ga *3d* and As *2p* lines where the measurements were performed at TOA = 45°. These XPS lines offer information about the presence of Gallium and Phosphorous in bulk GaP and in native oxides namely Ga_2O_3 and P_2O_5. In this regard, the XPS spectra included Ga $3d^{5/2}$, 19.00 eV and Ga $3d^{3/2}$, 19.49 eV in GaP and Ga $3d^{5/2}$, 20.53 eV and Ga $3d^{3/2}$, 21.52 eV in Ga_2O_3. Related to P line there are present the peaks P $2p^{3/2}$, 128.63 eV and P $2p^{1/2}$, 129.58 eV arisen from GaP and P $2p^{3/2}$, 133.36 eV and P $2p^{1/2}$, 134.39 eV peaks arisen from P_2O_5. The XPS signals are presented for Ga *3d* in **Figure 7** and for P *2p* in **Figure 8**.

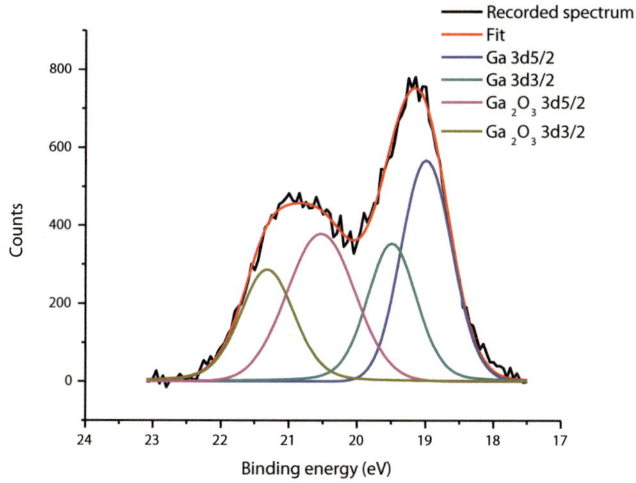

Figure 7. XPS spectra of Ga *3d* on n-GaP(111) oxidized surface.

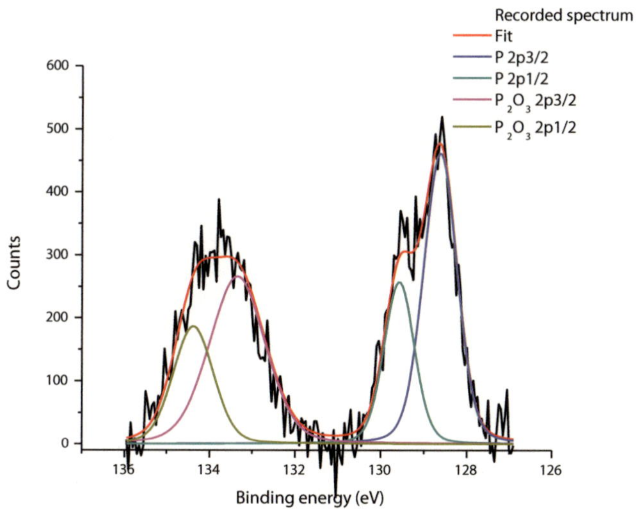

Figure 8. XPS spectra of P *2p* on n-GaP(111) oxidized surface.

The relative composition regarding GaP surface indicated a concentration of 52.7% Ga in GaP and 47.45 Ga in Ga_2O_3 and for P a percentage of 50.9% P in GaP and 49.1% P in P_2O_5. This surface structure indicates a relative balance in Ga and P, with a trend to stoichiometric value. The AFM image of GaP(111) naturally oxidized surface is presented in **Figure 10** and indicates a relative uniform distribution of Ga_2O_3 and P_2O_5.

The AFM image from **Figure 9** for n-GaP(111) surface indicates the existence of a conglomerate of Ga_2O_3 and P_2O_5 that determines on the surface roughness characteristic of RMS ~ 4.54 nm.

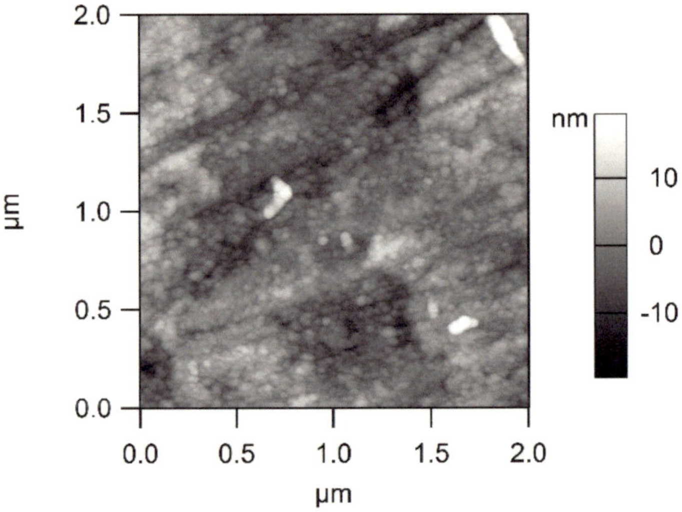

Figure 9. AFM image of naturally oxidized n-GaP(111) surface.

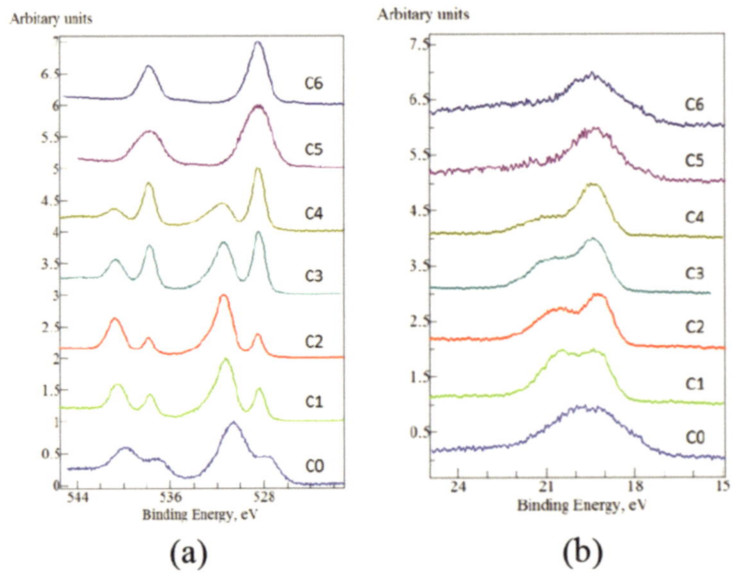

Figure 10. Superposition of XPS spectra for Sb *3d* (a) and Ga *3d* (b) at TOA: 90° (C0- sample as received; C1, C2, C3, C4, C5, C6).

4. Surface modification

As it was stated in Ref. [23], the implementation of GaSb-based alloys in device technology is depended on reproducible control of their surface properties. In this view, the real GaSb surfaces are modified by technological processes namely cutting, smoothing, or mechanical polishing of semiconductor wafer a fact that conduced to surface damage regions. Compared to GaAs or InP, the GaSb surface is more reactive and GaSb surface oxidized rapidly in ambient conditions and the resulted oxide layer is not self-limiting, stable nor abrupt [24]. Chemical reactivity of antimonides present them as a distinct category of materials [25]. The irreversibility character of the reactions implies that oxygen atoms, rather than molecules are ultimately involved in chemical bonding. It is important to note [23] that for oxygen involved there is a movement from the chemisorption concept (i.e., adsorbed layer structure is a function of underlying surface) to the complex formation concept characterized by a film with its own chemical and structural identity. Generally speaking the different structure of III-V semiconductor surface determined a basic influence on the surface electrical properties. This influence manifest itself by the generation of electronic states (i.e., surface and resonant states) that induced the Fermi level pinning on semiconductor surface. Fermi level pinning may conduced to density of surface states in the order $10^{12} \div 10^{13}$ cm^{-2}. Chemisorption and chemical reactions may conduced to a modification of surface states and to the generation of new donor and acceptor states in the band gap. Study of the preparation methods of semiconductor surfaces in order to obtain a surface with characteristic as: free of damage, contamination, and native oxides is essential for reproducible and efficient control of surface properties. It is observed [23] that in device technology practice contact with atmosphere is unavoidable and surface preparation usually takes place by etching processes. The surface of n-GaSb(100) Te doped is generally chemical prepared by a cleaning procedure in organic solvents, e.g., trichloroethylene, acetone, and afterwards is chemical etched HF:H_2O (DIW) (1:1) for 15 seconds, and HCl:H_2O(DIW) (1:1) for 15 seconds and then rinsed in DIW. In the following will be presented the effect of chemical etching in a mixture of HCl:H_2O(DIW):$C_6H_8O_7$ (citric acid), after an exposer at room temperature for an etching time in the range 35 ÷ 300 seconds. The evolution of native oxides was put into evidence by a XPS analysis at TOA = 90°, as is presented in **Figure 10**. It can be observed that the Sb signal from GaSb is increasing and the signal from oxide is decreasing (91.5 to 8.5% in composition table) and the signal from Ga in GaSb has a higher intensity as the O line in Ga oxide. This effect is a consequence of the surface chemical etching and to the fact that for this solution the Sb oxide and Ga oxide are both sensitive that is why this etchant can be used further in technology for a rapid cleaning of the n-GaSb(100) surface. The evolution of Sb native oxides and Ga native oxides on n-GaSb(100) as a result of chemical etching can be seen in the superposition presented in **Figure 10(a)** for Sb *3d* spectra in different stages of etching namely: C1 (t_1 ~ 30–35 seconds); C2 (t_2 ~ 65 seconds); C3 (t_3 ~ 90 seconds); C4 (t_4 ~ 120 seconds); C5 (t_5 ~ 180 seconds); C6 (t_6 ~ 300 seconds); (b) for Ga *3d* spectra in the same chemical etching stages. The signal arising from Ga *3d* shows the same trend for Ga oxides, the surface Ga oxides are removed after C5, which means that both Sb and Ga oxides had the same behavior, and the sample is prepared in these conditions of chemical etching for further processing.

Figure 11 presents the AFM image of n-GaSb(100) surface after 90 seconds of chemical etching. As can be observed there is a strong attack of the etching solution at the surface where the network aspect is characterized by valleys and hills and the roughness is high, i.e., RMS ~ 13.96 nm.

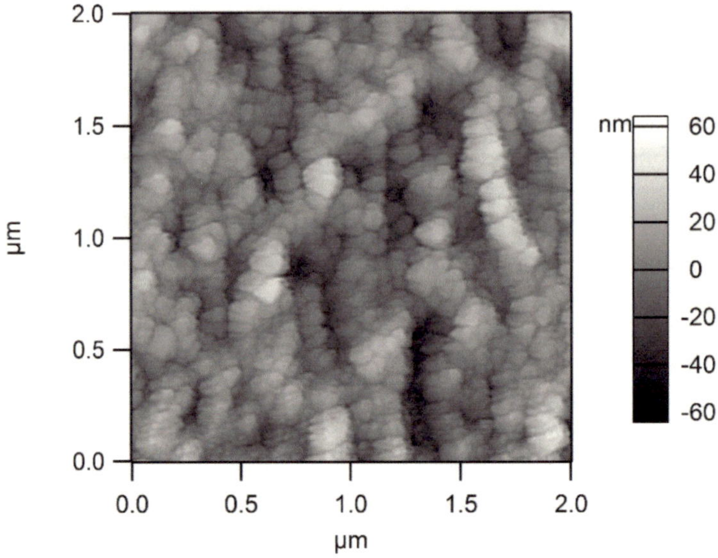

Figure 11. AFM image of n-GaSb(100) after 90-second chemical etching.

The effect of Ar^{+} ion etching on n-GaSb native oxides for different voltages and etching times can be observed from XPS spectra, i.e., C1 (0.5 kV, t_1 = 3 minutes); C2 (1 kV, t_2 = 3 minutes); C3 (1 kV, t_3 = 5 minutes); C4 (2 kV, t_4 = 10 minutes). **Figure 12** presents the XPS signal for a TOA = 20° for Sb *4d* line, a signal sensitive to the surface composition.

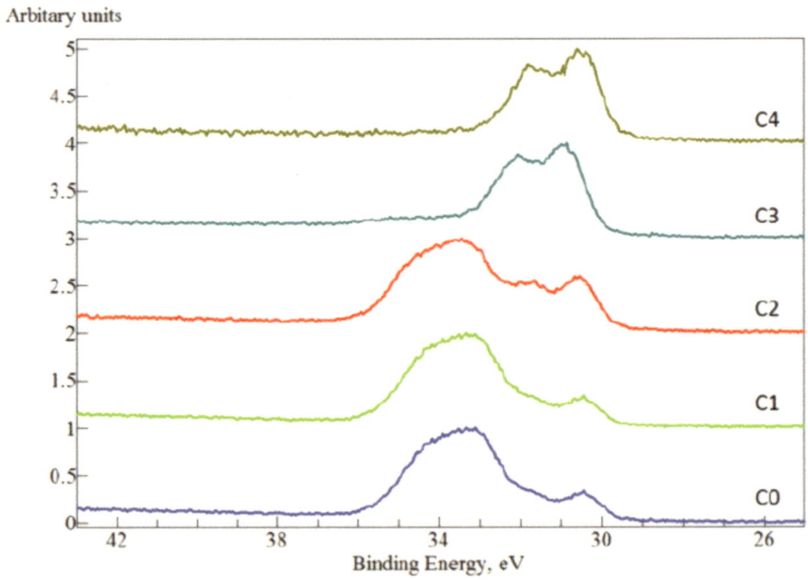

Figure 12. The evolution of Sb oxide after ion etching for Sb 4d signal C0- as received, C1, C2, C3, C4).

The signal for antimonide oxides present on the as-received sample (a sample covered with native oxides) starts to modify its intensity during etching process, and at the last Ar^+ etching (C4) it remains only the signal from Sb *4d* in the bulk of GaSb (~35 eV, ~32 eV). In the evolution of native oxides on GaSb during Ar^+ ion etching can be observed that the Sb oxides are the first one to be cleared off, and then are cleared off Ga oxides [26]. As the result of an ion beam cleaning process, the native oxides vanishes, e.g., Ga oxides signals are the first ones with a decreasing intensity until Ga *3d* signal from Ga_2O_3 disappears after C4 ion etching.

The effect surface passivation by chemical reaction on n-GaSb(100) with sulfur compounds (e.g., S_2Cl_2 and $(NH_4)_2S$) is presented in **Figure 13** as an AFM image.

Passivation by S_2Cl_2 chemical reaction and the presence of GaSb-S thin films exhibited moderate to high toxicity against gram-positive bacterial strain (*Staphylococcus aureus*), demonstrated by the formation of a clear inhabitation zone around the coated substrate. The presence of sulfur bond is putted into evidence by the S *2s* XPS as recorded signal observed in **Figure 14**.

We consider that the passivated n-GaSb studies open a gate for a possible active biological GaSb device obtained on a sulfur-passivated surface in a compatible technology of III-V compounds.

A real GaAs surface is covered with a relatively thick layer (a few nanometers) of native oxide, which is the origin of a high density of surface states $N_{SS} \sim 10^{12}$ cm^{-2} (under technological conditions) pinning the surface Fermi level within the band gap of the semiconductor. In the case of GaAs, as it was mentioned in Ref. [20], the GaAs(100) surfaces have a high surface energy and as a consequence they are very reactive and chemically unstable. Regarding the

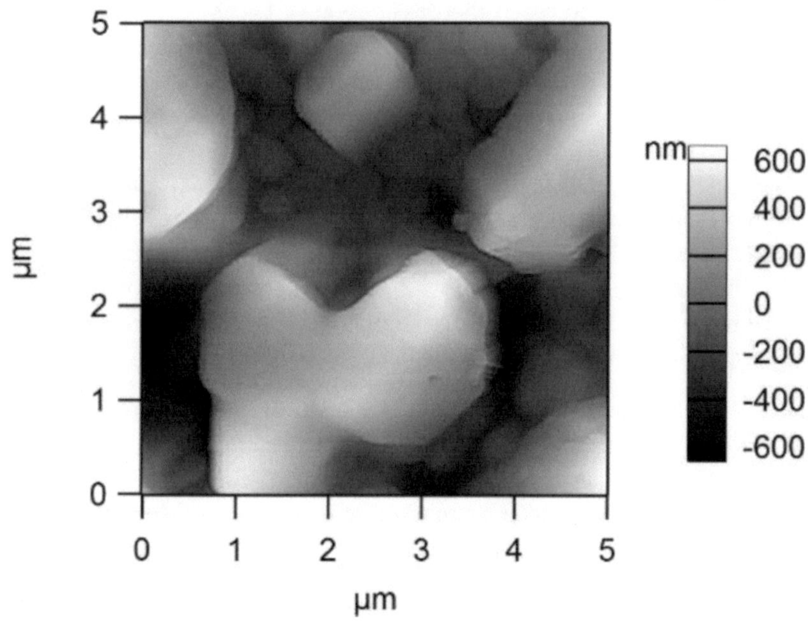

Figure 13. Aspect of n-GaSb(100) surface exposed to S_2Cl_2 action.

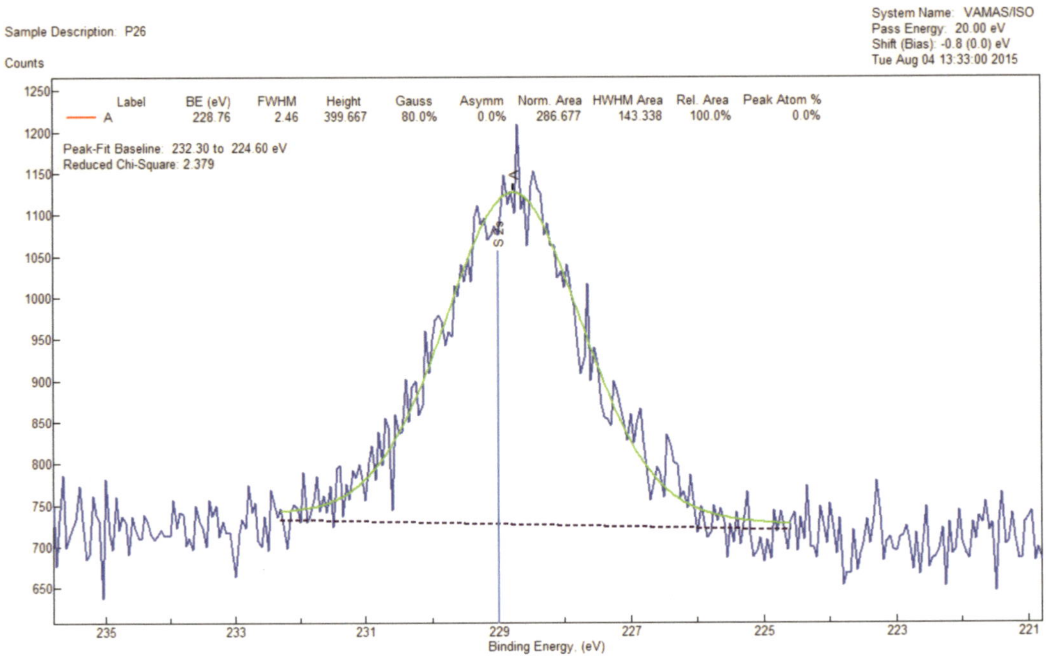

Figure 14. XPS recorded signal for S 2*s* line (linked to Sb) on n-GaSb passivated by S_2Cl_2.

surface analysis of GaAs native oxides in the case of XPS measurements there exist a great dispersion of XPS values [27]. In the perspective of native oxides reactivity, it is worth to understand that it is difficult to reach the passivation of GaAs surfaces. The aim of surface passivation is to reduce the surface state density, or preferably, to remove them from the band gap. The surface of n-GaAs(100) Te doped was prepared by a cleaning procedure at room temperature in organic solvents, i.e., trichloroethylene, acetone, and afterward it is generally chemical etched in $HCl:H_2O_2$ (1:1) for 30 seconds and then rinsed in DIW or chemical etched in $H_2SO_4:H_2O_2:H_2O$ (DIW) (1:1:1) for maximum 5 minutes and then rinsed in DIW. The As 2*p* XPS signal from the chemical etched n-GaAs(100) is observed in **Figure 15**. The signal derived from As in As_2O_3 after the chemical cleaning procedure is close to the background, a fact that suggested that the native oxide was removed. The same conclusion derived from the XPS Ga 2*p* signal from the chemical-etched GaAs(100) is observed in **Figure 16**. The Ga line from Ga_2O_3 after cleaning procedure is close to the background as a result of native oxide removal.

A variant of n-GaAs(100) surface preparation is the sulfur passivation by organic and inorganic compounds. The sulfur treatment of the GaAs surface in sulfide solutions results in the removal of the native oxide layer and the formation of As-S and Ga-S covalent bonds [28], where the processes are accompanied by a marked reduction of surface recombination velocity.

In this perspective in the case of GaAs the chemical treatment by sulfur tend to minimize the values of N_{SS} and v_s (surface recombination velocity) ~ 10^7 cms^{-1}. From a chemical point of view sulfur passivation involves a removal of native oxide layer and the formation of thin protective layer on n-GaAs in order to prevent its oxidation in atmosphere. From the electrical

http://dx.doi.org/10.5772/67916

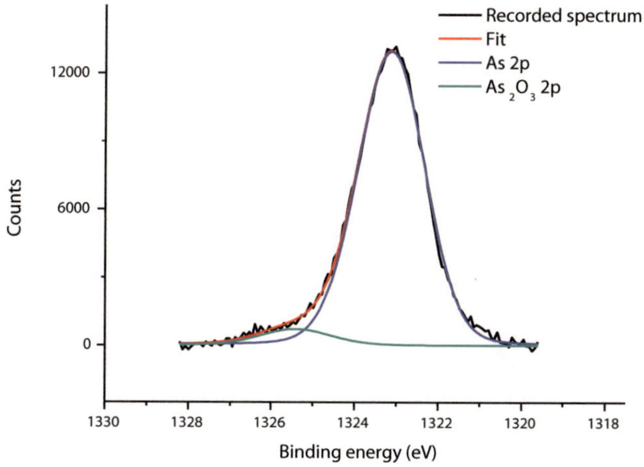

Figure 15. As *2p* XPS spectra from n-GaAs chemical etched surface.

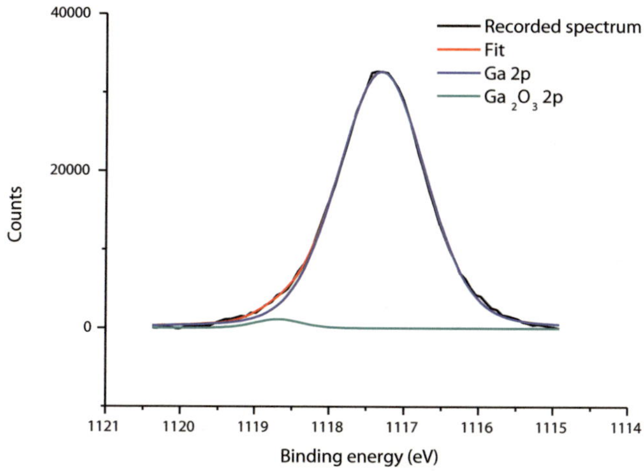

Figure 16. Ga *2p* XPS spectra from n-GaAs chemical etched surface.

point of view the passivation is related as was stated before with the reduction of surface state density and as a consequence is the change of surface potential barrier on GaAs. Prior to sulfur treatment the wafers were degreased in trichloroethylene (boiling for 2 minutes), rinsing in acetone and then chemical etched in HCl:H_2O (DIW) (1:1) for 100 seconds at room temperature. The n-GaAs wafers presented the aspect of an optical polished surface. The sulfide treatment was carried out in two different solutions namely: (a) pure ammonium sulfide $(NH_4)_2S$ (50% H_2O)—as stated an aqueous solution and (b) solution of sulfur monochloride (S_2Cl_2) in carbon tetrachloride (CCl_4) (1:10).The SEM micrograph on sulfur-treated n-GaAs(100) revealed a uniform deposition of sulfur compounds (S_2Cl_2 and $(NH_4)_2S$) on different regions on the sample area of over 1 cm^2 as presented in **Figure 17(a, b)**.

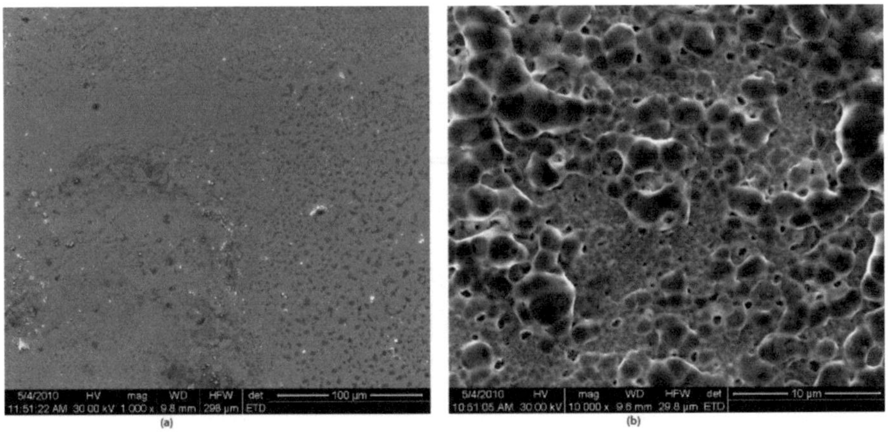

Figure 17. SEM images for (a) S_2Cl_2 and (b) $(NH_4)_2S$ deposition on n-GaAs(100).

In the followings, we present the results of a more complex procedure for n-GaAs(100) cleaning, namely combined chemical etching with controlled thermal treatment. The as-received GaAs wafers were firstly chemical etched in HCl:H_2O_2 (1:1) for 30 seconds and rinsed in H_2O (DIW) and afterwards exposed to thermal treatment at 200°C for t = 1 h in high vacuum (p ~ 10^{-9} torr). The XPS spectra of the as-prepared GaAs(100) surface recorded for As *2p* and Ga *2p* lines are presented in **Figure 18(a, b)**.

As can be observed the XPS signals for As and Ga in native oxides vanish and the only signals present are the ones for As and Ga in GaAs. The XPS extracted information indicate that the combined cleaning procedure used is effective, the n-GaAs(100) is prepared for further technological stages. Another possibility for preparing n-GaAs(100) surface is the cleaning by physical means, namely sputtering technique. The experiment of Ar^+ ion beam etching implied an accelerating voltage of U = 500 V, for medium time interval of 100 seconds.

Figure 18. XPS spectra for (a) As *2p* and (b) Ga *2p* for a complex treatment of n-GaAs(100) surface.

The result of the as-prepared n-GaAs(100) surface indicates that the ion beam etching is effective in cleaning the GaAs native oxides, but the exposed surface is damaged by ion beam. This damaging effect is observed in the XPS spectra for As *3d* and Ga *3d* line as an enlargement of main peak (FWHM-Full Width at Half Maximum). In **Figure 19(a, b)** can be observed the absence of As and Ga signals from native oxides (As_2O_3, Ga_2O_3), that indicated the presence of a cleaned surface, together with an enlarged signals of As and Ga arisen from GaAs ion damaged surface.

The naturally oxidized n-GaP(111) samples were cleaned by chemical etching in a solution of H_2SO_4:H_2O_2:H_2O (DIW) (3:1:1) for t = 5 minutes at a temperature of T ~ 39°C and then rinsed in DIW. As it has been observed in Ref. [8], the P/Ga atomic ratio on the semiconductor surface has a near stoichiometric value. For a complete preparation of GaP surface in order to be exposed to further technological procedures, the chemical etched wafer was exposed to a thermal treatment at about T = 200°C for a time range t = 2 h in high vacuum (p ~ 10^{-9}) torr conditions. The cleaning effect of chemical-thermal treatment can be observed in the Ga *3d* and P *2p* XPS surface recorded spectra as they are presented in **Figure 20**. The effect of surface

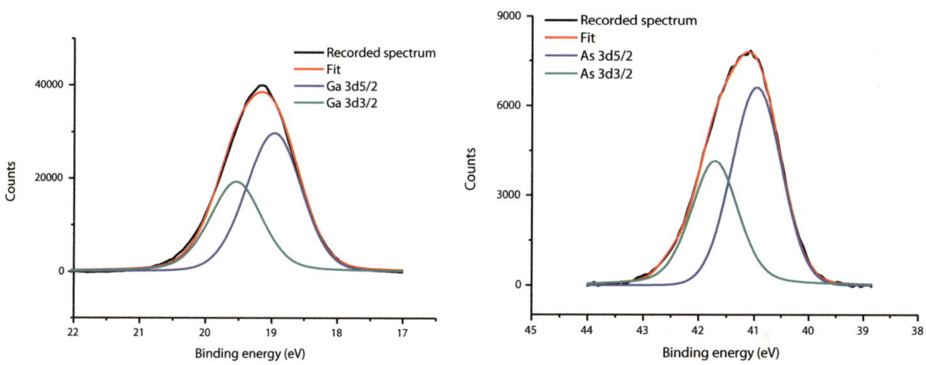

(a) **(b)**

Figure 19. XPS spectra of (a) Ga *3d* and (b) As *3d* lines on an ion beam etched n-GaAs(100) surface.

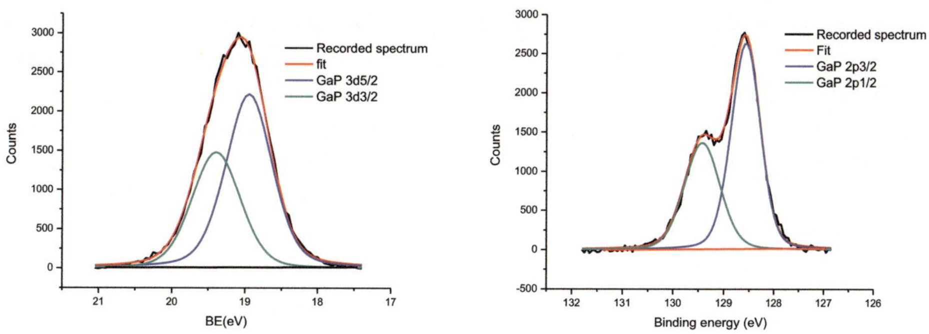

(a) **(b)**

Figure 20. XPS spectra of (a) Ga *3d* and (b) P *2p* for as-prepared n-GaP(111) surface.

preparation is characterized firstly by the vanishing of native oxide signals (Ga_2O_3 and P_2O_5) and secondly by the spectral line characteristics of a FWHM ~ 0.8 eV that indicates a well reconstructed surface without surface damages and with a Ga/P ~ 1 ratio, that means a stoichiometric aspect. The simple GaP wet chemical etching can remove the native oxide layer but as soon as the sample is exposed to air, the wafer is again naturally oxidized.

The Ar^+ ion sputtering GaP prepared surface, in the energy limit of 1 ÷ 5 keV and sputter time t = 5 minutes, indicates [8] a decrease of the XPS signal arising from oxides. The oxide overlayer was removed by sputtering and the changes in both core-level signals of Ga and P indicate that sputtering process causes P depletion in n-GaP surface. The Ar^+ ion bombardment generates a near-surface layer depleted of "volatile atoms" (Ar^+ sputter rates for GaP, GaAs are relatively low, i.e., <250 nm/min [29]) and the compositional changes generated by preferential sputtering is propagating into bulk by diffusion processes. In this view, the distribution of Ga and P atoms in outer layers can be explained by the competition between preferential sputtering of P versus Ga and radiation-enhanced surface diffusion Gibbson segregation by the analogy with the distribution mechanism for Ga and As [30, 31]. Due to the preferential sputtering, the resulted surface composition is considerably different from the stoichiometric bulk composition and the observed depletion of P species from the surface is related to a sublimation energy of P smaller than of Ga [30]. This preferential sputtering of P and P_2O_5 in the outer layer is responsible for the increase of Ga/P ratio to 1.5 and in consequence to a surface enrichment in Ga atoms. The Ar^+ ion sputtering of n-GaP(111) surface is efficient in removing the native oxides, but the remaining etched surface is intensely nonstoichiometric and the reconstruction procedure is more complicated.

5. Aspects of metal deposition on prepared surface of III-V compounds

As it was presented, gallium antimonide (GaSb) is of special interest as a substrate material for device applications as laser diodes with low threshold voltage, photodetectors with high efficiency, high frequency devices, or high efficiency thermophotovoltaic (TPV) cells. Fabrication of PV discrete devices generally begins with defining the active structure by a Schottky photosensitive contact or the growth of MBE (molecular beam epitaxy) heteroepitaxial layers. One of the most important parts of fabrication procedure is obtaining low resistivity ohmic contacts on broad area of photosensitive surface in order to realize a competitive serial resistance (R_S). For a PV device, a good R_s assures an optimal fill factor (ff) related to an extended spectral response of photocurrent (I_{ph}). The problems connected to the formation of ohmic contacts on p- or n-GaSb represent a technological skill considered in general a real state-of-art in obtaining semiconductor devices [3]. The investigation of the characteristics for low resistance Au/Ge/Ni nanometric contact layers deposited on n-GaSb(100) is a part of a general technological effort in order to define conditions for obtaining a PV competitive device. The Au/Ge/Ni with characteristic, i.e., Au (139 nm)/Ge (72 nm)/Ni (14 nm), contact layers were deposited in medium vacuum p ~ 8×10^{-5} torr) and then annealed in low vacuum (p ~ 7×10^{-2} torr) low temperature (T ~ 300°C) for t = 3 minutes. In this contact technology, the role of Au is to facilitate Ga diffusion from the surface, the role of Ge is to occupy the Ga vacancies

http://dx.doi.org/10.5772/67916

in GaSb and to dope the first semiconductor layers in a n^+ fashion, the consequence being the decrease the ohmic characteristic for the contact. In this conjuncture the Ni part is to assure the wettability of the contact and the thermal stability [32]. In order to analyze the as-mentioned characteristics of Au/Ge/Ni deposition the depth profiling measurement of AuGeNi contact was obtained in SPECS system by XPS analysis after controlled plasma etching (Ar^+ ions at energy E = 3 KeV, pressure p ~ 10^{-5} torr, ion current I_{ic} = 10 μA). In **Figure 21**(**a–c**), the as-recorded…" is appropriate.

Figure 21(**a–c**), the as-recorded XPS spectra for Au *4f*, Ge *2p*, and Ni *2p* arisen from the initial metal deposition n-GaSb surface are presented.

The effect of Ar^+ ion beam sputtering in the experiment of depth profiling can be observed in **Figure 22**(**a–c**) from the XPS recorded after the 16th controlled etching.

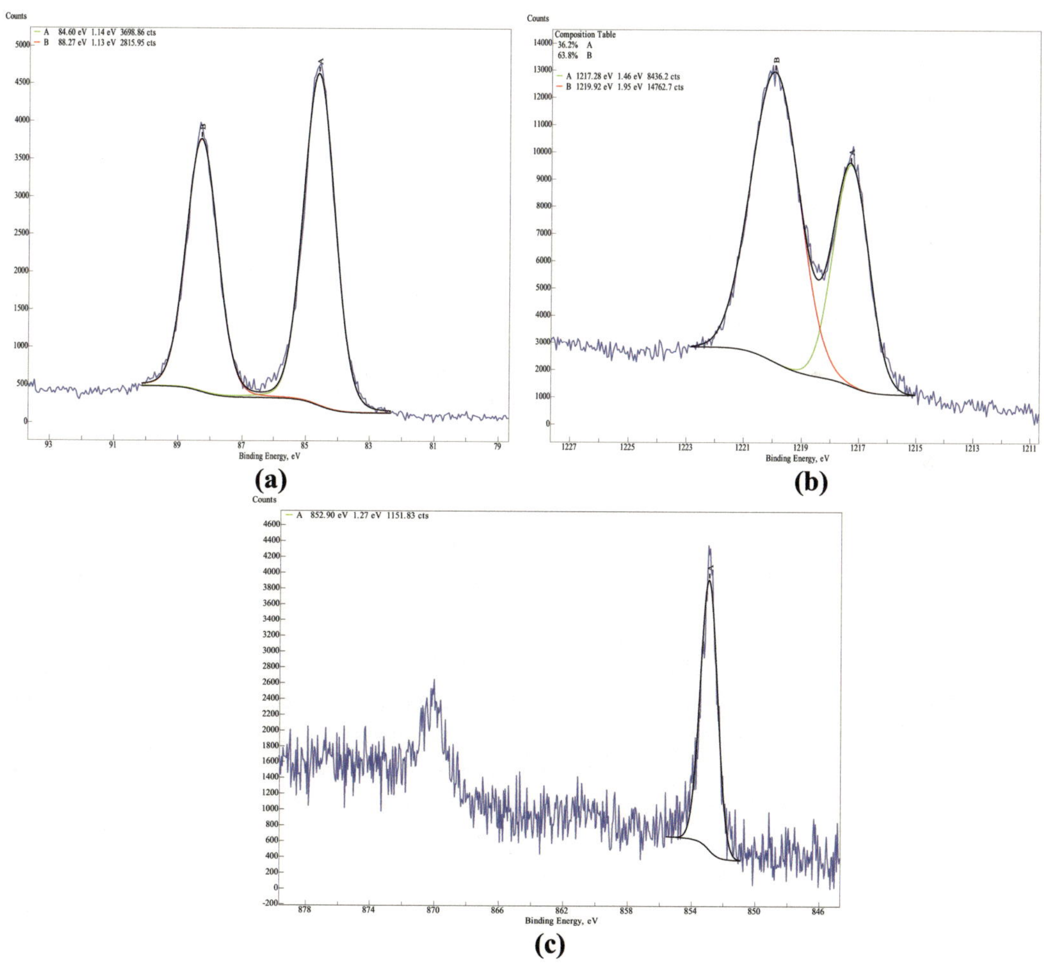

Figure 21. XPS spectra of (a) Au *4f*, (b) Ge *2p*, and (c) Ni *2p* deposited on n-GaSb(100).

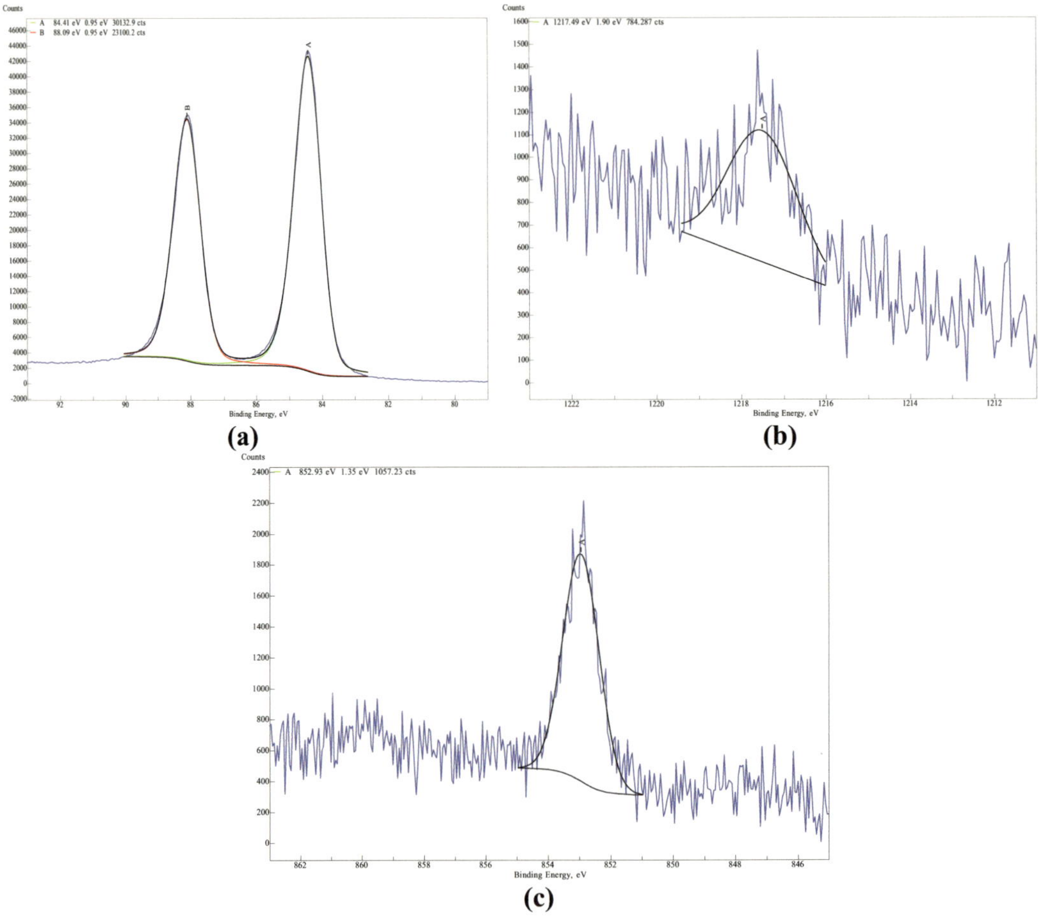

Figure 22. XPS spectra of (a) Au *4f*, (b) Ge *2p*, and (c) Ni *2p* after 16th ion beam etching.

The AFM aspect of the AuGeNi surface after controlled ion beam etched can be observed in **Figure 23**.

The elemental composition of the sample namely AuGeNi/n-GaSb was also investigated by Rutherford Backscattering Spectrometry (RBS), as a surface sensitive analysis technique. The depth profiles of the elements contained in the deposited contact layer are presented in **Figure 24**. The scattering profile of Au appear at an energy of 2769 keV. The thickness was determined to be 1500×10^{15} at/cm^2. The concentration of Au is increasing from 30 at. % to 70 at. % from the surface to a depth of 500×10^{15} at/cm^2. The depth profile of Ge is rather similar. The concentration of Ge is also increasing from 15 at. % to 30 at. % from the surface to a depth of 500×10^{15} at/cm^2. Ni was present only in the first 500×10^{15} at/cm^2 part of the layer, with a decreasing concentration starting from 55 down to 15 at. %.

The initial defining of AuGeNi as an ohmic contact on n-GaSb can be remarked in **Figure 25** presenting the I-V characteristics of a MBE Ni/GaSb Schottky diode (Ni contact thickness ~ 2 nm).

http://dx.doi.org/10.5772/67916

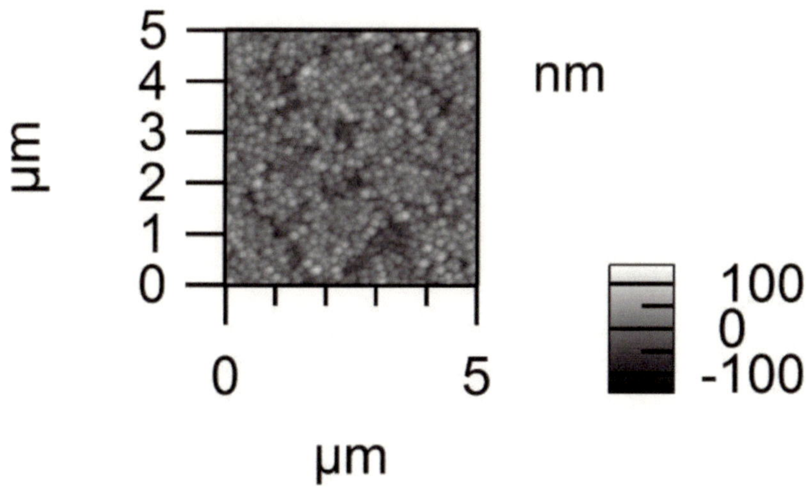

Figure 23. AFM characteristics of AuGeNi surface after Ar^+ ion beam sputtering.

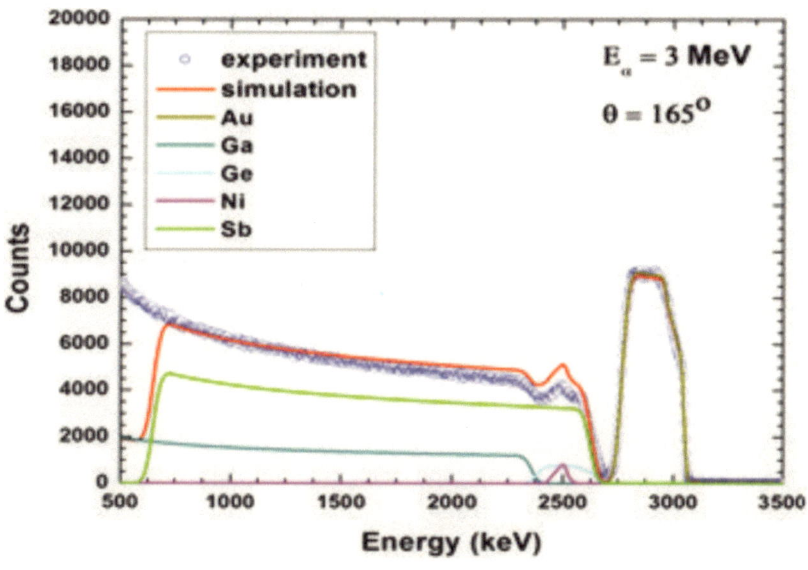

Figure 24. RBS spectrum of a AuGeNi deposited on n-GaSb substrate and a theoretical simulation for the elements present in the sample.

Gallium arsenide has been long time considered as one of the most important semiconductor material besides silicon, due to his attractive intrinsic electrical properties: direct energy and wider bandgap, higher carrier mobility, high power and operating temperature [33]. A key element in GaAs devices technology consists in the obtaining of high quality ohmic contacts because the surface of this compound is often covered with a chemically unstable oxide layer. Developed on a trial-and-error basis, the AuGeNi alloyed ohmic contacts remain the most widely used for n-GaAs devices manufacturing. This metallization system is characterized

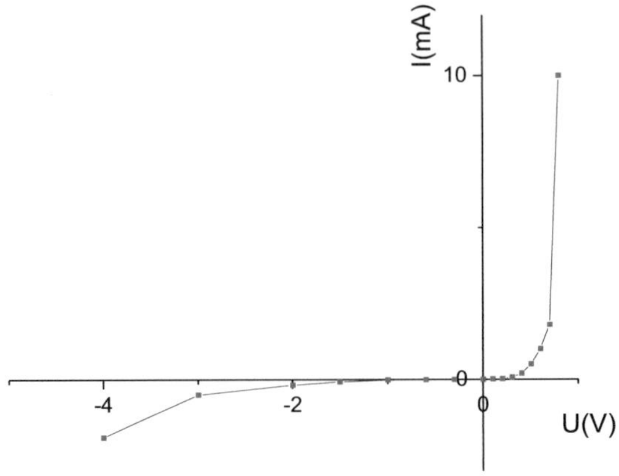

Figure 25. I-V characteristics of a MBE Ni/GaSb Schottky contact (back side ohmic contact is AuGeNi).

by low contact resistance, good thermal stability both during device fabrication and device operation, strong adhesion, low metal sheet resistance [34]. In this system, namely AuGeNi/n-GaAs(110) [35], Au acts as a base metal and his role is to promote Ga vacancies in n-GaAs due to high solubility of Ga in Au. Ge acts as a dopant element, diffusing into the lattice sites vacated by Ga, creating in this way a heavily doped n-type intermediate semiconductor layer that allows the tunneling mechanism which leads to the ohmic behavior. Ni acts as a catalyst during the alloying procedure to improve the uniformity, thermal and mechanical stability, and a decrease of surface roughness [32]. **Figure 26** presents the depth profile of AuGeNi layer and the interface with the n-GaAs semiconductor. From this figure, it can be seen that while the As concentration tends to grow steadily, the Ga concentration varies in a different manner: it is greater in the first layers, then starts to decrease to a minimum of 35% and then is growing once more. This last growing is explained through the decrease of Au concentration which is more aggressively removed from the surface by ion sputtering. This indicates that both Ni and Ge diffuse deeper in the n-GaAs semiconductor. As presented in **Figure 27(a, b)**, XPS spectra for Ge *3d*, Au *4f* before sputtering, firstly for the Ge *3d*, the spectrum is composed from two peaks: one situated at 29 eV binding energy and the other situated at 32.2 eV. The first component, representing 32% from total peak area, is attributed to elemental Ge, while the second component, containing the rest of 68% from peak area, is attributed to Ge_2O_3 specie. Even after the first sputtering session, the oxide completely disappears, remaining only the Ge metal component. The attempting to fit the Ge *3d* XPS recorded spectra with more than one peak provided unsatisfactory results and as a consequence the formation of other compounds, e.g., $GeAs_2$ or Ni_3Ge assigned to XPS BE of 29.7 and 29.1 eV is not realistic. From the XPS spectra Ni *2p* it is observed that Ni is missing in the first layers of the surface and the signal is arising from a metallic state.

Absence of shake up satellites indicates once more that Ni oxide is present in the layer. The evolution of Ni peak shape (e.g., close to the interface FWHM varies from 1.6 to 1.9 eV) is explained by the fact that a portion of Ni interacts directly with GaAs substrate in the semiconductor "wetting" process for which Ni role is intended. The presence of Ni even at

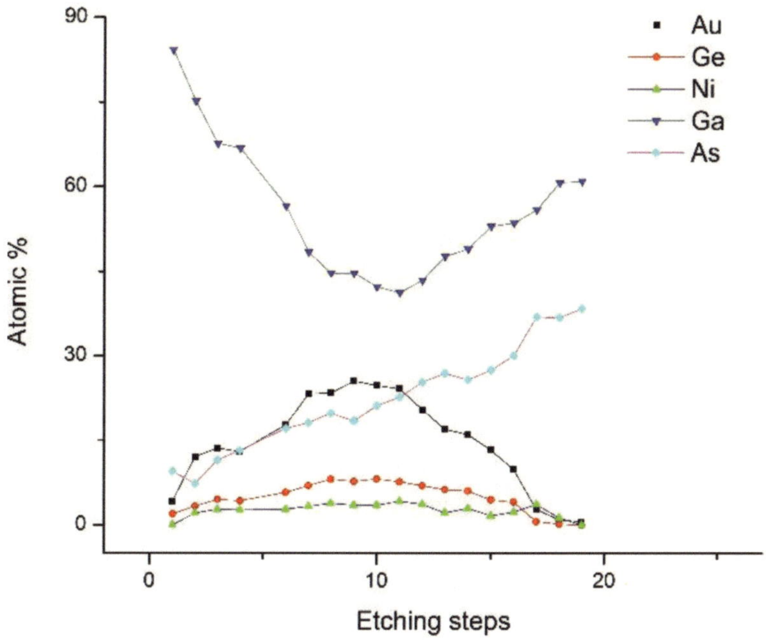

Figure 26. AuGeNi/n-GaAs profiling of chemical composition.

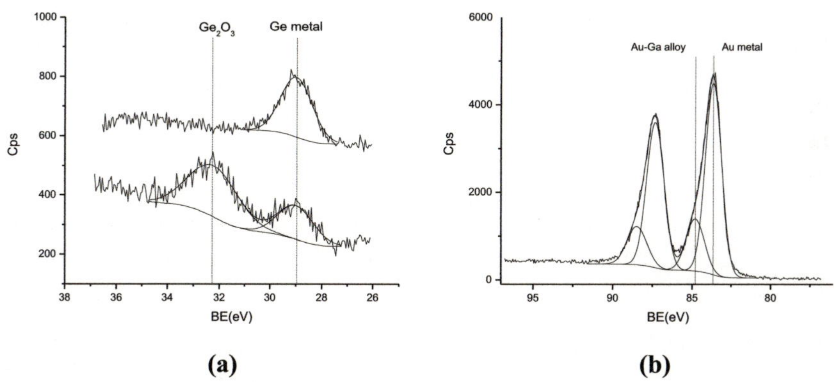

Figure 27. XPS spectra of (a) Ge *3d* and (b) Au *4f* at metal/semiconductor interface.

the interface is beneficial for the ohmic character of contacts in the sense that is reducing the lateral diffusion of Au and holds the Au-Ge melt in intimate contact with GaAs substrate during thermal annealing. From Au *4f* XPS doublet recorded, it is remarked the gold from surface to the interface. Spectra fitting of $4f^{7/2}$ line was made with two peaks, where the first one is attributed to the metallic gold Au $4f^{7/2}$, while the second is attributed to an alloy of Ga and Au [35], and the area of the second peak is less than a quarter of the total area. The results indicated that the chemical shift of Au $4f^{7/2}$ tends to be growing with the proportion of Ga, from 0.4 to 1.25 eV for AuGa, and 1.55 eV for $AuGa_2$ [36]. The chemical shift of 1.15 eV found in our experiment indicate more Au than Ga in the alloy but with close to 1 Ga/Au ratio. As a general remark Au, Ge and Ni are uniformly distributed in the metallic layer AuGeNi/n-GaAs(110) and both Ga and As diffuse to the surface, with a more pronounced Ga diffusion.

The as-prepared surface n-GaAs by controlled Ar^+ ion beam sputtering reveal the absence of native oxides and the metal deposition technique for obtaining good ohmic contacts is convenient and competitive to GaAs device technology.

6. Conclusions

Surface preparation of semiconductors, in particular, for III-V compounds is a necessary requirement in device technology due to the existence of surface impurities and the presence of native oxides. The impurities can affect the adherence of ohmic and Schottky contacts and due to the thermal decomposition of native oxides (e.g., GaSb) it also affect the interface metal/semiconductor. The practical experience reveals that the simple preparation of a surface is a nonrealistic expectation, i.e., surface preparation is a result of combined treatments. In this view, the ion beam sputtering surface preparation is an efficient technique that removes rapidly and in *in situ* the native oxides. The effect of major cleaning of semiconductor surface is accomplished, however, by surface damaging. The removal of surface damage is obtained by cycles of thermal annealing at a relative high temperature. It is worth to mention GaSb or GaAs III-V compounds, the ion beam etching leaves a surface enriched in Ga atoms, that means the presence of a nonstoichiometric characteristic. In the case of chemical etching preparation, the etchants generally specific to the semiconductor are selective in respect to the crystallographic plane (i.e., the etch rate different). The chemical etching is an effective and rapid *ex situ* preparation technique that can leave a slightly nonstoichiometric surface (e.g., there exists a depletion in Ga atoms). Also, the chemical etching leaves traces of native oxide. In this stage, the preparation includes thermal treatments, but the annealing implies a temperature lower that in case of ion beam cleaning.

The efficient technological method for III-V compounds surface preparation is considered to be the combination between chemical etching and controlled thermal treatment. The effect of surface preparation can be observed by XPS for the surface chemical quality, by AFM for surface morphology, and by LEED for the structural characteristics.

Acknowledgements

The authors thank for the financial support provided by the Ministry of Education from Romania-UEFISCDI Project No. 68/2014-Partnerships.

Author details

Rodica V. Ghita*, Constantin Logofatu, Constantin-Catalin Negrila, Lucian Trupina and Costel Cotirlan-Simioniuc

*Address all correspondence to: ghitar@infim.ro

National Institute of Materials Physics, Bucharest, Romania

References

[1] I. Vurgaftman, J.R. Meyer, and L.R. Ram-Mohan, J. Appl. Phys., 89, 5815 (2001).

[2] K. Nakahara et al., Electron. Lett., 32, 1585 (1996).

[3] P.S. Dutta, H.L. Bhat, and V. Kumar, J. Appl. Phys., 81, 5821 (1997).

[4] G. Motosugi and T. Kagawa, Jpn. J. Appl. Phys., 19, 2303 (1980).

[5] O. Hildebrand, W. Kuebant, and M.H. Pilkuhn, Appl. Phys. Lett. 37, 80 (1980).

[6] C. Hilsum and H.D. Rees, Electron. Lett., 6, 277 (1970).

[7] L.M. Fraas, G.R. Girard, J.E. Avery, B.A. Arau, V.S. Sundaram, A.G. Thomson, and J.M. Gee, J. Appl. Phys., 66, 3866 (1989).

[8] C. Cotirlan, C. Logofatu, C.C. Negrila, R.V. Ghita, A.S. Manea, and M.F. Lazarescu, J. Optoelectron Adv. Mater., 11(4), 386 (2009).

[9] J.W. Faust Jr. in "The Surface Chemistry of Metals and Semiconductors" (H.C. Gatos ed), Wiley, New York, 1960.

[10] H.C. Gatos and M.C. Lavine, Prog. Semicond., 9, 1 (1965).

[11] P.R. Camp, J. Electrochem. Soc., 102, 586 (1955).

[12] A.M. Morgan and I. Dalins, J. Vac. Sci. Technol., 10, 523 (1973).

[13] D. Klinov, B. Dwir, E. Kapon, N. Borovok,T. Molotsky, and A. Kotlyar, High-resolution atomic force microscopy of duplex and triplex DNA molecules, Nanotechnology, 18 (2007) 225102 (8pp) doi:10.1088/0957-4484/18/22/225102.

[14] L.R. Doolittle, NIMB, 15, 227 (1986).

[15] J. Makela, M. Tuominen, M. Yasir, M. Kuzmin, J. Dahl, M.P.J. Punkkinen, P. Laukkanen, K. Kokko, and R.M. Wallace, Appl. Phys. Lett., 107, 061601 (2015).

[16] Hartman et al. United States Patent, No. 4227975 (Oct. 14, 1980).

[17] O. El-Atwani, J.P. Allain and A. Suslova, Appl. Phys. Lett, 101, 251606 (2012).

[18] G.P. Schwartz, G.P. Gualtieri, J.E. Griffiths, C.D. Thurmond, and B. Schwartz, J. Electrochem. Soc. 127, 2488 (1980).

[19] O. El-Atwani, J.P. Allain, A. Cimarroli, A. Suslova, and S. Ortoleva, J. Appl. Phys. 110, 07301 (2011).

[20] Freiberger. General Specifications, Issue 2000 (www.fem-semicond.com/pdf/gen.spec.pdf).

[21] C.C. Negrila, C. Logofatu, R.V. Ghita, C. Cotirlan, F. Ungureanu, A.S. Manea, and M.F. Lazarescu, J. Cryst. Growth, 310, 1576 (2008).

[22] R.V. Ghita, C. Negrila, A.S. Manea, C. Logofatu, M. Cernea, and M.F. Lazarescu, J. Optoelectron. Adv. Mater., 5(4), 859 (2003).

[23] E. Papis-Polakowska, Electron Technology-Internet Journal, 37/38 1 (2005/2006).

[24] W.E. Spicer, I. Lindau, P. Pianetta, P.W. Chye, and C.M. Garner, Thin Solid Films, 56, 1 (1978).

[25] H. Iwasaki, Y. Mizokawa, and S. Nakamura, Jpn. J. Appl. Phys., 18, 1525 (1979).

[26] C. Cotirlan, R.V. Ghita, C.C. Negrila, C. Logofatu, F. Frumosu, and G.A. Lungu, Appl. Surf. Sci., 363, 83 (2016).

[27] C.C. Surdu-Bob, S.O. Saied, and J.L. Sullivan, Appl. Surf. Sci., 183, 126 (2001).

[28] V.N. Bessolov, M.V. Lebedev, N.M. Binh, M. Friedrich, and D.R.T. Zahn, Semicond. Sci. Technol., 13, 611 (1998).

[29] J.L. Sullivan, W. Yu, and S.O. Saied, Sur. Interface. Anal., 22, 515 (1994).

[30] J.S. Pan, A.T.S. Wee, C.H.A. Huan, H.S. Tan, and K.L. Tan, J. Phys. D: Appl. Phys, 30, 2514 (1997).

[31] J.S. Williams, Rep. Prog. Phys, 49, 491 (1986).

[32] M. Murakami, Science and Technology of Advanced Materials, 3, 1, (2002).

[33] M. R. Brozell and C.E. Stillman, Properties of Gallium Arsenide, INSPEC, Institution of Electrical Engineers, University of Michigan, Third Edition, (1996).

[34] A. Baca, F. Ren, J. Zolper, R. Briggs, and S. Pearton, Thin Solid Films, 308–309, 599 (1997).

[35] C.C. Negrila, M.F. Lazarescu, C. Logofatu. C. Cotirlan, R.V. Ghita, F. Frumosu, and L. Trupina, J. Nanomater., 2016, ID 7574526 (2016).

[36] D.T. Jayne, N.S. Fatemi, and V.G. Weizer., J. Vac. Sci. Technol. A., 10, 2802 (1992).

Section 3

Physical, Chemical, Biological and Environmental Properties and Applications

Chapter 6

Nanoscaled Fluorescent Films and Layers for Detection of Environmental Pollutants

Meizhen Yin and Chendong Ji

Additional information is available at the end of the chapter

http://dx.doi.org/10.5772/67869

Abstract

Hazardous gas and ion pollutants are the most serious environmental problems around the world. It is of great importance to develop devices for easy detection of these hazardous substances. Fluorescence technology with high resolution and operational simplicity has attracted a lot of attention in recent years. Organic fluorescent dyes absorb/emit lights within a broad wavelength range, which is suitable for various demands. Chromophores, such as perylene, cyanine dyes, spiropyran, and so on, are widely studied as fluorescent probes for gases and ions. The dyes could respond to external stimuli through structural changes of the conjugated chromophore itself or the attached functional groups, leading to detectable spectral changes. Organic dyes are incorporated into nanoscaled films and layers, which are portable and durable for effective sensing in complex environments. In this chapter, preparation and application of fluorescent films and layers (FFL) for gaseous/ionic detection are reviewed. We discuss the response mechanism of fluorescent dyes, the fabrication of nanoscaled FFL, and some examples of FFL for the detection of gas and ion pollutants.

Keywords: environmental pollutants, detection, fluorescence, nanoscaled films and layers, gases/ions sensor

1. Introduction

Human activities have introduced increasingly hazardous pollutants into the environment since the growth of industry, agriculture, and livestock. These pollutants including toxic gases and trace element ions caused pollutions of atmosphere, water, and soils, which severely harm the existence and development of human beings [1]. Thus, effective and inexpensive systems for detection and quantification of environmental pollutants have been progressively more important. Currently, conventional detection methods are based on chromatography-mass spectrometry (GC-MS), electrochemical systems, and spectrophotometers [2–4]. In these processes, large stationary instruments and complex components are required, which limit their use in resource-limited fields. Thus, there is currently a global effort to develop new technologies to evaluate and detect these environmental pollutants.

In recent years, fluorescence technology (FT) with high resolution and operational simplicity has attracted a lot of attention [5]. Taking advantage of FT, sensors based on chromophores have been considered promising alternatives for environmental measurements due to their simplicity, high sensitivity, and inexpensive nature. Particularly, organic fluorescent dyes are widely studied because they absorb/emit lights with a broad wavelength range, which could fulfill a variety of demands [6, 7]. Functionalized organic dyes could respond to external stimuli through structural or morphological change, which leads to the change of their apparent colors and fluorescence signals. It's worth mentioning that the response is easy to check by naked eye or using ultraviolet (UV) lamp. Therefore, design strategies to chemosensors for effective detection of environmental pollutants have attracted significant attention over the past few years.

On the other hand, recent developments in the field of nanotechnology have triggered increased interest in using nanomaterials for environmental applications [8]. Nanoscaled films and layers possess unique properties that can be used to improve the performance of existing sensors. Organic dyes are incorporated into self-supported films and layers as fluorescent films and layers (FFL) that are portable and durable for effective sensing in complex environments. Furthermore, taking advantage of their nanoscaled dimensions and controllable surface, FFL perform high sensitivity and selectivity to the analyte, which could be regarded as a promising sensing material especially for hazardous pollutants.

This chapter focuses on the recent developments of nanoscaled FFL for the detection of environmental pollutants. We first begin with the sensing mechanism of synthetic dyes and the fabrication of FFL based on these dyes. Then, some examples of FFL for the detection of gaseous pollutants (such as CO_2, NH_3, and HCl gas) and metal ions (such as Hg^{2+}, Pb^{2+}, and Cd^{2+}) is discussed.

1.1. Sensing mechanism of synthetic dyes for environmental applications

The main hurdle for fluorescent techniques lies in the construction of fluorescent sensors with high selectivity, sensitivity, and stability. Taking advantage of their structural designability, easy synthesis/modification, and high fluorescence quantum yields [9], organic sensors have become a new multidisciplinary research field involving organic synthesis, analytical chemistry, and environmentology. For environmental applications, different chromophores with various response mechanisms have been developed for the detection of hazardous pollutants. Herein, we present a survey on recent progress in sensing strategies of organic chromophore for the detection of hazardous gases and ions.

Organic dyes with aromatic structure, such as rylene dye and cyanine dye, emit strong and stable fluorescence but easily aggregate due to strong π-π stacking interaction [10]. For the

http://dx.doi.org/10.5772/67869

modification of these dyes, gas/ion responsive groups are attached to the periphery of the conjugation structure. The structural changes of functional groups usually affect the electron distribution or aggregate state of the fluorophores, resulting in distinct changes in the absorption/fluorescence spectrum. Based on organic dyes, a number of photophysical signaling mechanisms for sensing have been developed, such as photoinduced electron transfer (PET) [11, 12], intramolecular charge transfer (ICT) [11, 12], fluorescence resonance energy transfer (FRET) [13], and aggregation-disaggregation effect. In addition, different mechanisms of aggregation-induced signal change including aggregation-caused quenching (ACQ) [14] and aggregation-induced emission (AIE) [15] are studied for sensing environmental pollutants. The sensing strategies of fluorescent dyes are concluded in **Table 1**.

PET with "fluorophore-spacer-receptor" format is the most commonly exploited approach for designing fluorescent sensors and switches. Herman et al. reported an approach for sensing carbon dioxide based on PET (**Figure 1**) [16]. In this case, sensor **1** is nonfluorescent or weakly fluorescent due to the quenching by amine group through PET process. Upon reaction with CO_2, the basic amine group forms a carbamate salt and a positively charged ammonium group, resulting in prevention of the PET process, after which a turn-on fluorescent response is predictable.

Methods based on energy transfer (ET), such as FRET, hold great promise for pollutant detection, allowing precise and quantitative analysis and imaging even in complicated systems. Kim et al. reported a calix[4]crown chemosensor **2** based on dual sensing probes revealing high selectivity for Pb^{2+} ion over other metal ions (**Figure 2**) [17]. Sensors containing calixcrown molecules with a diazo group giving a visual color change and a pyrenyl group providing a fluorescence change were constructed. By Pb^{2+} ion complexation, the sensor performed fluorescence enhancement due to diminished FRET effect.

ACQ and AIE dyes are both developed as sensing materials with "turn-off" or "turn-on" models. Yin et al. reported a selective fluorescent chemosensor based on ACQ dye **3** for recyclable

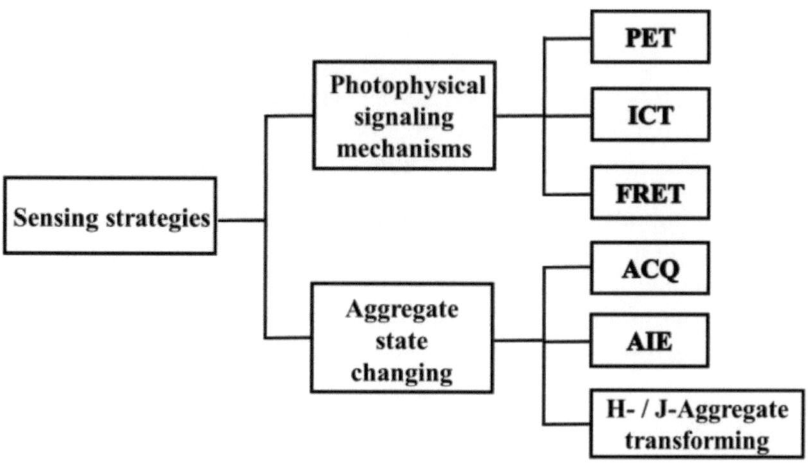

Table 1. Sensing strategies of the chromophores containing responsive groups.

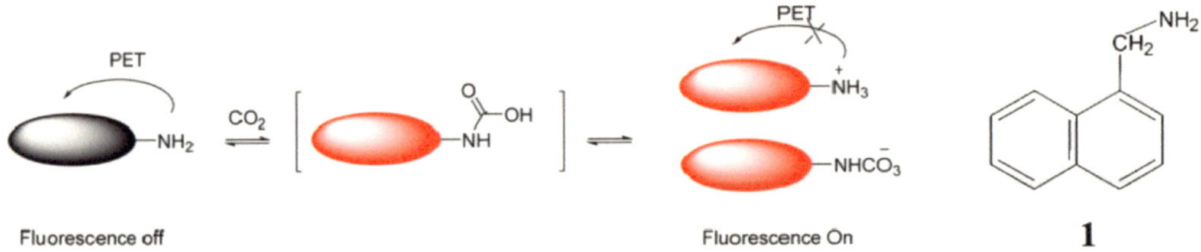

1

Figure 1. Strategy of PET-based chemosensor for CO_2 and chemical structure of **1**.

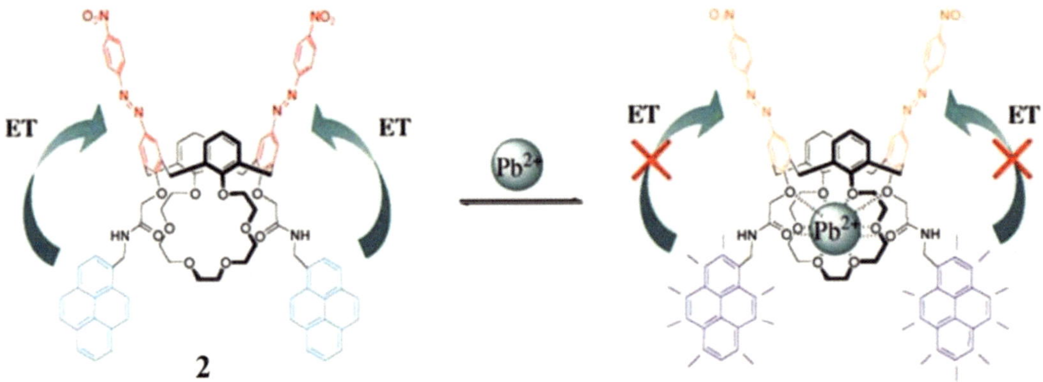

2

Figure 2. Sensing mechanism of probe **2** with Pb^{2+}. Copyright 2005 Elsevier B.V. All rights reserved. The aggregate state of dyes exerts a tremendous influence on their spectral properties.

Hg^{2+} detection with high sensitivity (**Figure 3**) [18]. With the addition of 20 mM of Hg^{2+}, the fluorescence of **3** (5 mM) is quenched completely.

Tang et al. reported a simple approach for quantitative detection of CO_2 gas with an AIE chromophore hexaphenylsilole (**4**, HPS) [19]. **4** is nonradiative in the very dilute dipropylamine (DPA) solution (~37 μM) due to the restriction of intramolecular motion (RIM) mechanism (**Figure 4a**). Bubbling DPA liquid with CO_2 gas results in the formation of a viscous and polar carbamate ionic liquid (CIL) with poor solvating power toward **4** (**Figure 4b**). The molecule **4** aggregates in CIL and exhibits the "turn-on" fluorescence as a CO_2 sensor.

Another strategy taking advantage of the self-association of dyes is also investigated. The aggregation of dyes also strongly affects their spectroscopic characteristics, and these spectral changes can be attributed to aggregation of the dye molecules in water to form dimers and higher order aggregates in the "J-" or "H-" type aggregation state [20]. Yin et al. developed a self-assembled fluorescent film by carboxylic acid group functionalized squarylium cyanine dyes **5** [21]. **5** in the solid state showed selective response to ammonia gas due to its unique molecular aggregation and size effect of ammonia molecule. The aggregated **5** could transit from J-aggregation to H-aggregation upon ammonia treatment and easily recovered by the treatment with HCl gas (**Figure 5**).

http://dx.doi.org/10.5772/67869

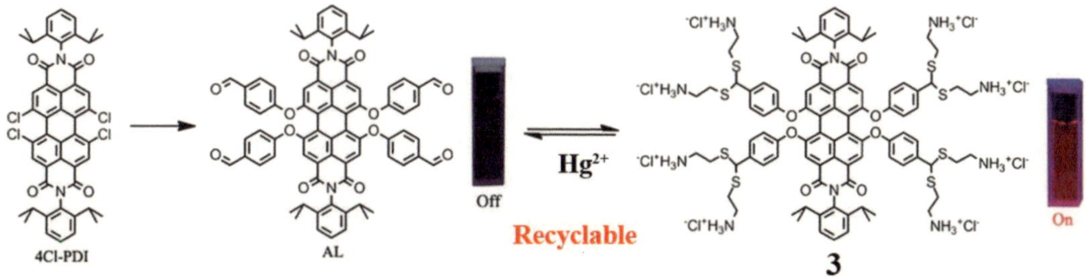

Figure 3. Synthesis approach and sensing mechanism of chemsensor **3**.

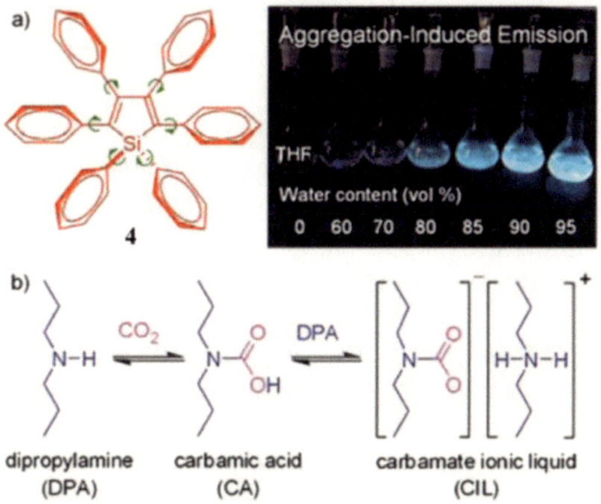

Figure 4. (a) Dye **4** is nonemissive when dissolved in THF but becomes strongly fluorescent when aggregated in THF/water mixtures with high water contents. (b) Formation of CIL by bubbling CO_2 gas through DPA liquid. Copyright 2007 American Chemical Society.

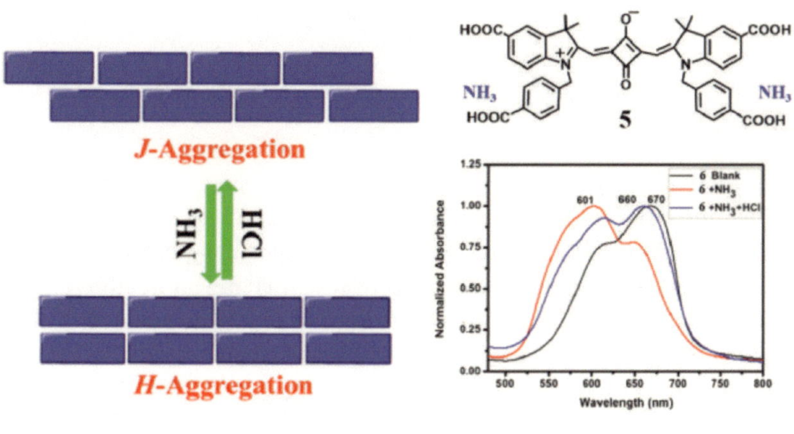

Figure 5. The mechanism of concentration-dependent self-assembly of dye **5** in the solid state for selective ammonia gas sensing.

1.2. Fabrication of films with organic dyes

Pollutant detection is challenging in atmosphere, water, and soil because of the complex components and rigorous conditions. Although organic dyes are sensitive to specific analytes, their applications in real environment are still limited due to their unstable nature. Nanomaterials have many excellent properties, such as strong adsorption, fine and tunable nanostructure, and stability [22]. The combination of organic dye and nanotechnology could maximize the functions of dye, while resulting in functional materials with higher performance. Nanomaterials with various architectures, such as particle, film, and layer, have been widely applied as sensors for environmental monitoring and pollutant detection. Film sensors have attracted enormous attention due to easy preparation, good applicability, and portable properties. In the chemosensory processes of films, the physical or chemical interactions occur at the surface or interface of the materials. Thus, controlling of the specific surface area, surface energy, and surface chemistry of films and layers through nanotechnology is of great importance. For example, compared with bulk materials, nanofibers with high surface area to volume ratio possess higher functional group density. Porous films also have high surface area and ordered porosity. These unique characters lead to higher sensitivity of the nanoscaled films and layers [23].

Advances in nanoscaled FFL are providing unprecedented opportunities to treat pollutants in environment. During the fabrication of FFL for pollutant detection, the following four precedent conditions should be met: (1) environment security, (2) reuse of sensors, (3) low cost, and (4) high detection efficiency. Based on these concerns, polymer films with dyes doped or embedded were widely used as sensing fluorescent films.

1.2.1. Casting films

Casting films were prepared by doping organic dyes into matrix polymers directly. The solvent mixture of dye and polymer was spread into a thin layer and then air-dried to prepare the film. The film is conveniently manufactured into devices and easily tailored according to practical need, which is widely adopted in the field of environment monitoring [24–27].

1.2.2. Cross-linking films

As polymer is an ideal supporting martial for organic sensing molecules, efforts have been made to further improve its stability and mechanical strength. Polymer networks could achieve stronger intermolecular interactions, closer molecular packing, and reduced polymer mobility by means of cross-linking using physical or chemical treatments [28]. Therefore, cross-linking has been explored as a viable method to improve the properties of polymer films as well as to give a stable environment for the embedded dyes.

Yin et al. reported a series of cross-linked fluorescent polyvinyl pyrrolidone (PVP) films covalently attached with different perylene derivatives (PDA, 4Cl-PDA, and 4Cl-PDI) (**Figure 6**) [29–31]. The ring-opening reaction of anhydride end groups and aromatic nucleophilic substitution

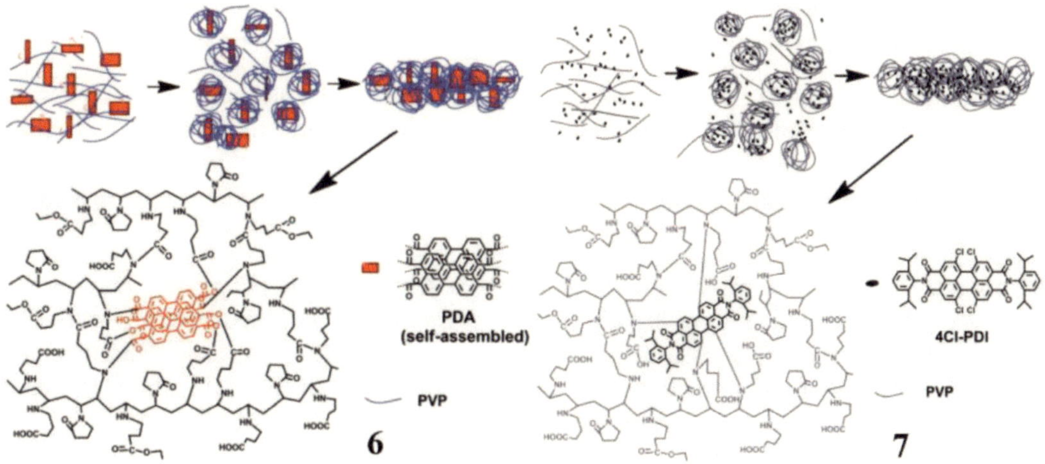

Figure 6. Formation of the fluorescent PVP films **6** and **7** based on perylene derivatives.

between the chlorine atom and the secondary amine in PVP both contributed to the PVP-chromophore conjugation. The obtained fluorescent film **6** and **7** showed highly stable optical performance originated from perylene chromophore.

1.2.3. Layer-by-layer (LbL) assembled films or layers

The LbL method has been reported to be a useful method for the preparation of molecularly assembled films. It was introduced in 1966 by Iler [32], and subsequently extended by Decher and coworkers to encompass the alternate adsorption of polycations and polyanions onto a surface [33]. A versatile approach for preparing nanoscaled thin film by LbL method is shown in **Figure 7** [34]. A substrate with inherent charge is first exposed to an oppositely charged polyelectrolyte,

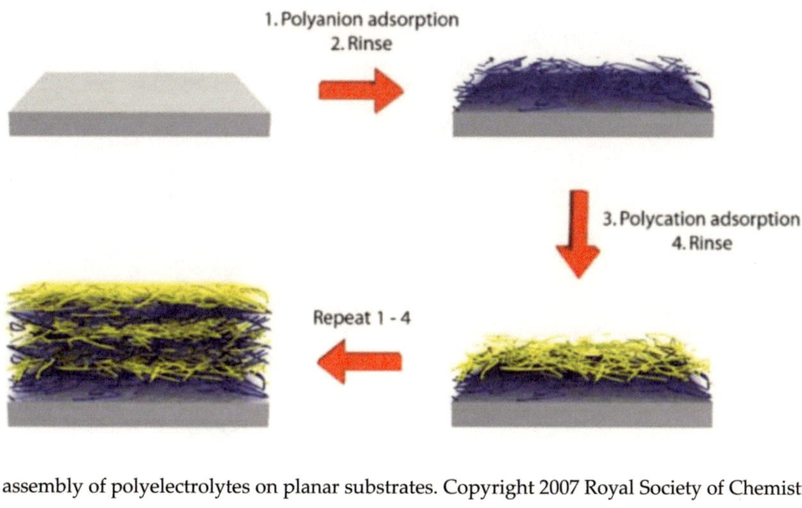

Figure 7. LbL assembly of polyelectrolytes on planar substrates. Copyright 2007 Royal Society of Chemistry.

followed by thorough rinsing. Reversal of the surface charge then facilitates further adsorption steps. The process is continued until the desired layer number (or thickness) is achieved.

The technique is still expanding its potential because of its versatility for the fabrication of ordered multilayers. Molecular self-assembly of macromolecules and nanoreactors is controlled by the interplay of intra- and intermolecular as well as interfacial interactions. Through tuning the intra- and intermolecular interactions, films and layers with varied surface properties could be obtained.

Serhiy et al. reported an LbL thin film **8** for ammonia sensing where porphyrin was adopted as a responsive element (**Figure 8**) [35]. It showed that the sensing sensitivity was greatly dependent on the film thicknesses. Upon ammonia exposure, the optical spectrum of film **8** obtained by LbL assembling for 15 cycles had a response time less than 30 s and it showed a linear relation with ammonia concentrations (0–100 ppm).

1.2.4. Electrospinning films

Ultrathin films, such as nanofiber films with different structures and functions, have been developed for environmental applications in recent years. Electrospinning was an easy but versatile method for various polymers to fabricate ultrathin films. Differing from normal fibers, the electrospun ultrathin fibers show nanoscale diameters, high lengths, and large surface area to volume ratios [36]. By controlling the processing parameters, different fibers could be obtained with various morphologies, such as core-shell, hollow, and porous structures [37, 38]. Generally, the sensitivity of a sensor will increase with a growing surface area

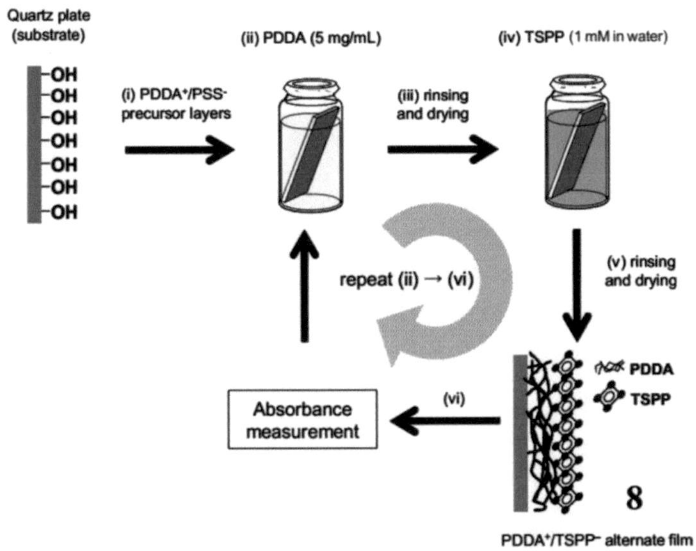

Figure 8. Preparation of film **8** via electrostatic LbL adsorption. Copyright 2006 Collegium Basilea & AMSI.

per unit mass, due to the detection carried out on the sensor surface [39]. Therefore, the large surface area to volume ratio and good interconnectivity make electrospun ultrathin fibers highly attractive for sensor applications. It is worth noting that a porous nanostructured fiber can be generated directly by controlling phase separation during electrospinning. The enhanced surface to volume ratio of the porous nanofibers could improve the sensitivity of the sensors.

There are two main classifications of fabrication processes of ultrathin fiber sensors. One is to incorporate sensing substances onto the outer surface of electrospun fibers. Sensors fabricated by this method usually have higher sensitivity but are unstable in external environment. Lee et al. reported a dots-on-fibers (DoF) hybrid nanostructure **9** via simple electrospinning and subsequent immobilization processes (**Figure 9**) [40]. The conjugated polymer dots (CPdots) was immobilized on the surface of poly(vinyl alcohol) (PVA)-silica nanofibers. The CPdots and amine-functionalized electrospun PVA-silica nanofibers interacted via electrostatic interaction.

Another strategy is to fabricate nanofiber sensors directly by adding sensing elements into the spinning solution. Ding et al. reported an approach for colorimetric quantitative detection of Pb^{2+} by embedding newly synthesized polydiacetylenes (PDA) into the electrospun polyacrylonitrile nanofibrous membrane (PAN NFM, **Figure 10**) [41]. Compared with the casting film-based PAN NFM, the electrospun film **10** displayed much higher sensitivity. Moreover, the morphology and porous structure of **10** could be further regulated by tuning the concentration of PAN, which displayed profound influence on sensing performance.

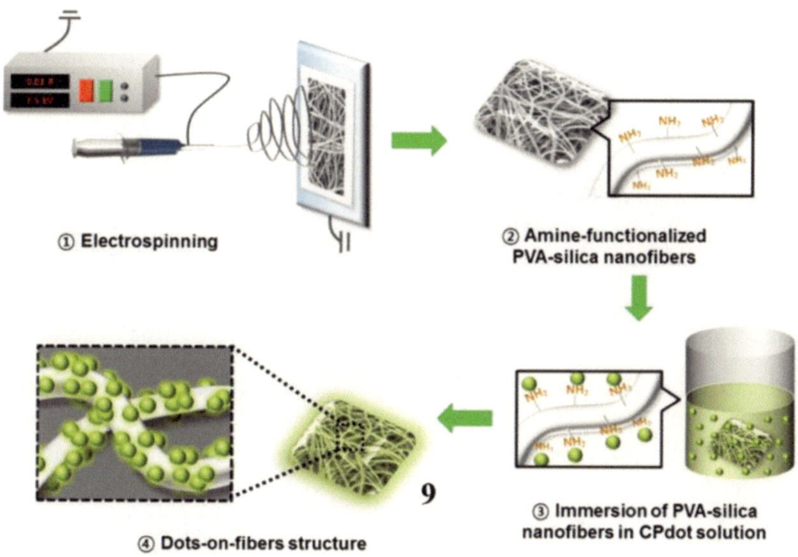

Figure 9. Preparation of DoF structure **9** by electrospinning and subsequently surface modification of the obtained fiber. Copyright 2014 American Chemical Society.

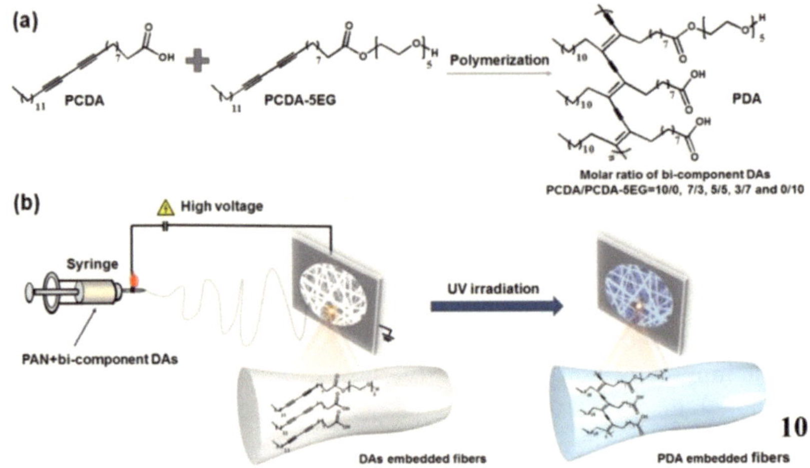

Figure 10. (a) Preparation of PDA polymers. (b) Schematic representation of the preparation of the PDA embedded film **10**. Copyright 2014 Royal Society of Chemistry.

2. FLL for the detection of environmental pollutants

2.1. Gaseous pollutant detection

2.1.1. Carbon dioxide (CO_2) detection

Carbon dioxide is a major public concern with widespread discussion because of its role in global greenhouse warming. Moreover, CO_2 is also quite critical to the modern agricultural, food, environmental, oil, and chemical industries. However, it is dangerous for living beings to stay at high concentration levels of CO_2 [42]. Therefore, the importance and prevalence of such a gas (CO_2) provides an incentive for development of new methods that can be used for the rapid and selective detection and monitoring of this relatively inert gas both in gaseous and liquid phases. Some traditional detection methods, such as electrochemical and infrared spectroscopic techniques, are usually employed for CO_2 sensing and detection; however, these methods are believed to be expensive and time consuming. Hence, the development of chemosensors for determining the concentration level of CO_2 is of great interest.

The use of pH indicators is one strategy in the construction of fluorescent sensors for CO_2 detection. The most famous pH indicator used for sensing CO_2 is 1-hydroxypyrene-3,6,8-trisulfonate (HPTS, **11**), as shown in **Figure 11a**. For instance, Sampathkumaran et al. reported the development of a sol-gel fluorescent sensor for detecting CO_2 with a detection limit below the amount of CO_2 in the atmosphere (ca. 387 ppm) (**Figure 11b**) [43]. In their work, a sol-gel fluorescent sensor was modified into silica-doped matrix with 11 by spin coating on glass slides. The fluorescent intensity of this indicator (513 nm) was quenched by CO_2 as a result of protonation of the anionic form.

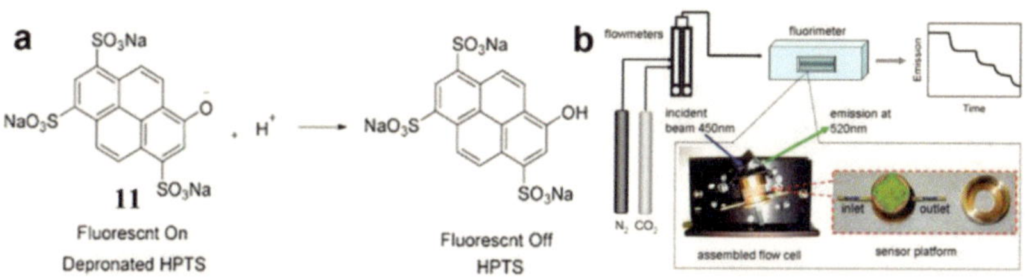

Figure 11. (a) Structures of pH indicator **11** as a CO_2 sensor. (b) Schematic diagram of the sensing system based on **11**-doped polymeric film. Copyright 2010 American Chemical Society.

Another strategy was developed based on chemical reaction between amine and CO_2. CO_2 is a weak electrophile that can react with an active basic amine to form corresponding carbamate salt and ammonium salt. Sijbesma et al. investigated the reaction between CO_2 and optically pure chiral diamines **12** ((R,R)-12 or (S,S)-12), **Figure 12**) [44]. This strategy for monitoring CO_2 is also based on the reversible carbamate formation-decomposition mechanism. The corresponding diamine/diol complex was utilized for reacting with CO_2 and results in apparent color change as well as in a helical twisting power. Moreover, this sensing platform showed a continuous reversible response for CO_2 monitoring based on the equilibrium complexation of **12** with CO_2.

2.1.2. Ammonia gas detection

Ammonia (NH_3) is widely employed in industrial and agricultural systems such as refrigeration, stock farming, fertilizer production, and food processing [45]. In addition to its broad applications, ammonia is also toxic, irritating, and corrosive to human skin, eyes, and respiratory system [46]. Recently, fluorescent sensors for ammonia detection have shown great promise with their low cost, easy use, and high sensitivity.

Gu and Huang reported the polyaniline **13** deposited filter paper as a reversible colorimetric sensor for gaseous ammonia (**Figure 13**) [47]. In this research, the authors utilized conducting

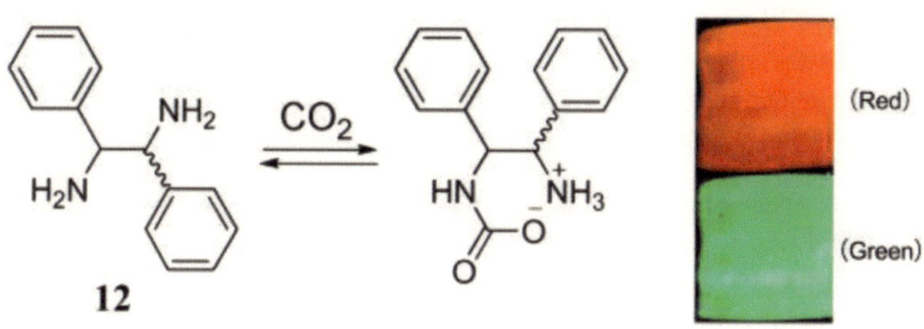

Figure 12. (Left) Structures of CO_2 responsive chiral dopant **12**. (Right) Cholesteric liquid crystals (CLC) film before (top) and after (bottom) exposure to CO_2 for 1 hour. Copyright 2010 American Chemical Society.

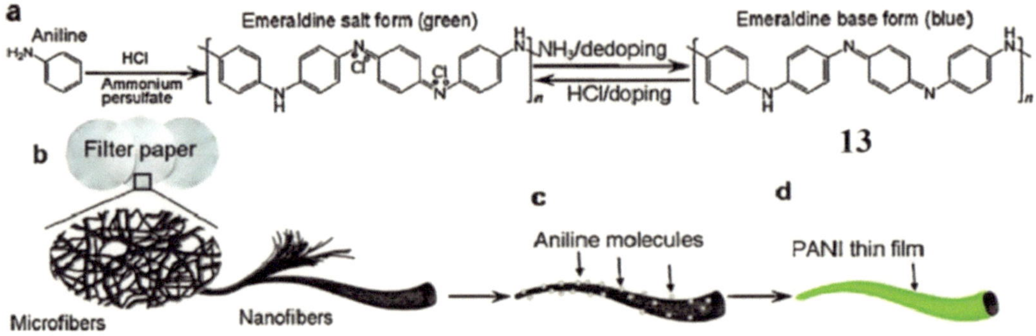

Figure 13. Scheme for the color transition of **13** (a) and the fabrication process of the **13**-deposited filter paper (b–d).

polymer, which has various advantages, such as tunable conductivity, a reversible doping/dedoping mechanism, and high affinity for ammonia. Nanometer-thick films were uniformly coated on cellulose nanofiber via in situ polymerization of anilines. Upon protonation by hydrochloride, the emeraldine salt type of **13** displayed green color, while the emeraldine base underwent deprotonation upon the addition of ammonia, resulting in color change into blue. The detection limits of this film by the naked eye were 100 ppm for the gaseous state ammonia under nitrogen atmosphere and 10 ppm for the evaporated ammonia aqueous solution. Furthermore, these color transitions were fully reversible.

Colored sensing films have slight color variations especially under low concentrations of ammonia, which are not easily perceived by visual inspection. Yin et al. reported a flexible naked eye ammonia sensor mat **14** by blending a hydrophobic cyanine dye into polylactic acid (PLA) porous fibers [48]. Upon exposure to ammonia vapor, the sensing mat experienced perceivable color change from white to blue. The fibrous mat **14** exhibited high selectivity for ammonia gas with a detection limit of 3.3 ppm. The white to blue color changing process was attributed to ionization-induced deaggregation of cyanine dyes (**Figure 14**).

2.1.3. Hydrogen chloride gas (HCl) detection

Sensitive detection and monitoring of HCl gas have attracted a lot of attention [49]. HCl is mainly released in the mass production of halogenated polymers. Therefore, it is of great importance to sense and regulate HCl concentration in the adsorbing towers of these factories, which benefits the emission monitoring and air quality control.

As shown in **Figure 15**, Wu and coworkers reported an HCl gas sensor **15** with rapid response-recovery manner, which was fabricated from the nanofibrous membrane of porphyrinated polyimide (PPI) [50]. Exposing **15** to HCl gas induced absorption change within 10 s and color varied in response to different concentrations of HCl gas. Apparent binding affinity constant calculated from surface plasmon resonance (SPR) analysis is found to be $(1.05 \pm 0.23) \times 10^4$ L mol^{-1}. The superior thermal stability of PPI makes this nanofibrous membrane sensor promising for monitoring HCl gas emission in incinerators for burning domestic, clinical, or industrial wastes.

http://dx.doi.org/10.5772/67869

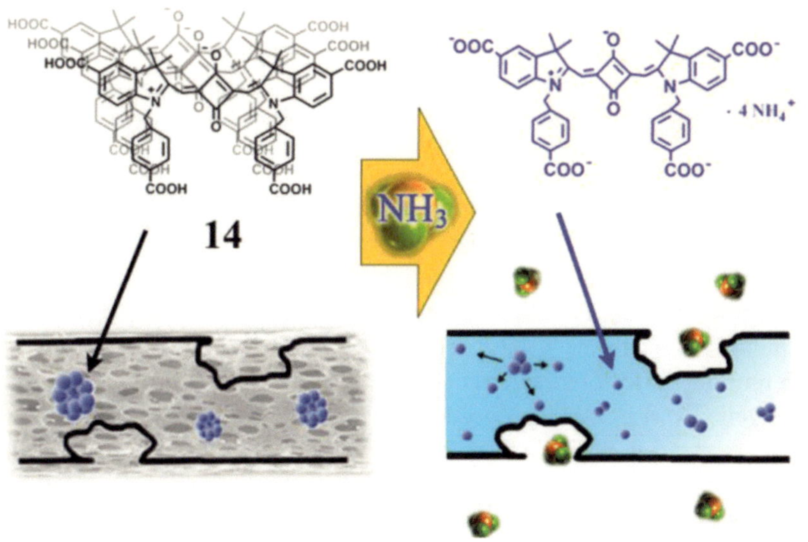

Figure 14. Structure of the sensor **14** and proposed sensing mechanism of **14**-doped electrospun fiber. Copyright 2016 WILEY-VCH Verlag GmbH & Co. KGaA, Weinheim.

2.1.4. Volatile organic compound (VOC) detection

The detection of volatile organic compounds is an important aspect considering that VOCs are continuously released into the environment by different sources like industrial processes, transportation, agriculture, etc. [51]. These VOCs not only cause environmental pollution but also directly affect human health. For example, alcohols and aromatic hydrocarbons

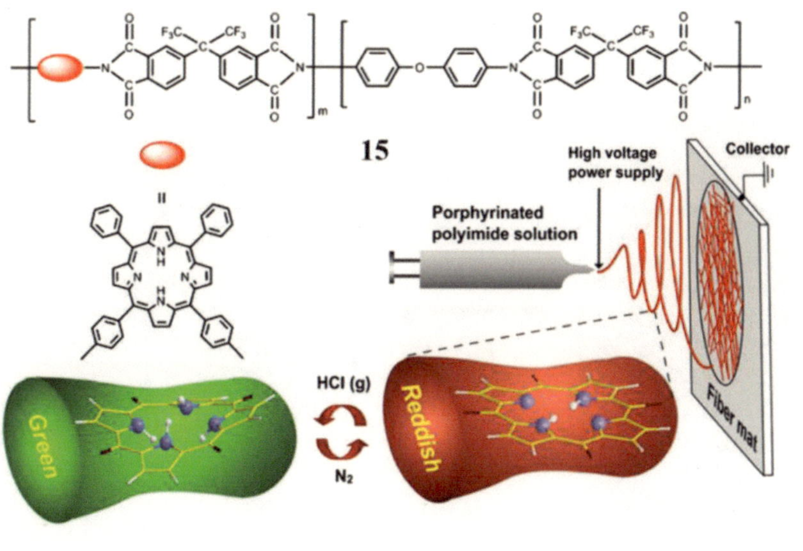

Figure 15. Molecular structure of **15** and the schematic illustration for the fluorescent detection of HCl gas. Copyright 2010 Elsevier B.V. All rights reserved.

are potentially hazardous to human health due to their capabilities to stimulate the mucous membranes and upper respiratory tracts [52]. Various detection methods have been proposed based on changes in electrochemical, conducting, and chromism properties of the corresponding sensor matrices. Colorimetric sensor systems are of particular interest thanks to their effectiveness and simplicity.

Formaldehyde could be slowly released from many building materials such as paints, adhesives, wallboards, and ceiling tiles, which irritates the mucous membranes and can make a person irritated and uncomfortable. The reaction of primary amine with aldehyde is often utilized to detect formaldehyde. Suslick and coworkers utilized simple pH indicators to detect the change in basicity as a simple colorimetric system for gaseous formaldehyde (**Figure 16**) [53]. Six different pH indicators, such as methyl red, bromocresol purple, 4-nitrophenol, alizarin, nitrazine yellow, and bromoxylenol blue, were introduced into a poly(ethylene glycol) (PEG) polymer containing five different amounts (0.24, 0.48, 0.58, 0.82, and 1.03 wt%) of PEG with amine termination. This polymer film **16** displayed distinct colorimetric changes with different concentrations of formaldehyde.

Synthetic amines are produced in millions of tons each year and have broad applications in agricultural, pharmaceutical, and food industries. The excess of organic amines in air would seriously damage the ecological environment and pose severe threats to human health. Selective detection of trace amines in the vapor phase has attracted intense attention because of the increasing concerns regarding air pollution monitoring, quality control of food, and even medical diagnosis. Lin and coworkers reported an AIE dye **17**-based fluorescence sensing film for alkylamines (**Figure 17**) [54]. The fluorescence intensity of **17** enhanced by 30- to 50-fold toward ammonia, hydrazine, and alkyl amines (benzylamine, $EtNH_2$, Et_2NH, NMe_3, Et_3N). The light-up ratios (I/I_0) indicate that these amine vapors can efficiently cleave the *O*-acetyl bond to generate highly emissive product through aminolysis reaction. However, for aromatic amines (aniline, 2-methylaniline), biogenic amines such as putrescine, cadaverine, histamine, and organic solvent vapors such as hexane, CH_2Cl_2, EtOAc, THF, MeCN, MeOH, and EtOH, almost no light-up response was observed.

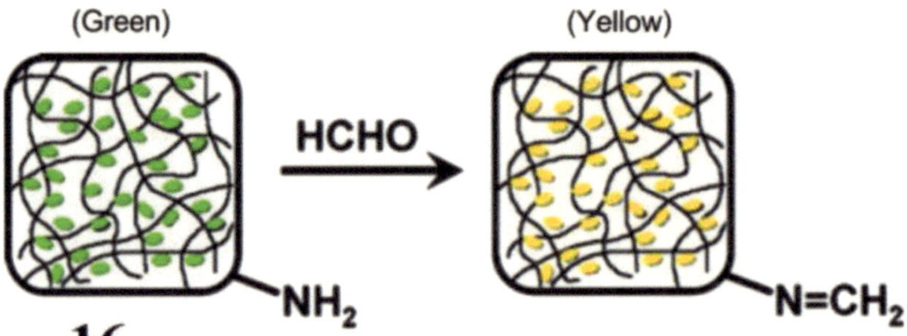

16

Figure 16. Schematic presentation of sensing film **16** for formaldehyde. Copyright 2010 American Chemical Society.

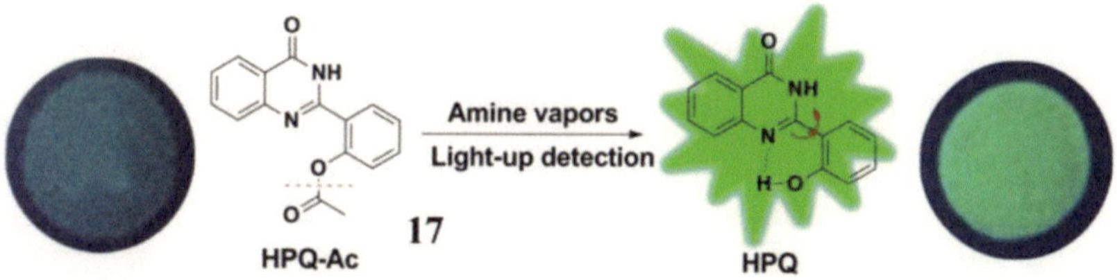

Figure 17. Chemical structure of **17** and its light-up response to amine vapors. Copyright 2016 American Chemical Society.

2.2. Metal ion detection

During the last decades, increased contributions of contaminants caused drastic changes in ecosystems. Soil and water pollution with metals especially heavy metals is very frequently encountered. Metals are not biodegradable, and therefore they remain in ecological systems and in the food chain indefinitely, exposing top-level predators to very high levels of pollution. These interactions metals have toxicological and carcinogenic effects such as those affecting the central nervous system (Hg^{2+}, Pb^{2+}, As^{2+}); the kidneys or liver (Cu^{2+}, Cd^{2+}, Hg^{2+}, Pb^{2+}); or skin, bones, or teeth (Ni^{2+}, Cu^{2+}, Cd^{2+}, Cr^{2+}) [55]. Therefore, World Health Organization (WHO) and Environmental Protection Agency (EPA) have strictly defined the concentration limits of metal ions that are allowed in the drinking water. Especially, lead, cadmium, and mercury ions are banned in electrical and electronic equipment by the European Union's Restriction on Hazardous Substances (RoHS) [56]. As is well known, standard techniques for trace heavy-metal analysis (even in the ppt and ppq range) require sophisticated analytical techniques such as atomic absorption spectroscopy (AAS), inductively coupled plasma-mass spectrometry (ICPMS), mass spectroscopy (MS), X-ray fluorescence spectroscopy (R-FS), and potentiometric methods and specialized personnel to carry out the operational procedures [55]. Efforts are ongoing to prepare rapid and inexpensive techniques for heavy-metal detection. Here, optical detections for heavy-metal ions based on fluorescent films are reviewed.

2.2.1. Mercury ion (Hg^{2+}) detection

Mercury is widespread in air, water, and soil and is notoriously known for its high toxicity to human beings such as severe damage to the central nervous system. Moreover, accumulation of mercury in human body could lead to various kinds of cognitive and motor disorders as well as Minamata disease [57]. Many sources are to blame for the generation of mercury, for example, coal plants, gold production, measuring instrument, and mercury lamps [58]. Mercury is mainly uptaken by human beings through daily diet such as fish, and it is vital that considerable efforts should focus on the development and evolution of sensitive and selective detection methods.

Yin et al. reported a highly sensitive and selective fluorescent nanofibrous membrane **18** for the detection of mercuric ions (II) (**Figure 18**) [59]. **18** was prepared via electrospinning and

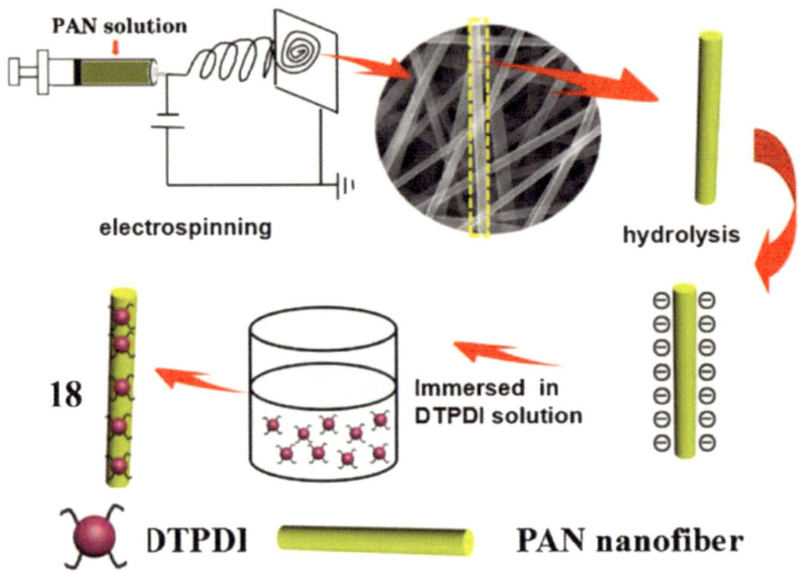

Figure 18. Schematic illustration for the fabrication sensing film **18** for Hg^{2+}.

subsequent immobilization. In this process, fluorescent chemosensor dithioacetal-modified perylenediimide (DTPDI) was introduced on the surface of PAN nanofibers with high stability under mechanical force by electrostatic interaction. DTPDI could detach from PAN nanofibers due to the hydrolysis of dithioacetals in the presence of Hg^{2+} and form an oil-soluble fluorescence dye AL. According to the linear correlation between AL and Hg^{2+}, the obtained **18** could be employed for the detection of Hg^{2+}. The limit of detection for Hg^{2+} can reach as low as 1 ppb. In addition, the strong fluorescence of **18** still can be observed even after the repeated use for seven times. **18** can be developed as a rapid, portable, and stable sensor for the detection of Hg^{2+}.

2.2.2. Lead ion (Pb^{2+}) detection

As an abundant but toxic metal, lead is often left in the environment due to its application in batteries, gasoline, pigments, etc. [60]. Lead pollution causes long-term damage to both human health and the environment, as most of the mined lead, 300 million tons to date, is still going back to soil and groundwater finally. Trace amounts of lead may cause neurological, reproductive, cardiovascular, and developmental disorders. Particularly, lead poisoning will introduce serious problems in children including slow motor responses, decreased IQs, and hypertension.

Kim et al. reported a rhodamine-based chemosensor **19** for the selective fluorescent sensing of Pb^{2+} (**Figure 19**) [61]. **19** was synthesized and self-assembled onto glass surfaces as a film sensor. The immobilized chemosensor showed a fluorescent response that was turned-on with Pb^{2+} in CH_3CN, selectively over various metal ions. T selective fluorescent change of sensor

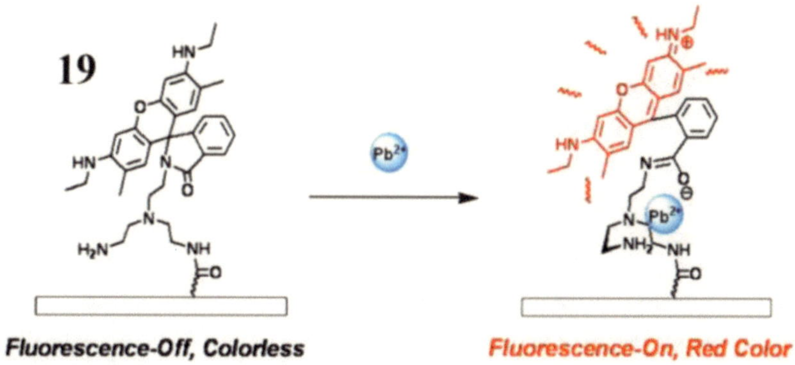

Figure 19. Rhodamine-based chemosensor **19** and its fluorescent "turn-on" response for Pb^{2+}. Copyright 2011 Elsevier B.V. All rights reserved.

was 1223 ± 166 counts only for Pb^{2+} over other cations. The detection limit of Pb^{2+} (S/N = 3) was approximately 10^{-4} M. The Pb^{2+} selective fluorescent switch of the immobilized chemosensor was also reversible, allowing for repeated use for Pb^{2+} detection.

2.2.3. *Cadmium ion (Cd^{2+}) detection*

Cadmium is a kind of heavy metal with severe toxicity and carcinogenicity. It is commonly seen in electric batteries, pigments in plastics, electroplated steel, and so on [62]. Human being tends to uptake cadmium mainly through smoking and daily diet. Exposure to high concentration of cadmium would result in increased risk in cardiovascular diseases and cancer, and it may also cause damage to liver and kidneys [63].

Wei et al. reported a fluorescence layered double hydroxide (LDH) ultrathin films **20** fabricated by alternate assembly of 2,2′-azino-bis(3-ethylbenzothiazoline-6-sulfonate) (ABTS) and ZnAl LDH nanosheets on quartz substrates using the LbL deposition technique (**Figure 20**) [64]. **20** performed as a good fluorescence chemosensors for Cd^{2+}. The fluorescent sensor with

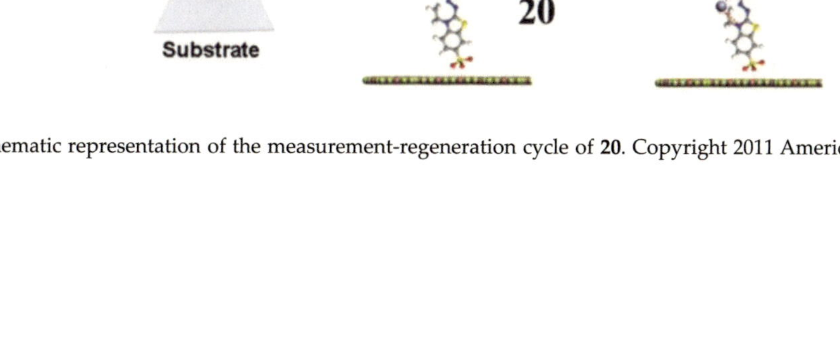

Figure 20. Schematic representation of the measurement-regeneration cycle of **20**. Copyright 2011 American Chemical Society.

20 bilayers showed a low detection limit of 9.5 nM for Cd^{2+}, good selectivity toward Cd^{2+} from Zn^{2+} at pH 6.5 to 7.5 and excellent stability. The regeneration of **20** can be achieved during a repeat procedure by using ethylenediamine tetraacetic (EDTA).

3. Conclusion

This chapter reports the preparation of nanoscaled FFL and the detection of environmental pollutants. Several sensing strategies are brought up based on the different dyes. Nanoscaled FFL act as a promising tool for hazardous pollutant detection, including gaseous pollutants (such as CO_2, NH_3, HCl gas, and VOCs) and metal ions (such as Hg^{2+}, Pb^{2+}, Cd^{2+}). However, although numerous chemosensors for gases/ions have been widely developed and applied, a big challenge remains to construct those that display high selectivities, sensitivities, and reusabilities.

Acknowledgements

This work was financially supported by the National Natural Science Foundation of China (21574009 and 51521062), Beijing collaborative innovative research center for cardiovascular diseases, and the Higher Education and High-quality and World-class Universities (PY201605).

Author details

Meizhen Yin* and Chendong Ji

*Address all correspondence to: yinmz@mail.buct.edu.cn

State Key Laboratory of Chemical Resource Engineering, Beijing Laboratory of Biomedical Materials, Beijing University of Chemical Technology, Beijing, China

References

[1] Lubchenco J. Entering the century of the environment: a new social contract for science. Science 1998; **279**: 491–497.

[2] Ji X, Banks CE, Silvester DS, et al. Electrochemical ammonia gas sensing in nonaqueous systems: a comparison of propylene carbonate with room temperature ionic liquids. Electroanalysis 2007; **19**: 2194–2201.

[3] Moujahid W, Eichelmann-Daly P, Strutwolf J, et al. Microelectrochemical systems on silicon chips for the detection of pollutants in seawater. Electroanalysis 2011; **23**: 147–155.

[4] Balakumar V, Prakash P, Muthupandi K, et al. Nanosilver for selective and sensitive sensing of saturnism. Sens. Actuators B. Chem. 2017; **241**: 814–820.

[5] Wu J, Kwon B, Liu W, et al. Chromogenic/fluorogenic ensemble chemosensing systems. Chem. Rev. 2015; **115**: 7893–7943.

[6] Sinkeldam RW, Greco NJ, Tor Y. Fluorescent analogs of biomolecular building blocks: design, properties, and applications. Chem. Rev. 2010; **110**: 2579–2619.

[7] Chen M, Yin M. Design and development of fluorescent nanostructures for bioimaging. Prog. Polym. Sci. 2014; **39**: 365–395.

[8] Perreault F, de Faria AF, Elimelech M. Environmental applications of graphene-based nanomaterials. Chem. Soc. Rev. 2015; **44**: 5861–5896.

[9] Fölling J, Bossi M, Bock H, et al. Fluorescence nanoscopy by ground-state depletion and single-molecule return. Nat. methods 2008; **5**: 943–945.

[10] Lü B, You S, Li P, et al. Kinetically trapped supramolecular assembly of perylene dianhydride derivative in methanol: optical spectra, morphology, and mechanisms. Chem. Eur. J. 2016; **22**: 1–6.

[11] De Silva AP, Gunaratne HQN, Gunnlaugsson T, et al. Signaling recognition events with fluorescent sensors and switches. Chem. Rev. 1997; **97**: 1515–1566.

[12] Martínez-Manez R, Sancenon F. Fluorogenic and chromogenic chemosensors and reagents for anions. Chem. Rev. 2003; **103**: 4419–4476.

[13] Sapsford KE, Berti L, Medintz IL. Materials for fluorescence resonance energy transfer analysis: beyond traditional donor–acceptor combinations. Angew. Chem. Int. Ed. 2006; **45**: 4562–4589.

[14] Zhang J, Liu K, Müllen K, et al. Self-assemblies of amphiphilic homopolymers: synthesis, morphology studies and biomedical applications. Chem. Commun. 2015; **51**: 11541–11555.

[15] Hong Y, Lam JWY, Tang BZ. Aggregation-induced emission: phenomenon, mechanism and applications. Chem. Commun. 2009; **29**: 4332–4353.

[16] Herman P, Murtaza Z, Lakowicz JR. Sensing of carbon dioxide by a decrease in photoinduced electron transfer quenching. Anal. Biochem. 1999; **272**: 87–93.

[17] Lee SH, Kim SK, Bok JH, et al. Calix[4]crown in dual sensing functions with FRET. Tetrahedron Lett. 2005; **46**: 8163–8167.

[18] Liu K, Xu Z, Yin M, et al. A multifunctional perylenediimide derivative (DTPDI) can be used as a recyclable specific Hg^{2+} ion sensor and an efficient DNA delivery carrier. J. Mater. Chem. B. 2014; **2**:2093–2096.

[19] Liu Y, Tang Y, Barashkov NN, et al. Fluorescent chemosensor for detection and quantitation of carbon dioxide gas. J. Am. Chem. Soc. 2010; 132: 13951–13953.

[20] Yagai S, Seki T, Karatsu T, et al. Transformation from H-to J-aggregated perylene bisimide dyes by complexation with cyanurates. Angew. Chem. Int. Ed. 2008; **47**: 3367–3371.

[21] Li J, Lv B, Yan D, et al. Tunable self-assembled micro/nanostructures of carboxyl-functionalized squarylium cyanine for ammonia sensing. Adv. Funct. Mater. 2015; **25**: 7442–7449.

[22] Zhang Q, Uchaker E, Candelaria SL, et al. Nanomaterials for energy conversion and storage. Chem. Soc. Rev. 2013; **42**: 3127–3171.

[23] Innocenzi P, Malfatti L. Mesoporous thin films: properties and applications. Chem. Soc. Rev. 2013; **42**: 4198–4216.

[24] Jayasekara R, Harding I, Bowater I, et al. Preparation, surface modification and characterisation of solution cast starch PVA blended films. Polym. Test. 2004; **23**: 17–27.

[25] Sasaki DY, Singh S, Cox JD, et al. Fluorescence detection of nitrogen dioxide with perylene/PMMA thin films. Sens. Actuators B. 2001; **72**: 51–55.

[26] Liu Y, Wang KR, Guo DS, et al. Supramolecular assembly of perylene bisimide with b-cyclodextrin grafts as a solid-state fluorescence sensor for vapor detection. Adv. Funct. Mater. 2009; **19**: 2230.

[27] Jiang BP, Guo DS, Liu Y. Self-assembly of amphiphilic perylene-cyclodextrin conjugate and vapor sensing for organic amines. The J. Org. Chem. 2010; **75**: 7258–7264.

[28] Remunan-Lopez C, Bodmeier R. Mechanical, water uptake and permeability properties of crosslinked chitosan glutamate and alginate films. J. Org. Chem. 1997; **44**: 215–225.

[29] Yin M, Ye Y, Sun M, et al. Facile one-pot synthesis of a polyvinylpyrrolidone-based self-crosslinked fluorescent film. Macromol. Rapid Commun. 2013; **34**: 616–620.

[30] Sun M, He Y, Ye Y, et al. Nucleophilic substitution of tetrachloroperylene diimide in fluorescent polyvinylpyrrolidone film. Macromol. Chem. Phys. 2014; 215: 493–498.

[31] Sun M, He Y, Yang W, et al. A fluorescent perylene-assembled polyvinylpyrrolidone film: synthesis, morphology and nanostructure. Soft Matter, 2014; **10**: 3426–3431.

[32] Iler RK. Multilayers of colloidal particles. J. Colloid Interface Sci. 1966; **21**: 569–594.

[33] Decher G, Hong JD. Buildup of ultrathin multilayer films by a self-assembly process, 1 consecutive adsorption of anionic and cationic bipolar amphiphiles on charged surfaces. Macromol. Symp. 1991; **46**: 321–327.

[34] Quinn JF, Johnston APR, Such GK, et al. Next generation, sequentially assembled ultrathin films: beyond electrostatics. Chem. Soc. Rev. 2007; **36**: 707–718.

[35] Korposh SO, Takahara N, Ramsden JJ, et al. Nano-assembled thin film gas sensors. 1. Ammonia detection by a porphyrin-based multilayer film. J. Biol. Phys. Chem. 2006; **6**: 125–132.

[36] Ding B, Wang M, Wang X, et al. Electrospun nanomaterials for ultrasensitive sensors. Mater. Today, 2010; **13**: 16–27.

[37] Zhao Y, Cao X, Jiang L. Bio-mimic multichannel microtubes by a facile method. J. Am. Chem. Soc. 2007; **129**: 764–765.

[38] Li D, Xia Y. Direct fabrication of composite and ceramic hollow nanofibers by electrospinning. Nano Lett. 2004; **4**: 933–938.

[39] Wang X, Drew C, Lee SH, et al. Electrospun nanofibrous membranes for highly sensitive optical sensors. Nano Lett. 2002; **2**: 1273–1275.

[40] Jo S, Kim J, Noh J, et al. Conjugated polymer dots-on-electrospun fibers as a fluorescent nanofibrous sensor for nerve gas stimulant. ACS Appl. Mater. Interfaces 2014; **6**: 22884–22893.

[41] Li Y, Wang L, Yin X, et al. Colorimetric strips for visual lead ion recognition utilizing polydiacetylene embedded nanofibers. J. Mater. Chem. A. 2014; **2**: 18304–18312.

[42] Tian T, Chen X, Li H, et al. Amidine-based fluorescent chemosensor with high applicability for detection of CO_2: a facile way to "see" CO_2. Analyst 2013, **138**: 991–994.

[43] Dansby-Sparks RN, Jin J, Mechery SJ, Sampathkumaran U. et al. Fluorescent-dye-doped sol–gel sensor for highly sensitive carbon dioxide gas detection below atmospheric concentrations. Anal. Chem. 2009; **82**: 593–600.

[44] Han Y, Pacheco K, Bastiaansen CWM, Broer DJ and Sijbesma RP. Optical monitoring of gases with cholesteric liquid crystals. J. Am. Chem. Soc. 2010; **132**: 2961–2967.

[45] Timmer B, Olthuis W, Van Den Berg A. Ammonia sensors and their applications—a review. Sens. Actuators B. 2005; **107**: 666–677.

[46] Max A. Ammonia: Principles and Industrial Practice. Vch Verlagsgesellschaft Mbh, 1999.

[47] Gu Y, Huang J. Colorimetric detection of gaseous ammonia by polyaniline nanocoating of natural cellulose substances. J. Colloids Surf. A: Physicochem. Eng. Aspects 2013; **433**: 166–172.

[48] Ji C, Ma L, Yin M, et al. A cyanine dye encapsulated porous fibrous mat for naked-eye ammonia sensing. Chem. Asian J. 2016; **11**: 2316–2321.

[49] Matsuguchi M, Kadowaki Y. Poly (acrylamide) derivatives for QCM-based HCl gas sensor applications. Sens. Actuators B. 2008; **130**: 842–847.

[50] Lv YY, Wu J, Xu ZK. Colorimetric and fluorescent sensor constructing from the nanofibrous membrane of porphyrinated polyimide for the detection of hydrogen chloride gas. Sens. Actuators B. 2010; **148**: 233–239.

[51] Adgate JL, Goldstein BD, McKenzie LM. Potential public health hazards, exposures and health effects from unconventional natural gas development. Environ. Sci. Technol. 2014; **48**: 8307–8320.

[52] Room R, Babor T, Rehm J. Alcohol and public health. Lancet 2005; **365**: 519–530.

[53] Feng L, Musto CJ, Suslick KS. A simple and highly sensitive colorimetric detection method for gaseous formaldehyde. J. Am. Chem. Soc. 2010; **132**: 4046–4047.

[54] Gao M, Li S, Lin Y, et al. Fluorescent light-up detection of amine vapors based on aggregation-induced emission. ACS Sens. 2015; **1**: 179–184.

[55] Aragay G, Pons J, Merkoçi A. Recent trends in macro-, micro-, and nanomaterial-based tools and strategies for heavy-metal detection. Chem. Rev. 2011; **111**: 3433–3458.

[56] Kim HN, Ren WX, Kim JS, et al. Fluorescent and colorimetric sensors for detection of lead, cadmium, and mercury ions. Chem. Soc. Rev. 2012, **41**: 3210–3244.

[57] Carvalho CML, Chew EH, Hashemy SI, et al. Inhibition of the human thioredoxin system a molecular mechanism of mercury toxicity. J. Biol. Chem. 2008; **283**: 11913–11923.

[58] Di Natale F, Lancia A, Molino A, et al. Capture of mercury ions by natural and industrial materials. J. Hazard. Mater. 2006; **132**: 220–225.

[59] Ma L, Liu K, Yin M, et al. Fluorescent nanofibrous membrane (FNFM) for the detection of mercuric ion (II) with high sensitivity and selectivity. Sens. Actuators B. 2017; **238**: 120–127.

[60] Flegal AR, Smith DR. Current needs for increased accuracy and precision in measurements of low levels of lead in blood. Environ. Res. 1992; **58**: 125–133.

[61] Ju H, Lee MH, Kim J, et al. Rhodamine-based chemosensing monolayers on glass as a facile fluorescent "turn-on" sensing film for selective detection of Pb^{2+}. Talanta 2011; **83**: 1359–1363.

[62] Nordberg GF, Herber RFM, Alessio L. Cadmium in the Human Environment, Oxford University Press, Oxford, UK, 1992.

[63] McFarland CN, Bendell-Young LI, Guglielmo C, et al. Kidney, liver and bone cadmium content in the Western Sandpiper in relation to migration. J. Environ. Monit. 2002; **4**: 791–795.

[64] Shi W, Ji X, Zhang S, et al. Fluorescence chemosensory ultrathin films for Cd^{2+} based on the assembly of benzothiazole and layered double hydroxide. J. Phys. Chem. C. 2011; **115**: 20433–20441.

Chapter 7

Mechanical Nanoprocessing and Nanoviscoelasticity of Surface-Modified Polycarbonate

Shojiro Miyake and Mei Wang

Additional information is available at the end of the chapter

http://dx.doi.org/10.5772/67512

Abstract

To clarify their potential as atomic force microscope (AFM) memory media, the nanometer-scale mechanical processing properties of untreated and fluorocarbon plasma-treated polycarbonate samples were determined via the sliding of an AFM tip. The surface energy of the polycarbonate was reduced by the fluorocarbon plasma treatment, as well as the force necessary for processing. Nanometer-scale precise processing of the polycarbonate was realized after the fluorocarbon plasma treatment, and the interval pitch in the formation of lines, spaces, and nanometer-scale fine dots was minimized to 60 nm with these samples. The viscoelastic properties of the fluorinated polycarbonate were evaluated using an AFM in force modulation mode. The fluorocarbon plasma treatment reduced the friction force of the polycarbonate sample and improved its wear resistance, which caused the friction durability corresponding to the reliability of data reproduction to be markedly improved. These results show that high-density recording can be realized by nanometer-scale processing of fluorocarbon plasma-treated polycarbonate samples.

Keywords: nanometer-scale processing, viscoelasticity, atomic force microscopy, polycarbonate, fluorocarbon plasma treatment, high-density memory

1. Introduction

Based on the application of scanning probe microscopy in nanoprocessing technologies, mechanical nanoprocessing technology at the atomic scale is attracting great attention for nanomachines and nanodevices. Scanning tunneling microscopy and related techniques are promising for the development of the manufacture of atomic-scale writing by means of tip scanning using a piezoelectric element [1]. In nanoprocessing, the semiconductor manufacturing process comprising etching, deposition, and lithography has mainly been applied. In future nanotechnology processes, however, it is expected that various machine elements such as bearings and sliders will be manufactured by manipulating atoms or molecules. On the one hand, microtribology has been studied using atomic force microscopy (AFM), whereby tribological processing is performed by scanning the AFM tip on a sample surface. On the other hand, wear tests have been carried out on diamond-like carbon (DLC), ion-implanted DLC, and cubic boron nitride films [2–4] whereupon a $1 \times 1\ \mu m^2$ square with a depth of 5–10 nm was formed by wearing and the maximum roughness depth at the bottom of a processed groove was about 1 nm [3].

The atomic wear of a material with a layered crystal structure has been evaluated in previous studies based on the topographic changes that occur because of sliding. Materials with layered crystal structures such as muscovite mica have basal planes that interact weakly, so the surface of these materials is easily cleaved and can be observed on the atomic scale. Wear on the mica surface mainly occurs to a depth of about 1 nm, which corresponds to the distance from one cleavage plane surface to the one immediately beneath it [4]. By exploiting this wear mechanism, high-precision processing can be performed whose units correspond to the period of the layers in a multilayered crystal material that has periodic weak van der Waals bonds [5]. Processing begins at a certain critical load, and the processing depth increases discretely with the load. Fracture easily occurs at the two cleavage planes of the SiO_4-K and K-SiO_4 interfaces of muscovite mica, upon which 1- and 0.8-nm-deep mechanical processing has been performed by AFM in a previous study [6]. Because the processing of nanometer-scale shapes to a layer-period unit depth is possible, the mechanical processing of the standard ruler and the application to high-density recording technology has been proposed [7, 8]. To test the molecular bearing supported by the van der Waals force, a nanometer-scale slider has been proposed by applying AFM mechanical processing. Very fine dots composed of clusters of atoms supported by the van der Waals force were processed, called nanosliders, and the movement of these nanosliders has subsequently been evaluated [9, 10].

Polymer storage media play a crucial role in storage systems, and polycarbonate (PC) has attracted great attention as an AFM memory media. Although PC materials have been applied to light discs in large-scale applications owing to their smooth surface property and low cost, line-scratched PC surfaces exhibit plastic deformations at the nanometer scale [11] because PC has a low mechanical strength and a high molecular weight. In this way, the plastic deformation causes the development of large-sized surface damage and wear when attempting fine processing and reproduction at low loads. Furthermore, PC lacks wear-resistant properties to prevent the development of deformations. However, when considering their microtribological properties, polymer materials easily generate the deformation of lumps in thermal processing and typically have an inferior wear-resistant property. Because PC lacks wear resistance, a head-disc system made with this material is easily damaged owing to the contact between the head and the tip. Therefore, various protective films have been prepared to cover these materials.

Presently, the application of PC materials for pit fabrication, reproduction, and ultrahigh-density recording is being investigated [12] using the fabrication and formation techniques of field evaporation and electron kinetic energy [13–15]. In field evaporation, a voltage is applied between the probe and the substrate, causing local heating and evaporation to occur. This evaporation makes it possible to remove and deposit an individual atom or a cluster, while the heating creates fine holes on the surface and changes the electronic properties of the surface [13–15]. If this heating technique was applied to a PC substrate for ultrahigh-density recording by AFM, these effects are expected to enable fabrication of very fine patterns and to permit recording densities of more than 60 Gbits/in^2. Polycarbonate substrates, therefore, are a promising material for Millipede memory recordings that combine ultrahigh density, terabit capacity, a small form factor, and a high data rate.

The facilitation and precision of micromachining are indispensable to media recording, and this high-precision recording at the nanometer scale must be performed at a high speed. In addition, it is important to reproduce recording marks by tip scanning without inducing any change via surface deformation of the media; that is, it is necessary to control surface wear deformation at the nanometer scale. It is known that a contact start/stop cycle during recording operation is influenced by the micromachining stiction, which is that occurs between the head and the media. The capillary condensation of liquids from vapor, and especially water, can have additional effects upon the whole physical state of the contact zone, so the effect of absorbed water upon the tribological properties of the media surface is considered. For example, if the surfaces contain adhesive water molecules, these will increase the frictional force and atomic force between the tip and media surface. Conversely, this absorbed water also acts as a protective lubricating film on the media. It is difficult to control the effects of the absorbed water upon the recording, which can be reproduced using an AFM tip at a low load, and the absorbed water results in accidental errors in the recording because of water adhesion between the tip and media. For these reasons, it has been proposed to process the DLC film using a fluorocarbon plasma treatment and to characterize and optimize the treated film to improve its wear properties. However, PC materials are a promising ideal medium that should be easily deformable for bit writing, and yet the writing bits should be stable against tip wear and thermal degradation. It has also been proposed to treat the PC specimens with fluorocarbon plasma and subsequently coat them with an ultrathin DLC film. The surface-free energy of these modified PC specimens is thereby reduced and the interaction between the head and medium is thus decreased, which enables fine nanoprocessing to be realized.

In this study, to realize high-density memory based on PC materials as a recording media and exploit its unique tribological and viscoelastic properties, PC specimens were subjected to fluorocarbon plasma treatment. The untreated polycarbonate (untreated PC) and fluorocarbon plasma-treated polycarbonate (CF_4-treated PC) samples were characterized by measuring the surface-free energy, nanoindentation hardness, and the wear properties using the liquid-drop method and AFM techniques. High-density recording was realized by nanometer-scale processing of the untreated and CF_4-treated PC samples, and the processing properties of these samples were evaluated and characterized.

2. Experimental methods

2.1. Sample preparation

Polycarbonate was used as specimens in our study. A PC sample was treated by radio frequency (RF) magnetron sputtering with carbon tetrafluoride (CF_4) gas. The PC sample surface was subjected to a high-frequency (13.56 MHz) voltage for fluorocarbon plasma treatment. All experiments have been described elsewhere [16]. The experiments were performed under the conditions of an RF power of 0.25 W on the sample, a bias voltage of 0.8 kV, a vacuum pressure of 2×10^{-2} Torr, and a sputtering time of 30 min [16]. To investigate the effect of fluorocarbon plasma on PC samples, liquid-drop method was used to measure the surface-free energy of the samples [16]. A shape of a liquid drop deposited on the sample surface was evaluated to measure the contact angle between the drop and the surface, from which the surface-free energy was obtained based on the extended Fowkes theory of Ref. [3]. Drops of refined water, methyl iodide, and n-hexadecane were used in this study, and the hydrophobicity of these drops was determined by the contact angle of the droplet to the specimen surface. The surface-free energy (γs) is given by the sum of the dispersion force (γsd), the dipole force (γsp), and the hydrogen bond (γsh), such that

$$\gamma s = \gamma sd + \gamma sp + \gamma sh. \quad (1)$$

2.2. Nanoindentation evaluation

To study the mechanical properties of untreated and CF_4-treated PC samples, an AFM in combination with a dynamic stiffness measurement system (Digital Instruments Nanoscope III, Hysitron Inc.) was used as shown in **Figure 1** [16]. The nanoindentation measurement has been described elsewhere [16]. The indentation experiment was performed at an 80-μN load, with both the loading and unloading times being 5.0 s. A diamond tip with an approximately 100-nm radius was used [17].

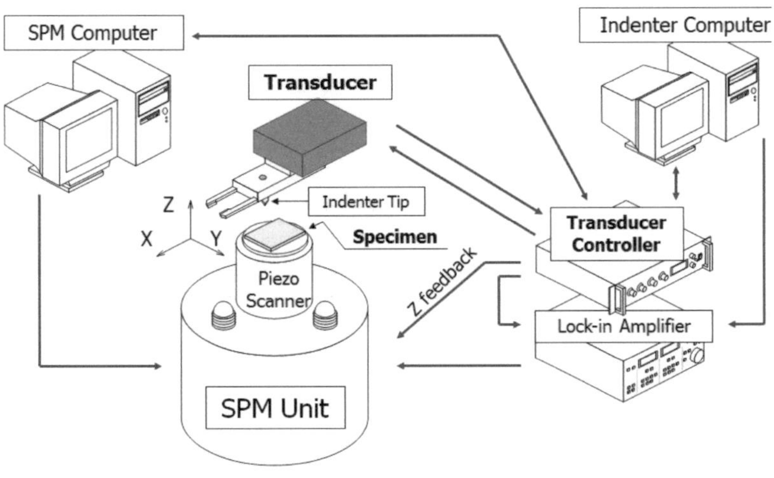

Figure 1. Schematic diagram of the scanning probe microscope and dynamic stiffness measurement system.

http://dx.doi.org/10.5772/67512

2.3. Microwear evaluation

The wear properties of the untreated and CF_4-treated PC samples were evaluated by a microwear test technique using AFM, where a cantilever with a pyramidal diamond tip (radius ~50 nm, Hysitron Inc.) was used. A schematic of the wear test is shown in **Figure 2**, wherein the diamond tip was slid against the sample surface. The wear test was conducted by scanning an area of 1 × 1 μm² and by applying varying loads of 0.5, 1.0, and 2.0 μN. The observation area was of 3 × 3 μm² [16].

2.4. Nanoprocessing of lines and dots

Line nanoprocessing was performed on untreated and CF_4-treated PC samples by using a diamond tip with a radius less than 50 nm (SPI 3800N, Seiko Instrument Inc. SII) [16]. The spring constant of the cantilever and tip was 45-51 N/m. An applied load was 150 nN. As shown in **Figure 3** [16], 500-nm-long lines and spaces were produced at 66- and 62-nm intervals on untreated and CF_4-treated PC samples, respectively, in an 800 × 800 Nm² scanning area. Second, dots were produced at intervals of 45 and 40 nm on the untreated and CF_4-treated PC samples, respectively, by applying a load of 150 nN.

2.5. Evaluation of viscoelastic properties of processed area

Viscoelastic properties were evaluated using the force modulation method of scanning probe microscopy (SPM) as shown in **Figure 4** [18, 19]. This apparatus was added to a nanoindentation system along with a lock-in amplifier. The Berkovich-type diamond indenter, where

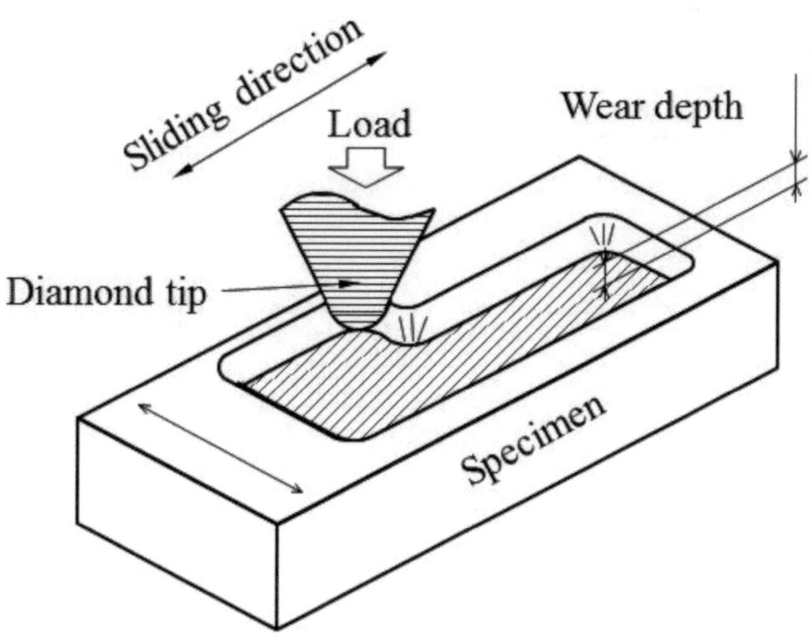

Figure 2. Schematic image of the micro-wear test by atomic force microscopy.

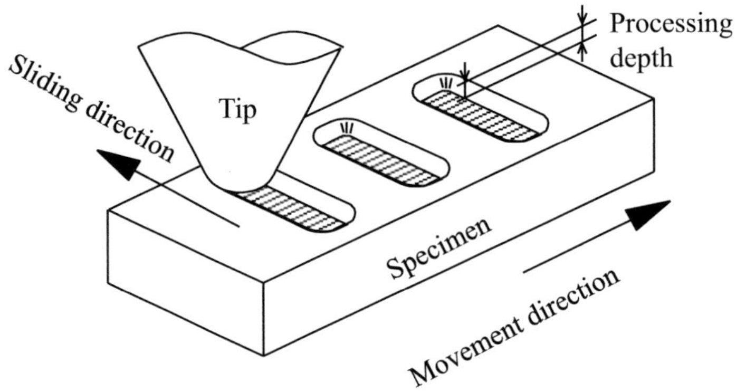

Figure 3. Schematic image of line processing by atomic force microscopy.

a vibration in the vertical direction was applied, was conducted to scan on the sample. As the phase lag and displacement of the tip cantilever were measured via the response of the tip indenter, which possessed a transducer controller, the viscoelastic properties such as the storage modulus, loss modulus, and tanδ were obtained from the force modulation [18]. The method in detail was reported in our previous study [16]. With a 100-nm-radius equilateral-triangle pyramidal diamond indenter, these tests were performed under loads varying from 5 to 35 μN, a load amplitude of 5 μN, and a frequency of 400 Hz [16]. After the tests, the sample was scanned using the same tip under a load of 2 nN with a vibration of a frequency of 50 Hz. Viscoelastic properties such as storage modulus, loss modulus, and tanδ were evaluated by means of atomic force microscopy [18].

To evaluate the viscoelastic properties, the dynamic model and force modulus of the AFM system were used (**Figure 4**). In this method, the dynamic mechanical properties (e.g., storage and loss moduli) were measured when the scanner was vibrated in the *z*-axis direction, which was possible when the dynamic response of the indenter of the AFM was well calibrated and modeled. **Figure 5** shows the correlated sinusoidal force and displacement curves, from which the mechanical properties can be calculated using well-established models. The amplitude and phase shift can be used to calculate the contact stiffness using a dynamic model [18], whose standard analytical solution is given below.

The amplitude of the displacement signal (X_0) and the phase shift between the force and displacement (Φ) are given as

$$X_0 = \frac{F_0}{\sqrt{(k - m\omega^2)^2 + [(C_i + C_s)\omega]^2}}, \tag{2}$$

$$\Phi = \tan^{-1}\frac{(C_i + C_s)\omega}{k - m\omega^2} \tag{3}$$

where F_0 is the ac force amplitude, m is the indenter mass, ω is the frequency (rad/s), and C_s and C_i are the damping coefficients of the specimen and the air gap in the capacitive displacement sensor, respectively. The combined stiffness, k, is given by

$$k = K_s + K_i, \tag{4}$$

http://dx.doi.org/10.5772/67512

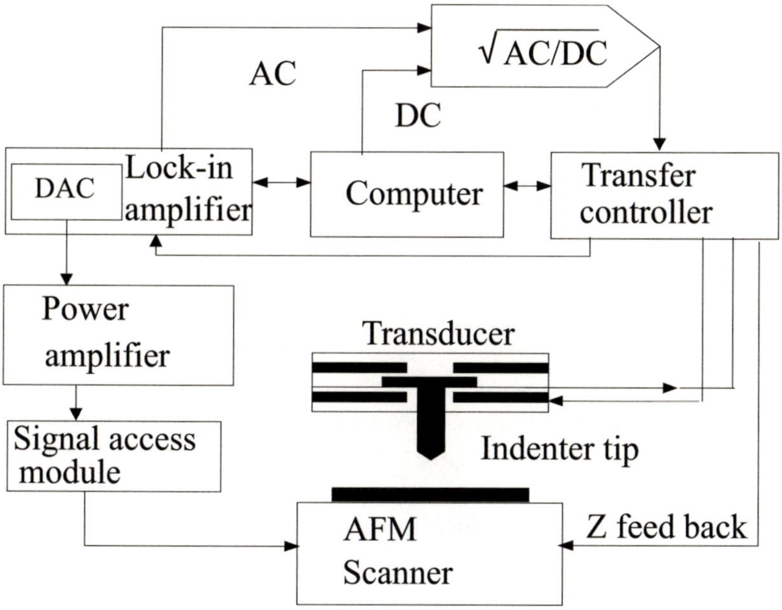

Figure 4. Schematic diagram of the atomic force microscopy system for evaluating the viscoelastic properties.

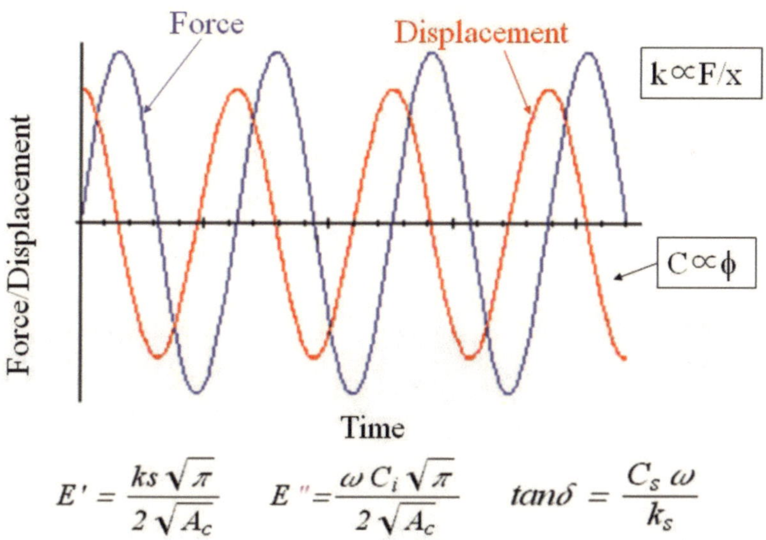

Figure 5. Example of the sinusoidal force/displacement data.

where K_s is the contact stiffness and K_i is the spring constant of the leaf springs holding the indenter shaft. The storage modulus (E'), loss modulus (E''), and tanδ can now be found as

$$E' = \frac{k_S\sqrt{\pi}}{2\sqrt{A_c}}, \tag{5}$$

$$E'' = \frac{\omega\, C_S\sqrt{\pi}}{2\sqrt{A_c}}, \tag{6}$$

$$\tan\delta = \frac{C_S\,\omega}{k_S}, \tag{7}$$

where δ is the phase lag between the force and displacement. Eqs. (5)–(7) are used to calculate the storage and loss moduli and to study the viscoelastic properties of polymeric materials [20].

3. Result and discussion

3.1. Measurement of surface-free energy

The effects of the fluorocarbon plasma treatment upon PC samples were evaluated by measuring surface contact angle and surface-free energy. Distilled water, methyl iodide, and n-hexadecane were used in this study. **Figure 6** shows the drop-profile images of the samples, and the measured values of the contact angle and surface-free energy are listed in **Table 1** [16]. **Figure 6** shows that the contact angles of the CF_4-treated PC are greater than those of the untreated PC, while the measured surface-free energies are 19.2 mN/m for the untreated PC and 16.1 mN/m for the CF_4-treated PC. These results show that the surface tension of the PC material is reduced after fluorocarbon plasma treatment. It is possible that the C-F compound, which is a peculiarity of polymer materials and has a weak C-F bond interaction, was formed on the PC sample by the fluorocarbon plasma treatment [16, 19, 21, 22].

3.2. Nanoindentation properties

Figure 7 shows the load displacement curves of untreated and CF_4-treated PC samples obtained in the nanoindentation test with a maximum applied load of 80 μN. The maximum

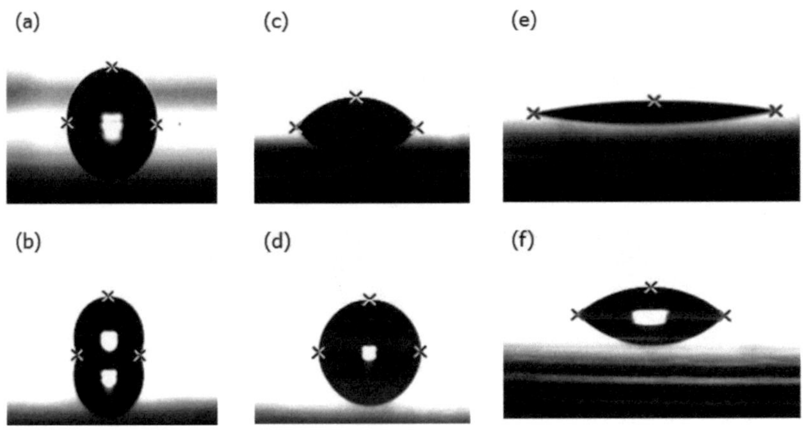

Figure 6. Drop profiles on the untreated and CF4-treated PC samples. The profiles show drops of refined water on the (a) untreated and (b) fluorinated samples; drops of methylene on the (c) untreated and (d) fluorinated samples; and drops of n-hexadecane on the (e) untreated and (f) fluorinated samples.

http://dx.doi.org/10.5772/67512

	Contact angle (°)			
	Refined water	**Methylene iodide**		***n*-hexadecane**
Untreated polycarbonate	95.8	84.2		36.6
CF_4-treated polycarbonate	103.5	89.5		44.8
	Surface-free energy (mN/m)			
	Dispersion force (γsd)	Dipole force (γsp)	Hydrogen bond (γsh)	Surface free energy (γs)
Untreated polycarbonate	16.5	0.0	2.7	19.2
CF_4-treated polycarbonate	14.8	0.0	1.3	16.1

Table 1. Values of drop-contact angles (upper) and surface-free energies (lower) for untreated (polycarbonate) and CF4-treated (fluorinated polycarbonate) samples.

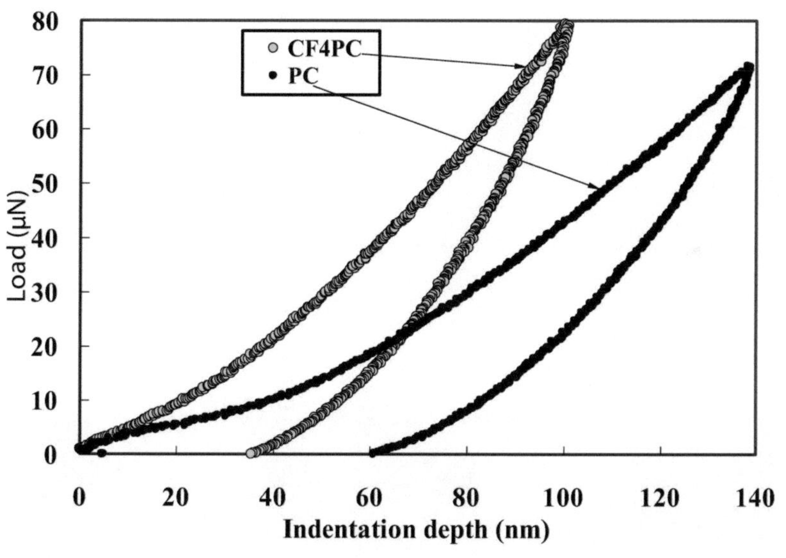

Figure 7. Load versus indentation depth (displacement) for (a) untreated (PC) and (b) CF_4-treated (CF_4-PC) PC samples at a peak load of 80 μN.

indentation depths are approximately 132 nm for the untreated PC and approximately 100 nm for the CF_4-treated PC [16]. It can be clearly seen that the residual depth (deformation) of the CF_4-treated PC is smaller than that of the untreated PC, indicating that more plastic deformation occurs on the untreated PC. Meanwhile, it is believed that the hardness of the PC sample is improved by the fluorocarbon plasma treatment [16].

3.3. Wear properties of untreated and CF_4-treated PC

Wear tests were performed by sliding a tip repeatedly over the same sample area under applied loads ranging from 0.5 to 35 μN. A sinusoidal signal with a frequency of 400 Hz was added to the tip. **Figures 8** and **9** show the profile AFM images of wear marks made at loads 0.5, 1, and 2 μN [16]. The square wear marks in the CF_4-treated PC are clearly formed compared with those in the untreated PC, whereas the wear depths of the former are smaller than those of the latter. Wear marks with a depth ranging from 2 to 13 nm were formed on the CF_4-treated PC, indicating that it is possible to perform nanoprocessing on it.

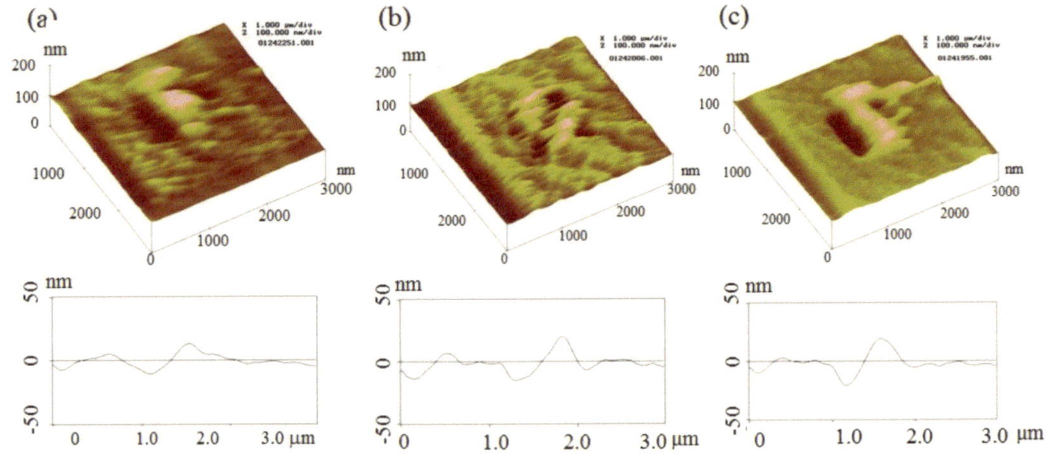

Figure 8. Atomic force microscope images (upper) and line-scan profiles (lower) of wear marks on untreated PC, created at loads of (a) 0.5, (b) 1, and (c) 2 μN.

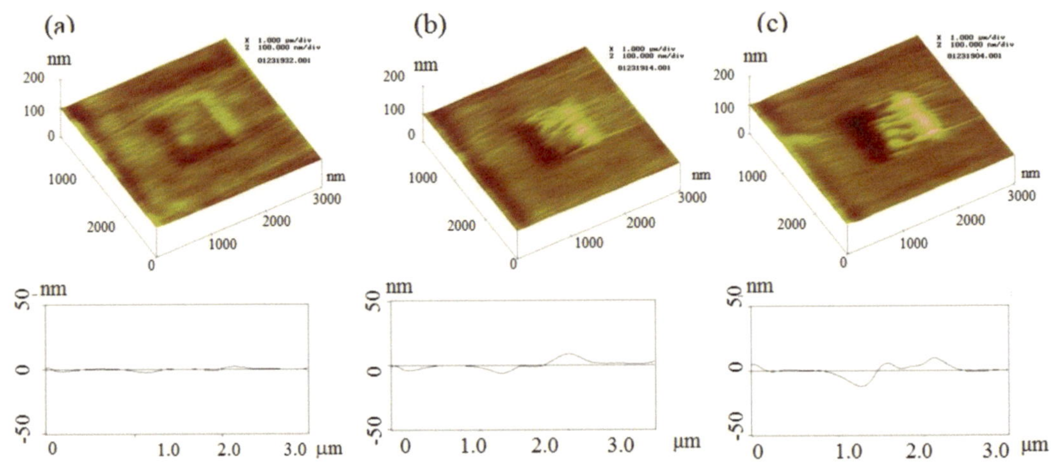

Figure 9. Atomic force microscope images (upper) and line-scan profiles (lower) of wear marks on CF_4-treated PC, created at loads of (a) 0.5, (b) 1, and (c) 2 μN.

Figure 10 shows a plot of wear depth versus load for the untreated and CF_4-treated PC. The wear tests were repeated five times and exhibited a high degree of repeatability with almost no changes in the observed trend [16]. It is found that the wear resistance of the PC sample is improved with fluorocarbon plasma treatment.

3.4. Load dependence of nanoprocessing properties

To clarify the nanometer-scale processing properties of the PC samples, these were investigated using load as a factor. **Figure 11** shows the horizontal waveform and magnitude of the friction force on the tip for varying loads. The friction force is measured by sliding the tip on the sample for one trace and a reciprocal retrace, where the tip is varyingly loaded with 0, 130, 190, or 250 nN. The untreated PC (**Figure 11a**) shows large and unstable friction forces, whereas the CF_4-treated PC (**Figure 11b**) shows small and stable friction forces at the same processing loads. **Figure 11c** shows the dependence of the magnitude of the friction force upon load for the untreated and CF_4-treated PC samples, where it is clear that the friction force of the CF_4-treated PC is smaller than that of the untreated PC.

Figure 12 shows the shape and depth images of grooves (traces) processed on the untreated (**Figure 12a**) and CF_4-treated (**Figure 12b**) PC samples. As shown in **Figure 12a**, the groove shapes in the untreated PC are not clearly visible, and lumps and processing residues are generated along the processing grooves, which indicate that the processed grooves generate deformations. By contrast, the grooves are clearly formed in the processed area of the CF_4-treated PC, as shown in **Figure 12b**.

Figure 12c shows the mean heights and depths of the deformation components (lumps and grooves) at the center of the processed grooves of the untreated PC for varying loads. The lump height (deformation height) increases with processing tip loads from 50 to 250 nN. However, with loads exceeding 300 nN, the mean height of the processed grooves decreases to negative values, which indicates that the grooves are formed by material removal. These

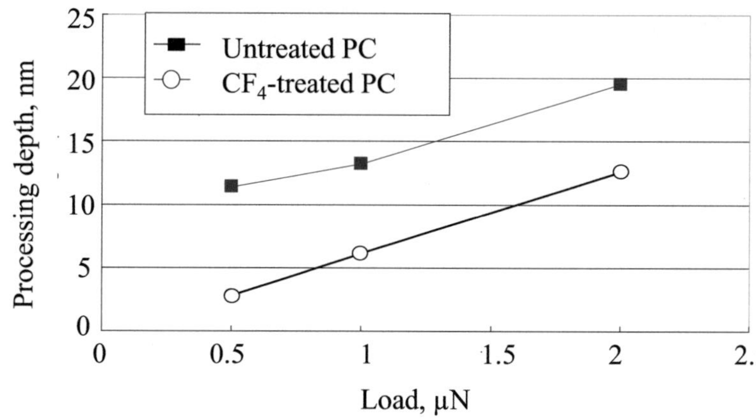

Figure 10. Wear depth as a function of load for untreated (filled squares) and CF4-treated (empty circles) PC samples.

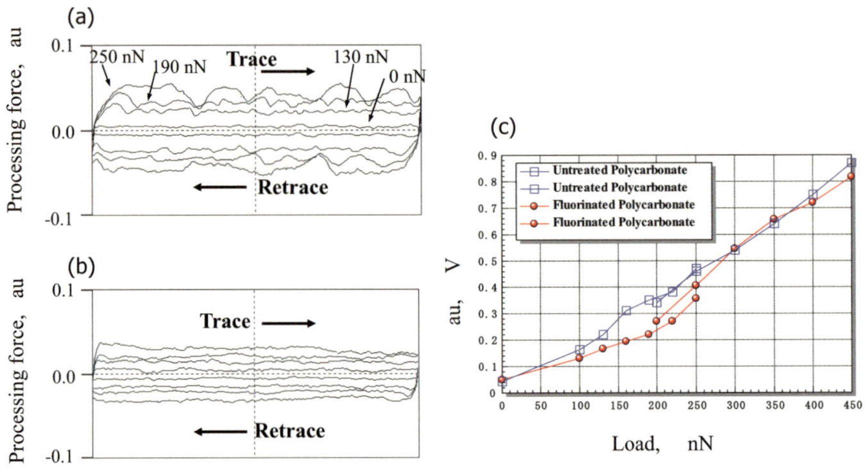

Figure 11. Friction force measured at varying loads for the (a) untreated and (b) CF4-treated PC samples, with the friction force waveforms arbitrarily scaled from the zero axis (dashed line) in the order of 0, 130, 190, and 250 nN for both the trace and retrace plots. (c) Magnitude of the friction force as a function of tip load for both samples.

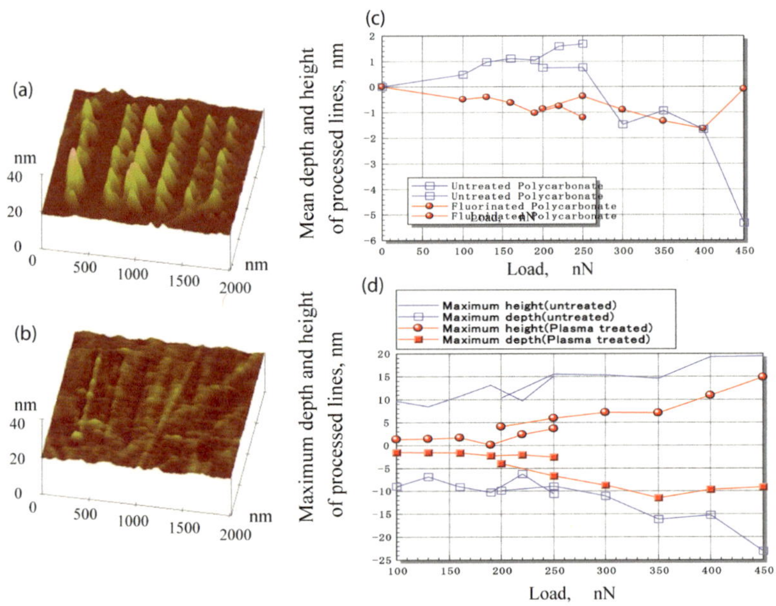

Figure 12. Atomic force microscope images of grooves processed on the (a) untreated and (b) CF4-treated PC samples. Depths of processed grooves for the samples, plotting the (c) mean depth and the (d) maximum and minimum depths of the groove centers.

results illustrate that molecular chains are dragged and deformed at loads less than 300 nN, while at loads greater than 300 nN there is removal of the PC. It is concluded that the molecular chains of the PC material are dragged by the friction force between the tip and processed surface at low loads, which results in expanded distances between the molecules.

As a result, lumps are formed. As the load increases further, the friction force increases to a level sufficient to remove part of the molecular chain, thereby forming a groove. By contrast, as shown in **Figure 12d**, the CF_4-treated PC exhibits very different results from that of the untreated PC. Removal processing in the treated sample begins from 100 nN loads, and the major restructuring of the processed surface is not observed. It is concluded that the friction force of the CF_4-treated PC is reduced by the fluorination treatment and that the molecular chain dragging that causes the lumps and surface roughening is suppressed.

3.5. Nanometer-scale dot and line processing

Dots with intervals of 45 nm were processed by sliding the tip over the untreated PC surface (**Figure 13a**), creating dots whose shapes are not uniform. Dots with inter-dot intervals of 40, 35, and 30 nm were also processed on the CF_4-treated PC and are, respectively, shown in **Figure 13b–d**, where uniformly shaped dots are clearly observed. The line profiles of these fine dots obtained from **Figure 13** AFM images are given in **Figure 14**, which clearly indicate that fine-dot processing is realized on the CF_4-treated PC sample. Because fine processing performed in a region of 25 × 25 μm^2 is thought necessary for the realization of 1 Tbit/inch^2 recording, these results demonstrate that this level of processing can be realized using a sharp tip.

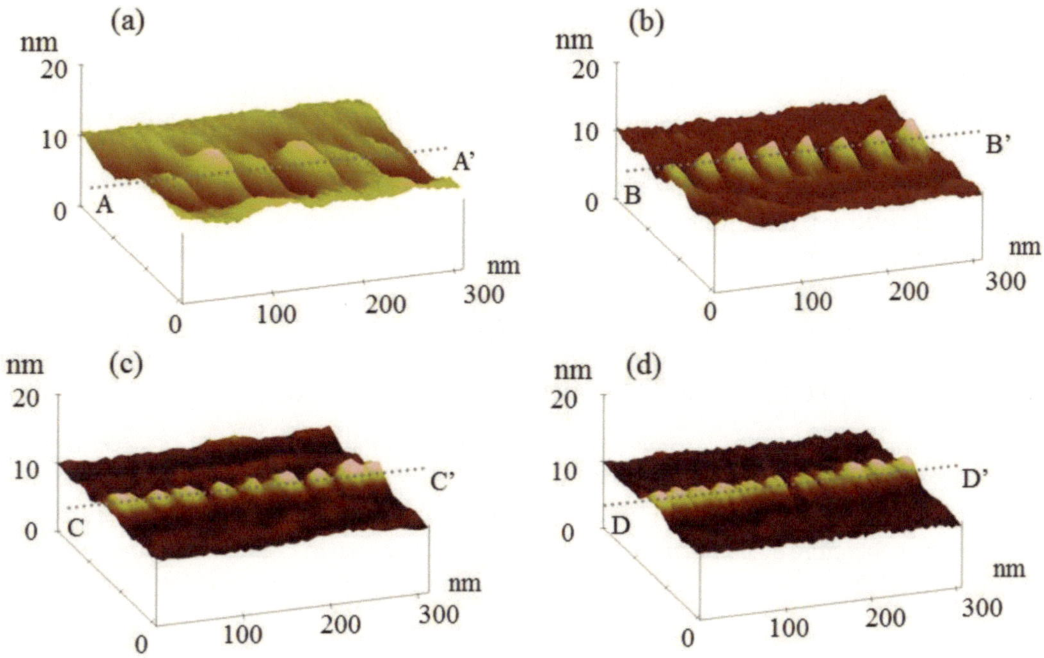

Figure 13. Atomic force microscope images of the fine dots processed with varying dot intervals on untreated and CF_4-treated PC samples. Shown are images of dots processed on (a) untreated PC (groove interval 45 nm), and CF4-treated PC with groove intervals of (b) 40, (c) 35, and (d) 30 nm. The lines (A–A′), (B–B′), (C–C′), and (D–D′) mark the line from which sectional profiles are obtained for **Figure 14**.

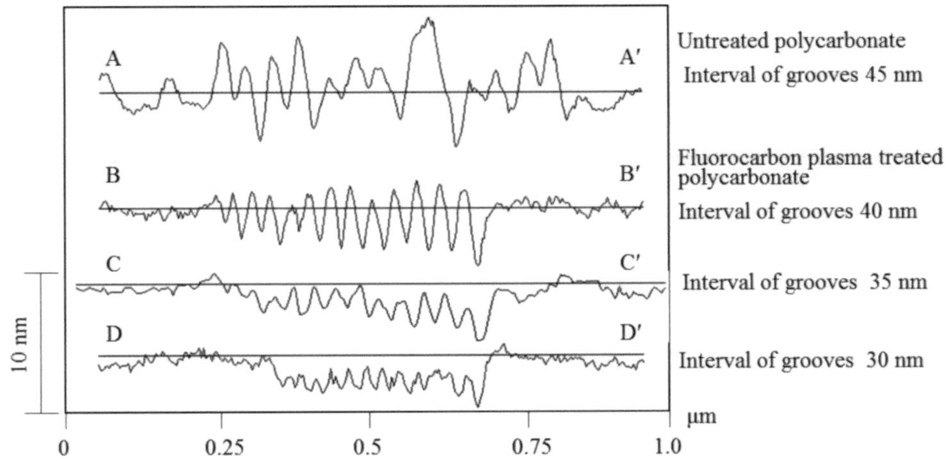

Figure 14. Atomic force microscope sectional profiles of fine dots processed at varying dot intervals on untreated and CF4-treated PC samples. Plotted are profiles of dots processed on untreated PC (groove interval 45 nm) (A–A′), and CF4-treated PC with groove intervals of 40 (B–B′), 35 (C–C′), and 30 (D–D′) nm.

Lines were processed on untreated PC by scanning a DLC-coated tip under a load of 150 nN at a 66-nm line interval. As shown in the AMF images and line profiles in **Figure 15a**, no obvious formation of precise lines are seen, but many lumps with heights approximately 3–5 nm are generated. As discussed earlier, these lumps are formed when the molecule chains of the polymer PC surface are dug up by tip scanning, causing plastic deformations in the processed area. It is clear that precise fabrication of fine lines is impossible on the untreated PC. **Figure 15b** shows AFM images and line profiles of the fine lines processed on the CF_4-treated PC by scanning a DLC-coated tip under a load of 150 nN at a 62-nm line interval [16]. Macroscopically, precision processing is realized. Contrasted with the roughness of the unprocessed area, the shapes of the formed lines are uniform and a change in depth can be observed. Furthermore, it is found that the roughness of the surface does not vary with the processing. This is possibly because that the processing resistance becomes constant, and any variation is virtually negligible during processing [16]. The processing properties of the CF_4-treated PC can be possibly improved because of a decrease in the PC friction coefficient of the PC with fluorocarbon plasma treatment.

Regarding the nanometer-scale processing characteristics of the CF_4-treated PC, because the surface-free energy is reduced with CF4 treatment, it is possible that fine dots are produced while scanning the tip at low loads. However, the processing is typically performed more than 10 times, and we therefore evaluated repetitive processing, as shown in **Figure 16**. On the untreated PC, shown in **Figure 16a**, there are quite large and irregular lumps generated at the peripheral areas of the scanned regions, which is thought to be because molecule chains are easily dug up on this sample even at low loading forces. Therefore, it is difficult to perform high-precision line processing at the nanometer scale on the untreated PC. For the CF_4-treated PC shown in **Figure 16b**, the interaction between atomic particles is weak owing to the C-F compound, and this produces a low frictional force during tip scanning that is insufficient to dig up the molecule chains. Therefore, fine processing at the nanometer scale can be successfully realized at low loads.

http://dx.doi.org/10.5772/67512

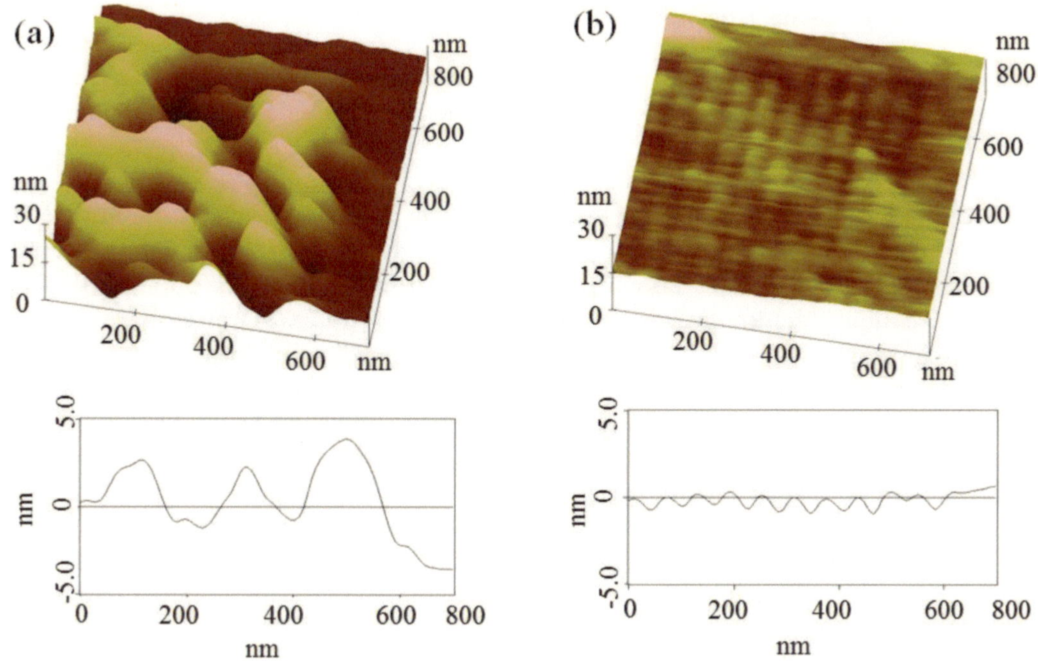

Figure 15. Atomic force microscope images (upper) and line-scan profiles (lower) of the lines and spaces processed at a load of 150 nN on (a) untreated PC with a 66-nm interval and (b) CF4-treated PC with a 62-nm interval.

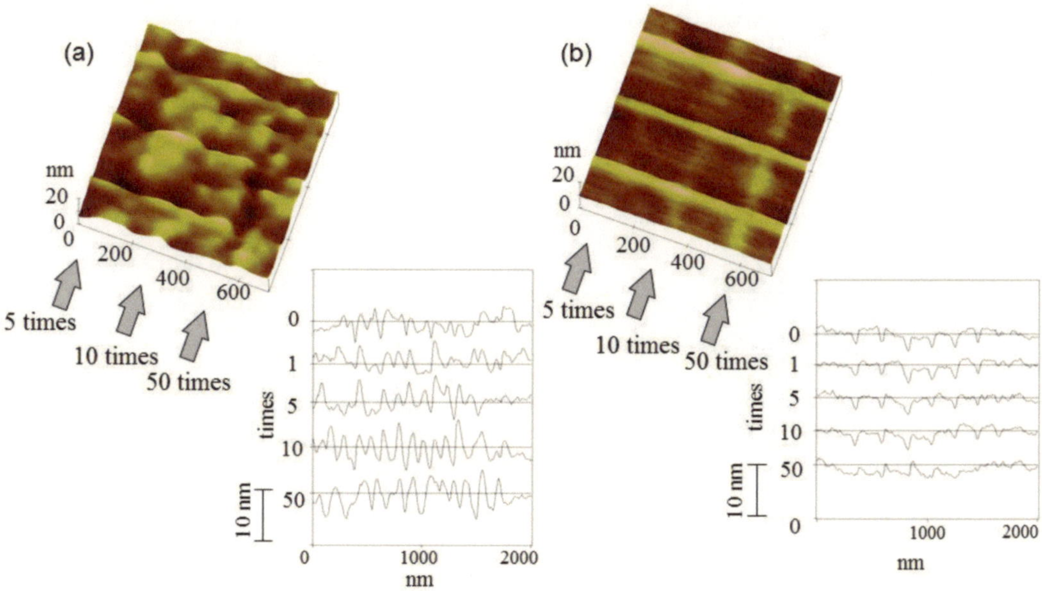

Figure 16. Atomic force microscope images (upper) and line-scan profiles (lower) of regions repetitively scanned 1, 5, 10, and 50 times on the (a) untreated and (b) CF4-treated PC samples.

3.6. Evaluation of viscoelastic properties

To investigate the effect of the fluorocarbon plasma upon precision processing, the viscoelastic properties were evaluated by a wear test on the untreated and CF_4-treated PC. **Figure 17** plots the storage (**Figure 17a**) and loss (**Figure 17b**) moduli of the worn (processed) and unworn (unprocessed) areas of the CF_4-treated PC and of the unprocessed and untreated PC. It can be seen in the plot that the storage modulus of the unprocessed CF_4-treated PC (i.e., CF_4-treated PC without wear) is lower than that of the untreated PC and the processed CF_4-treated PC (i.e., CF_4-treated PC with wear). It is also observed that the unprocessed CF_4-treated PC exhibits significantly smaller values of the loss modulus (as well as lower stiffness and damping values, c.f. **Figure 18**) compared with those of the untreated PC and the processed CF_4-treated PC. As shown in **Figure 17**, at loads ranging from 5 to 10 μN on the CF_4-treated PC, the value of the storage modulus is very small, while the loss modulus is almost at zero at a load of ~8 μN. Owing to the adhesion that exists between a tip and a surface, the interaction between the tip and the surface of the CF_4-treated PC sample is smaller than that of the untreated PC and processed CF_4-treated PC samples. When analyzing the CF_4-treated PC, it may be speculated that the energy loss of its interface is extremely small at low loads because the surface-free energy is low in the C-F compound structure and the interaction between a tip and sample surface is small when the tip is scanning the sample surface.

However, as the load increases, both *E'* (storage modulus) and *E"* (loss modulus) show an increasing trend, with the untreated PC and processed CF_4-treated PC increasing in a similar fashion. This similar tendency is shown in both the untreated PC, whose atomic interactions are strong, and the CF_4-treated PC, whose atomic interactions are weak, and the tendency is mainly affected by the adhesion between a tip and processing surface under a low load. It is observed in these samples that when the interaction between the tip and the sample surface is raised along with the gradually increasing load, the stiffness and damping are also seen to increase (**Figure 18**).

From **Figures 17b** and **18a**, the loss modulus and stiffness of the unprocessed CF_4-treated PC sample are significantly lower than those of the untreated PC and processed CF_4-treated PC

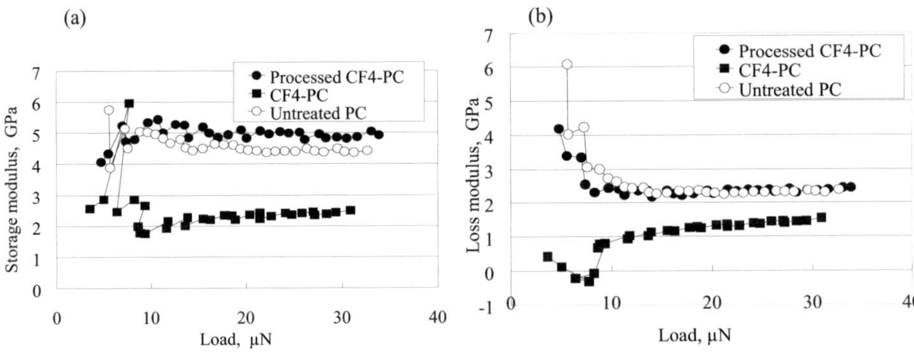

Figure 17. Plots of the viscoelastic properties at loads ranging from 5 to 35 μN, showing the (a) storage and (b) loss moduli of untreated PC (empty circles) and CF4-treated PC (filled-squares) samples with no wear induced on their surfaces, and a processed CF4-treated PC (filled-circle) sample, where "processed" indicates that wear is induced on its surface.

http://dx.doi.org/10.5772/67512

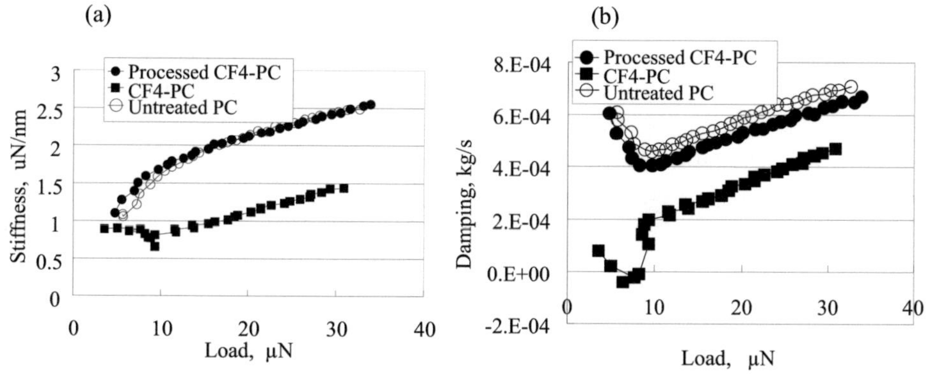

Figure 18. Plots of the viscoelastic properties at loads from 5 to 35 μN, showing the (a) stiffness and (b) damping of untreated PC (empty-circle) and CF4-treated PC (filled-square) samples with no wear induced on their surfaces, and a processed CF4-treated PC (filled-circle) sample, where "processed" indicates that wear is induced on its surface.

samples [16]. Compared with the untreated and processed CF_4-treated PC samples, since the unprocessed CF_4-treated PC has a fluorocarbon plasma layer on its surface, the interaction between the tip and its surface is weak during processing. As a result, the unprocessed CF_4-treated PC shows a lower loss modulus and stiffness [16, 19]. It is thought that C-F compound might be formed on the PC during the fluorocarbon plasma treatment, which results in a reduction of the shear resistance of the treated surface. **Figure 18b** shows that the damping of the CF_4-treated PC decreases owing to C-F bonds.

As shown in **Figure 19**, at a load ranging from 0 to 8 μN, the tanδ-value of the unprocessed CF_4-treated PC sample was much lower than those of the untreated PC and processed CF_4-treated PC samples. It was known that tanδ is the ratio of loss to the storage and is called damping [23]. The results showed that the unprocessed CF_4-treated PC sample has a small tanδ-value (damping value) at less than 10-μN load because the CF_4-plasma treatment layer on its surface still remained, so that the interaction between the tip and its surface is small

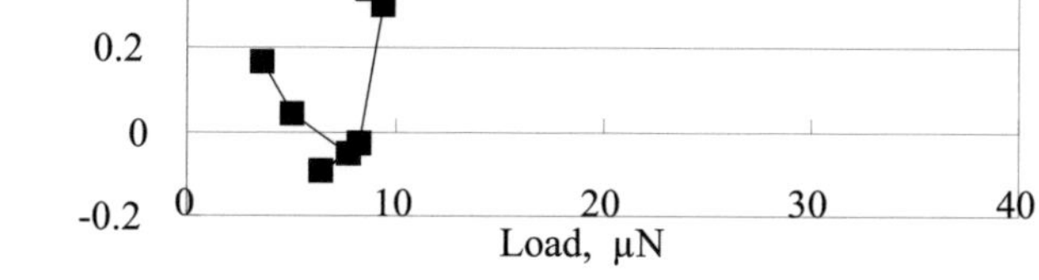

Figure 19. Plot of a viscoelastic property at loads ranging from 5 to 35 μN, showing the tan δ of untreated PC (empty circles) and CF4-treated PC (filled-square) samples with no wear induced on their surfaces, and a processed CF4-treated PC (filled-circle) sample, where "processed" indicates that wear is induced on its surface.

owing to the formed C-F bonds [16, 19]. However, the tanδ-value of the unprocessed CF_4-treated PC sample sharply increased to be close to those of the untreated and processed CF_4-treated PC samples at the load of 10 μN and it became even slightly higher than others when the load was more than 10 μN, indicating that the CF_4-plasma treatment layer on the unprocessed CF_4-treated PC sample was removed by tip scanning at the 10 μN and more load, resulting in the interaction being increased [16, 19].

4. Conclusion

Based on the surface modification and nanometer-scale mechanical processing of PC and the evaluation of its viscoelastic properties using AFM, it is feasible to realize high-density storage memory using modified PC as a media.

Our current study focuses upon the evaluation of various viscoelastic properties of untreated PC, processed CF_4-treated PC (i.e., wear induced on its surface), and unprocessed CF_4-treated PC (i.e., no wear induced on its surface), whereby the following points can be concluded:

1. A PC material was modified by fluorocarbon plasma treatment, whereby the surface-free energy of the PC was reduced. Nanometer-scale line and dot processing were realized owing to a subsequent reduction in the interaction force between the tip and the fluorinated sample.
2. The indentation evaluation result shows that the plastic deformation of the PC was improved by the fluorocarbon plasma treatment. The realization of nanometer-scale line and dot processing is possible because the plastic deformation caused the surface damage and wear when attempting fine processing at low loads was reduced. Furthermore, the CF_4-PC shows its super wear-resistant properties to prevent the development of deformations.
3. The viscoelastic properties of the CF_4-PC show the values of loss modulus, stiffness and tanδ lower than those of the untreated PC, which indicates that the shear resistance of PC was reduced. It is possible that the C-F combination was formed on the PC by the fluorocarbon plasma treatment, which would imply that the processing resistance decrease was owing to the weak interaction of the C-F bonds.
4. The storage and loss moduli, stiffness and damping of a processed area on the CF4-treated PC were practically the same as those of untreated PC because the CF4-treated layer of the processed area was removed by tip scanning. Therefore, the processed areas exhibited the same viscoelastic properties as the untreated PC.

Acknowledgements

This research was performed with the help of the graduate students at Nippon Institute of Technology. This research was partly supported by the Storage Research Consortium (SRC) and "High-Tech Research Center" Project for Private Universities: matching fund subsidy from MEXT (Ministry of Education, Culture, Sports, Science and Technology), Japan.

Author details

Shojiro Miyake[1] and Mei Wang[2]*

*Address all correspondence to: mwang@osg.co.jp

1 Department of Innovative System Engineering, Nippon Institute of Technology, Saitama, Japan

2 Department of Research and Development, OSG Corporation, Aichi, Japan

References

[1] Binnig G, Quate CF, Gerber Ch: Atomic force microscope. Phys. Rev. Lett. 1986;56: 930–933. DOI: https://doi.org/10.1103/PhysRevLett.56.930.

[2] Kaneko R, Oguti S, Miyamoto T, Andoh Y, Miyake S: Microtribology for magnetic recording. STLE Special Publication. 1990;SP-27:31–36.

[3] Miyake S, Kaneko R, Kikuya Y, Sugimoto I: Micro-tribological studies on fluorinated carbon film. Trans. ASME. J. Tribo. 1991;113:384–389. DOI:10.1115/1.2920633.

[4] Miyake S, Kaneko R, Miyamoto T: Micro wear increasing of polished chemically vapor deposited diamond film by nitrogen ion implantation by atomic force microscope. Nucl. Instrum. Meth. Phys. Res. Sect. B. 1996;108:70–74. DOI:10.1016/0168-583X(95)00873-X.

[5] Miyake S: 1 nm depth mechanical processing of muscovite mica by atomic force microscopy. Appl. Phys. Lett. 1995;67:25–2927. DOI: 10.1063/1.114844.

[6] Miyake S, Ishii M, Otake T, Tsushima N: Nanometer-scale mechanical processing of muscovite mica by atomic force microscope. J. Jpn. Soc. Precis. Eng. 1997;63-3:426–430. DOI: 10.2493/jjspe.63.426.

[7] Miyake S, Otake T, Asano M: Mechanical processing of standard rulers with one-nanometer depth of muscovite mica using an atomic force microscope. J. Jpn. Soc. Precis. Eng. 1999;65-4:570–574. DOI: 10.2493/jjspe.65.570.

[8] Miyake S, Kim J: Nano protuberance and groove processing of silicon by diamond tip sliding. The Institute of Electrical Engineers of Japan: Trans. Sensors Micromach. 2000, 120-E-7:350–356. DOI: 10.1541/ieejsmas.120.350.

[9] Miyake S, Asano M: Nanoprocessing by hard film coated tips and movement of processed nanoslider supported by van der Waals Forces. J. Jpn. Soc. Precise. Eng. 2000;66-8:1275–1279.

[10] Miyake S, Asano M: Nanometer scale mechanical processing of muscovite mica and application to high density memory. IEEJ Trans. Sensors Micromach. 2000;120-E-1:8 http://doi.org/10.1541/ieejsmas.120.8.

[11] Miyake S, Matuszaki K: Realization of nanometer-scale processing by fluorocarbon plasma treatment of polycarbonate and its potential application to high-density memory. IEEJ Trans. Sensors Micromach. 2001;121-E-10:564–570. http://doi.org/10.1541/ieejsmas.121.564.

[12] Mamin HJ, Rugar D: Thermomechanical writing with an atomic force microscope tip. Appl. Phys. Lett. 1992;61:1003–1005. http://dx.doi.org/10.1063/1.108460.

[13] Hosaka S, Koyanagi H, Kikukawa A, Miyamoto M, Imura R, Ushiyama J: Fabrication of nanometer-scale structure on insulators and in magnetic materials using a scanning probe microscope. J. Vac. Sci. Technol. 1995; B-13-3:1307–1311. http://dx.doi.org/10.1116/1.587843.

[14] Hosoki S, Hosaka S, Hasegawa T: Surface modifications of MoS_2 using an STM. Appl. Surf. Sci. 1992;60-61:643–647. DOI:10.1016/0169-4332(92)90489-K.

[15] Mamin HJ, Guethner PH, Rugar D: Atomic emission from a gold scanning-tunneling-microscope tip. Phys. Rev. Lett. 1990; 65-19:2418–2421. DOI: https://doi.org/10.1103/PhysRevLett.65.2418.

[16] Wang M, Miyake S: Application of nanometer-scale processing technique in high-density recording. Key Eng. Mater. 2005;291-292:407-412. DOI: 10.4028/www.scientific.net/KEM.291-292.407

[17] Bushan B: Boundary lubrication studies using AFM/FFM. In: Bushan B, editor. Handbook of micro/nanotribology, Boca Raton, FL: CRC Press, LLC, 1995:322–336.

[18] Syed Asif SA, Wahl KJ, Colton RJ: Nanoindentation and contact stiffness measurement using force modulation with capacitive load-displacement transducer. Rev. Sci. Instrum. 1999;70-5:2408–2413. DOI: 10.1063/1.1149769.

[19] Wang M, Watanabe S, Miyake S: Deposition of C-F thin films by sputtering and their micromechanical properties. New Diam. Front. Carbon Technol. 2005;15-1:29–35. DOI: 10.2979/ESJ.2005.4.1.29.

[20] Harris SJ, Weiner AM, Meng WJ: Tribology of metal-containing diamond-like carbon coatings. Wear. 1997;211-2:208–217. DOI: 10.1007/BF00156909.

[21] Prioli R, Jacobsohn LG, Maia da Costa MEH, Freire Jr FL: Nanotribological Properties of Amorphous Carbon-Fluorine Films. Tribol. Lett. 2003; 15-3:177–180. DOI: 10.1023/A:1024848816646.

[22] Leezenberg PB, Johnston WH, Tyndall GW: Chemical modification of sputtered amorphous-carbon surfaces. J. Appl. Phys. 2001;89:3498–3508. http://dx.doi.org/10.1063/1.1338989.

[23] Lakes RS: Viscoelastic measurement techniques. Rev. Sci. Instrum. 2004; 75-4:797–810. DOI: 10.1063/1.1651639.

Chapter 8

Green Intelligent Nanomaterials by Design (Using Nanoparticulate/2D-Materials Building Blocks) Current Developments and Future Trends

Dinesh Kumar and Shamim Ahmad

Additional information is available at the end of the chapter

http://dx.doi.org/10.5772/intechopen.68434

Abstract

Feasibility of designing and synthesizing 'smart' and 'intelligent' materials using nano-structured building blocks has been examined here based on the current status of the progress made in this context. The added advantages of using 2D layered/nonlayered materials along with phytosomal species derived from natural plants are highlighted with special reference to their better programmability along with minimum toxicity in biomedical applications. The current developments taking place in their upscaled productions are also included while assessing their upcoming industrial usages in diverse fields.

Keywords: materials by design, smart and intelligent materials, hybrid nanomaterials, targeted drug and gene deliveries, supramolecular complexes

1. Introduction

The study of crystalline materials, initiated since the beginning of the twentieth century, took almost 6–7 decades to mature in the form of microelectronics and microsystems technologies creating a wealth of information in form of industrial know-how (FCM, 2009). Subsequently, the discovery of the nanomaterials followed by extensive R&D efforts put in during the last few decades resulting in simultaneous developments of production/application technologies enabled them to be integrated into the biological systems and is currently offering valuable supports to the activities in the area of nanobiotechnology —an interdisciplinary field [1].

Finding inorganic/organic molecules biocompatible in hybrid nanomaterials syntheses, it was natural to explore mimicking the features of some simple living organisms in the pursuit of developing 'smart' and 'intelligent' materials responding to the environmental stimuli to start with. Consequently, attempts were made to identify the basic requirements of the materials to qualify for being 'smart' or 'intelligent.' Despite using these terms interchangeably earlier, some clarity started emerging particularly after several comprehensive deliberations held on this issue in different forums in form of two distinct levels of interactions between materials and the external stimuli. For instance, the materials that respond to the external stimuli by showing appreciable changes in their properties are termed as 'smart' leading to fabricate a variety of sensors and actuators. However, the material becomes 'intelligent' once it is endowed with the capability of reorganizing itself internally to take care of the changes due to external stimuli adaptively, and in such cases, a number of in-built features must be involved internally in a way similar to those in a conventional control system. For instance, the changes in material properties arising out of material-stimuli interactions must be communicated to a decision-making component of the material along with memory functions for taking decisions to initiate appropriate actions for countering the changes within a reasonable time period, which is an important parameter that may vary under different circumstances [2–5].

Coming back to the discussion of nanostructured materials, it may be noted that the modifications introduced into their physico-chemico-biological properties are considered as the result of the quantum confinement superimposed upon their bulk properties culminating into the morphology-specific features with enhanced activity arising from the exposure of the surface residing atomic species with sufficient unsaturated chemical bonds. These nanomaterials are subjected to still further modifications via their chemical conjugations involving strong/weak interactions in preparing the 1/2/3-dimensional nanobuilding blocks like nanowires, nanotubes, nanocoils, nanoropes, besides synthetic superlattices, and functional nanocomposites in addition to many other functional entities that are still being explored for their resultant structure-activity relationships (SAR) for developing newer materials. Further, the macromolecular species involving fullerenes, nanowires, nanotubes, and dendrimers prepared using different constituent materials are also being considered for new material discoveries owing to their chemical conjugations with a whole host of inorganic, organic, and biomolecular species [6–9].

The established links between the inorganic nanoparticles (NPs) and the biomolecular species using the biopolymeric compounds in different configurations possessing biocompatible, biodegradable, and low immunogenic features are currently being used in fabricating nanobiocarriers in drug/gene deliveries involving the polysaccharides, proteins, and nucleic acids to name a few [6]. For such applications, it is indeed imperative to control their morphology, surface charges, and the release profiles of the loaded therapeutic species. Subsequently, numerous bioactive nanomaterials were developed using silk proteins, collagen, gelatin, casein, albumin, protein-mimicking polypeptides, and polysaccharides like chitosan, alginate, pullulan, starch and heparin as typical examples. Protein engineered polymeric scaffolds, in addition, have been used in developing protein-polymer hybrids, where polymerization induces multifunctional properties leading to improved performances. Various kinds of supramolecular hydrogels with physicochemical properties for drug and gene deliveries owing to their features like good water-retention, better drug loading, biodegradability, biocompatibility, stability combined with multiple functionalities including optoelectronic properties, bioactivity, self-healing, and shape-memory effects were consequently explored and put to use. In addition, their stimuli responding gel/sol transitions (reversible) due to their noncovalent cross-linkages–based interactions were considered as promising bioscaffolds in theranostics. Various SAR aspects of these hydrogels with particular reference to their applications in bioimaging/detection, therapeutic delivery, and tissue engineering were reviewed recently [6–8, 10–18].

The phenomenon of self-assembly with special reference to the supramolecular assemblies using noncovalent intra-/intermolecular interactions has been invoked in producing the micro-/nanostructures including micelles, membranes, vesicles, and liquid crystals in the framework of crystal engineering. Molecular recognition-based 'host-guest' complexes are currently finding increasing applications in the development of molecular sensors and catalysis. It is important to note that the enhanced reactivity associated with the nanostructured materials has always been useful in molecular recognition-based self-assemblies providing environment for the ensuing chemical reactions. Noncovalent bonds between the reactants and 'template' holding the reactants close to the reactive sites provide the required environment for chemistry. Mechanically interlocked molecular architectures of topologically connected molecules involving noncovalent bonds in catenanes, rotaxanes, molecular knots, molecular rings and ravels are known to mimic the biological systems in form of photo-electro-chemical systems, catalytic systems, protein designs, and self-replications. A template molecule surrounded by functional monomers starts attaching them via intergroup interactions that helps in forming an imprinted matrix after polymerization. Subsequent removal of the template forms complementary cavities offering selective binding sites for new material synthesis [19].

While developing these numerous types of nanomaterials, a new class of condensed state materials was discovered exhibiting properties that surpassed those of the bulk and conventional nanomaterials. The theoretical studies of these atomically thin two-dimensional (2D) nanomaterials have, in the mean time, uncovered a number of novel features particularly arising out of the electron confinement in the third dimension without interlayer interactions (monolayer) resulting in extreme mechanical flexibility and optical transparency well suited for the fabrication of highly flexible and transparent electronic/optoelectronic devices, and the large surface to volume ratio making them appropriate for surface active applications. This fascinating field of graphene-like 2D layered nanomaterials (GLNs) includes a number of already explored materials like graphene, hexagonal boron nitride (h-BN), transition metal dichalcogenides (TMDCs), graphitic carbon nitride (g-C_3N_4), layered metal oxides, layered double hydroxides (LDHs) besides materials belonging to metal-organic frameworks (MOFs), covalent organic frameworks (COFs), polymers, metals, black phosphorus (BP), silicene, and MXenes. Driven by their extraordinary characteristic properties assessed theoretically and validated partly in some cases, a large number of synthetic methods including mechanical/chemical exfoliations, ion-intercalation and exfoliation, anion-exchange and exfoliation, chemical vapor deposition (CVD), and wet-chemical syntheses have been developed for preparing them for numerous applications. These nanomaterials are showing high promises for a variety of applications in electronics, optoelectronics, catalysis, energy storage and conversion, biomedicine, sensors, and many more [20]. Their other physicochemical properties like strong mechanical strength, unparalleled thermal conductivity, remarkable biocompatibility, and ease of surface functionalization make them highly useful in biochemical/medicinal applications particularly in biosensors, and nanomedicine comprising of electrochemical biosensors, optical biosensors, bioimaging, drug delivery and cancer therapy [21].

In particular, after the discovery of graphene, these 2D layered materials were exfoliated in the form of solvent dispersed crystalline few-/monolayers comprising the covalently bonded few atoms forming the crystalline stack, when bundled together with the van der Waals forces. For instance, heterostructures formed out of these monolayers of chalcogenides, graphene, and hBN are currently being examined as building blocks with tailored electronic band structures and associated physicochemical properties. Some of these predictions are yet to be realized experimentally as cited by many [22–38].

While studying the influence of structural features of these 2D material species, the characteristic features of the lamellar and 2D layered/nonlayered materials were found showing additional features worth use. It may particularly be noted that these 2D nanosheets not only exhibit novel optical and electronic properties due to the confinement of electron states along c-axis but also help in forming a variety of layered nano-/microsize entities involving differential stresses present in their bilayers to enforce numerous kinds of self-assemblies that are being explored in targeted deliveries. Based on these special features of 2D layered and nonlayered thin films along with the synthesis of lamellar nanoassemblies, an attempt has been made here in this chapter to highlight their syntheses along with their some novel applications already studied.

In the context of examining the usability of biopolymeric species in nanomedicines, the self-assembled liposomes were found offering special features that are quite useful in the targeted deliveries, where the hydrophilic/hydrophobic contents embedded in nano-/microsize double-layer enclosed spherical volumes with very effective protections from the enzymes, alkaline solutions, digestive juices, bile, and intestinal flora inside human body as well as free radicals. Accordingly, the liposomes are not only noted to check the oxidation and degradation of the embedded cores but also retain their double-layer barrier intact until the contents are delivered to the desired site. Discovered in the 1960s, liposomes are known to possess versatile features owing to their compositional variability and structural properties leading to a number of pharmaceutical, nutraceutical, and cosmeceutical applications, wherein, even the herbal extracts like flavonoids, glycosides, and terpenoids have been enclosed and transported from the hydrophilic to the lipophilic part of the membrane showing better bioavailability/efficacy, as noted in case of ginkgo biloba, grape seed, green tea, milk thistle, ginseng, and many other herbal families already explored for their applications in therapeutic formulations and dietary supplements [6, 7, 11, 18]. Liposomes are currently used in the pharmaceutical applications showing promises as intracellular delivery systems for antisense molecules, ribosomes, proteins/peptides, and DNA. Liposomes with enhanced drug delivery and long circulation times are finally getting clinically accepted as the liposomal drugs exhibiting reduced toxicities while retaining enhanced efficacy compared to their free complements [39–41].

Before putting these synthetic materials species to use in the form of nanomedicines, their toxicity features must be rigorously evaluated following the recommended standard procedures. Somehow, the data available in this context are inadequate as the associated toxicity is a complex function of the surface modification causing highly variable solubility of the inorganic nanomaterials. Accordingly, the associated toxicity scare is so strong that the regulatory authorities permit no relaxation in allowing for their human trials. Because of the mostly unknown nanotoxicological properties of these newer kinds of synthetic nanomaterials, even their nonbiological applications are presumably not considered safe, which, to a certain extent, has been creating hurdles in developing their further applications [42].

A relatively safer approach of mitigating these toxicity issues could be to use the benign nanomaterials particularly derived from the plants as phytochemicals. Such nanobuilding blocks of natural origin, already studied extensively in the recent past, are not only found adequately safe but are also compatible with numerous biomolecular species ensuring more benign interactions in contrast to those derived from purely synthetic materials of inorganic/organic origins. This approach is certainly green in nature while meeting various requirements of hybrid nanostructured materials species put to use in form of novel applications. A large variety of biomimetic designs are thus becoming feasible to invoke once green phytosomal building blocks are put to use in synthesizing new kinds of materials [11].

The recent developments in the field of liposomal encapsulations involving single/multiple bilayer nanosize enclosures and graphene-like few-/monolayered nanosheets, introduced above in very brief, do indicate toward the suitability of using these nanosize thin films as building blocks capable of imparting novel features for their applications that will be appreciated with wider impacts in the times to come. Keeping in view the importance of this growing field of nanosize thin film materials, an effort has been made here in this review by assessing the current status before attempting to foresee the trends from the angle of developing intelligent materials in due course of time by employing them in new materials discoveries.

2. Programmable physicochemical properties

The phenomenon of quantum confinement involved in preparing 1-, 2-, and 3D nanomaterials has already been validated experimentally before using them in newer applications while considering the electrons and the photons together in the form of diverse material building blocks designs as highlighted in the following.

2.1. Metal and semiconductor NPs

The plasmon resonance excited in metal NPs in the presence of a dielectric as a function of morphology and the metal used, and falling in the visible/IR region, have been exploited in electromagnetic enhancement resulting in Raman, fluorescence, and infrared absorption spectroscopies for single molecule detection; tip enhanced Raman spectroscopy, optical circuits, high efficiency LEDs, chemical/biochemical sensors, and efficient solar cells due to better light confinement in the photoactive material, or achieving resonant internal light scattering. In one of the highest efficiency organic solar cells, for instance (P3HT: PCBM bulk HJ), the efficiency limitations due to lower cutoff wavelength ($\sim$650 nm/$E_G$$\sim$2.1 eV) was taken care of by enhancing optical absorption as reported [43–55].

Semiconducting NPs, also called quantum dots (QDs), behave like quasi-atoms with electrons/holes possessing discrete energy levels, and exhibiting size-specific absorption and luminescence spectra characterized by the material used. QDs are very efficient light emitters with photoluminescence quantum efficiency ~80% and size-dependent emission characteristics. For instance, varying the size of CdSe QDs from 3 to 6.5 nm changed the emission from 470 to 630 nm. A number of optoelectronic devices using QDs are reported including biological tags, white LEDs, OLEDs, and photovoltaic solar cells (PVSCs) with efficiencies exceeding Shockley-Queisser limit [56–61].

2.2. Polymeric NPs

A variety of polymeric NPs involving either dispersion of preformed polymers or the polymerization of monomers have been reported using techniques like solvent evaporation, salting-out, dialysis, supercritical fluid technology, microemulsion, miniemulsion, surfactant-free emulsion, and interfacial polymerization, where the actual choice depends on a number of factors like particle size, size distribution, and the area of applications as discussed in a recent review [62].

Investigations have already correlated the physicochemical properties of the polymeric NPs with their biological responses, in which the morphology and surface charges on biodegradable entities were explored in designing various formulations in a recent review highlighting the challenges involved with in vivo trials [63]. Polymeric NPs are known to offer not only protection from environmental stimuli but also providing site-specific deliveries, particularly in case of charged NPs that are well protected. Parameters like uptake, bioavailability, and long-term therapeutic efficacies are possible to optimize by controlling their electrostatic interactions as mentioned below [64].

The study of drug release characteristics of diazepam loaded PLGA NPs confirmed that the parameters like sonication time, polymer content, surfactant, ratio of organic to aqueous phase, and the amount of drug—all influenced their sustained release [65]. Similarly, surface modifications of poly (ethylene glycol)-*b*-poly (ε-caprolactone) (PEG-*b*-PCL) NPs by 1, 4 and 8-residue-long oligoarginines caused substantial increase in cellular uptake highlighting the influence of surface functionalization of polymeric NPs in subcellular targeting [66]. In yet another study, core-shell type NPs, loaded with doxorubicin (DOX), were assessed for their in vitro cytotoxicity against breast cancer and human fibroblast cell lines in which AgNPs, Ag/PVA and Ag/PVP NPs were found more cytotoxic to MCF-7 cells than normal fibroblasts, as well as DOX-Ag, DOX-Ag/PVA, DOX-Ag/PEG and DOX-Ag/PVP nanocarriers exhibiting enhanced cytotoxicity to breast cancer cells [67]. The inorganic NPs embedded polymer matrix showed enhanced performance as discussed in a recent review dealing with preparation and characterization of cytocompatible multifunctional polymeric NPs by analyzing their fluorescence efficiency, the nature of the artificial cell-membrane structure, and their performance as in-cell devices [68]. In enzyme replacement therapy of lysosomal storage disorders (LSDs), using PLGA NPs modified with 7-aminoacid glycopeptide (g7) NPs along with high MW drug (FITC-albumin) improved the barrier crossing of albumin while delivering the drug to the brain [69].

2.3. Supramolecular systems

Molecular self-assembly is exploited in supramolecular formulations for preparing molecular assemblies through noncovalent intra-/intermolecular interactions resulting in the formations of micelles, membranes, vesicles, and liquid crystals. Molecular recognition-based host-guest complexes are now being exploited in molecular sensors and catalysis. It is noteworthy that reactive species are found essential for participating in such molecular-recognition–based self-assemblies providing environment for chemical reactions. Noncovalent bonds between the reactants and template holding the reactants near the reactive sites provide the environment for chemistry. Mechanically interlocked molecular architectures are formed consisting of topologically linked molecules involving noncovalent bonds in molecular architectures like catenanes, rotaxanes, molecular knots, rings and ravels. There are a number of these systems that mimic the biological processes through photo-electro-chemical/catalytic systems, protein designs, and self-replications. In another process of molecular imprinting, a host is constructed out of suitable molecules as a template that is subsequently removed leaving the guests stabilized through steric interactions besides incorporating hydrogen bonding and other interactions.

Numerous molecular systems that have been studied recently include mechanically interlocked systems employing π/π charge-transfer interactions of bipyridinium with dioxyarenes and diaminoarenes, crown ether binding with metal/ammonium cations, formations of carboxylic acid dimers and other hydrogen bonding interactions, bi/tri-pyridines combinations with ruthenium, silver, and other metal ions, and complexation of porphyrins and phthalocyanines around metal ions used by Nature in abundance. Similarly, there are macrocycles providing cavities surrounding guest species during chemical modifications for fine-tuning of their features, cyclodextrins, calixerenes, cucurbiturils, and crown ethers; cyclophanes, and cryptands; metallocycles with metal ions in the ring formed from angular and linear modules including triangles, squares, and pentagons, each bearing functional groups that connect the pieces via self-assembly; and metallo-macrocycles generated from fused chelate-rings that have been studied for their several possible applications. Further, for introducing suitable spacing and conformations relative to each other, a number of structural units have been employed including spacers, connecting groups out of polyether chains, bi/tri-phenyls, and simple alkyl chains; NPs, NRs, fullerenes, dendrimers offering nanometer-sized structure and encapsulation units, scaffolds on surfaces for interfacing electrochemical systems with electrodes. In addition, photochromic and photoisomerizable groups with ability to change their shapes and properties upon light exposure, TTF, and quinones capable of being switched with redox chemistry or electrochemistry, usage of benzidine derivatives, viologens groups and fullerenes in supramolecular electrochemical devices [70–80].

3. Nanomaterial synthesis

Knowing well about the advantages associated with quantum confinement in 1/2/3 dimensions in nanostructured material species during their preliminary investigations, it became imperative to explore the possibility of synthesizing them in quantity with controlled properties in stable form. Some of these aspects of nanostructured material species are examined here from the point of view of their industrial applications.

3.1. Synthesis of nanoparticulate materials

The basic scheme of chemical synthesis of metal NPs employs a suitable stabilizing agent added to a mixture of metal salt precursor and reducing agent chosen out of several including sodium citrate, sodium borohydride, and alcohols to convert metal ions into metal atoms that ultimately form NPs [81–83]. Silver colloidal solution prepared using citrate reduction reported in 1982 did contain 20–600 nm NPs, in which pH affected the morphology by changing from triangular to spherical/cylindrical shapes after changing the pH from 5.7 to 11.1 [84–87]. Similarly, polyol-based synthesis of Ag NPs was reported producing a wide range of NPs using precursor along with capping agent [88–91]. Using propylene, and 1,2-propylene glycols or 1,5-pentanediol as reducing agents and controlling temperature and precursor concentration was also found to influence the morphology of the final product [92].

Alternately, in a different route of photochemical synthesis, light irradiations helped in synthesizing metal NPs as seen in laser irradiated aqueous solution of metal salt + surfactant producing metal NPs besides using laser melting of metal nanospheres to produce nanoplates [93–100]. In contrast, green syntheses of metal NPs were reported from metallic electrodes and spray pyrolysis-based synthesis of 10 and 100 nm (average PS) Ag NPs, respectively [101, 102].

Seed crystals-mediated synthesis reported lately in 2010 produced Ag nanocubes from spherical/cubic single crystal seeds with the edges ranging from 30 to 200 nm [92, 103–105]. In another, very old process of silver mirror reaction, discovered way back in 1835 for depositing Ag metal on solid surfaces using $Ag(NH_3)_2OH$ reduction by sugar or any aldehyde containing compound, has been used for Ag NP synthesis [106].

Precisely controlled morphologies of metal NPs have been realized successfully using template-assisted synthesis already known to depend to follow the template features, in which the surfactant molecules behaved as soft templates in contrast to porous anodic aluminum oxide (AAO) membranes as hard templates [107, 108]. Soft template-assisted synthesis has been found producing metal nanowires, nanorods, hollow spheres, and nanoplates by exploiting a variety of surfactant compounds including ionic surfactants like cetyltrimethyl ammonium bromide, octadecyltrimethylammonium chloride, disodium (2-ethylhexyl) sulfosuccinate, and sodium dodecylsulfate; non ionic surfactants like oleic acid, oleylamine, trioctylphosphine and trioctylphosphine oxide; and polymer surfactants involving poly (vinyl pyrrolidone), poly(vinyl alcohol), and poly(ethylene oxide) as reported extensively [109–141]. Template-assisted synthesis is known to produce well-dispersed forms due to reduced particle aggregation along with mild reaction conditions [142–144].

Unlike above-mentioned chemical syntheses, lithographic patterning and deposition-based nanofabrication processes involving optical, e-beam, scanning probe, and multiphoton lithographies are known to produce precisely controlled nanostructured materials not limited to metal species alone [145–147].

Nanoparticulate material species are well-known building blocks to assist self-assembly processes forming micron size nanosheets and other structures. DNA-assisted self-assembly of metal NPs is another route explored for synthesizing plasmonic NPs into chain, triangular shape, 3D lattices, and Janus nanoclusters, in which the interaction of NPs in solutions involving attractive and repulsive forces comprising of hydrophobic, electrostatic, hydrogen bonding and biospecific interactions impacts the self-assembly to a large extent [148–161].

High-temperature hydrothermal synthesis of triangular shaped Ru nanoplates (~3 nm thick) was reported using $RuCl_3{\cdot}H_2O$ + HCHO + PVP @ 160°C, which changed to irregular shaped but with reduced thickness of 1.5nm subsequently by changing the concentration of Ru salt and PVP. In case of silver salt, it produced triangular Ag nanoplates with sharp and curved corners [162–164].

Recently, the potentials of living microorganisms including bacteria, fungi and plants have been examined for the synthesis of NPs like CdS, Ti/Ni, titanate, zirconia, Au, and Ag [165–171]. It is very important to note that using microorganisms is environmental friendly and benign synthesis route providing good control over size distribution of nanostructures. For example, Ag nanoparticulates were synthesized using bacteria with size less than 200 nm.

3.2. Physical and chemical syntheses

The physical/chemical methods of preparing NPs include lithography, laser ablation, high-energy irradiation, chemical reduction, electrochemistry, and photochemical reduction [172–179].

A number of process parameters that are important to consider in NP synthesis include temperature, concentrations, process kinetics describing interactions between metal ion precursors and the reducing agent, and adsorption kinetics involving the stabilizing agent and the NPs. Consequently, the current emphasis is on designing processes capable of ensuring adequately precise control of the size, shape, stability, and physicochemical properties of the NPs [180–183].

The conventional methods of NP syntheses, however, are known to involve chemical/physical processes that often use toxic materials like organic solvents, reducing agents, and stabilizers causing ultimately substantial environmental pollutions, cytotoxicity, and carcinogenicity in addition to the toxicity of some of the NPs due to their compositions, size, shape, and surface chemistry [184]. However, all these hazardous factors associated with NP syntheses are possible to mitigate using biologically mediated production schedules. There is a strong emerging interest in developing clean, reliable, biologically compatible, benign, and environment-friendly green processes to synthesize NPs for their numerous applications [185].

3.3. Green synthesis

Green synthesis of NPs involving microorganisms and plants is noted to be safe, inexpensive, and environment-friendly as they absorb and accumulate inorganic metal ions from their surroundings leading to an unexplored field of useful research [186–189]. A number of biotechnological applications including bioremediation and bioleaching are possible to implement by exploiting the efficacy of the microorganisms to interact with the environment via their lipid-based amphipathic membranes enabling a variety of oxidation-reduction mechanisms occurring and promoting these biochemical conversions [190–194]. It has been shown that unicellular/multicellular organisms are capable of synthesizing (extra/intracellular) inorganic micro-/nano-sized particulate materials in particular environment of their culture promoting coupled oxidation and reduction reactions that needs further investigations to understand the processes of nucleation, and subsequent NP growth kinetics and the interaction of these processes with metabolic processes of the microorganisms involved [192, 193, 195–198].

A similar situation is met in case of plants based synthesis of NPs with the advantage of plants over other eco-friendly biologically based systems such as bacteria and fungi that avoid using culture preparations and isolation techniques that are involved and expensive. Conversely, biosynthesis of NPs using plants/plant-based extracts is safe with relatively short production times, and having a lower cultivation cost compared to other biological systems [199].

Various methods explored in biologically synthesizing metal NPs include actinomycetes, algae, bacteria, fungus, plants, viruses, and yeast, where each entity has varying degrees of biochemical processing capabilities for preparing metallic or metal oxide NPs. Generally, biological entities with a potential to accumulate heavy metals offer better chance of synthesizing metal NPs [200]. In case of microorganisms, optimization of parameters like nutrients, light, pH, temperature, mixing speed, and buffer strength used in their culture could significantly enhance the enzyme activity [188, 201, 202].

3.4. Microemulsions synthesis

The most popular approach of synthesizing NPs from microemulsions uses a mix of two or more separate microemulsions of the required constituents that participate in nucleations on the micellar edges due to supersaturation of the reactants inside causing growth around the nucleation sites with the arrival of more reactants from intermicellar exchange. Examination of the process details clarified the NP growth starts at the interface, and subsequently moves on to the micellar cores with a intermicellar exchange specific rate limiting phenomenon occurring over a time duration longer than the times involved in reagents diffusions inside the polar domains causing large variation in reaction completion times compared to those observed in native aqueous solutions. Controlling this process by modifying the interfacial characteristics of the surfactant membrane was exploited subsequently as noted in case of $BaTiO_3$ using three separate microemulsions [10, 203–206]. NPs were also prepared from single microemulsion with stabilized reactant inside the reverse micelles after adding the reducing agent to produce the metal NPs. Recently, silver halide and cuprous oxide NPs were synthesized using direct reaction of solubilized silver metal with dioctyldimethylammonium halide counter-ion in reverse micelles, and gamma irradiation of copper nitrate micellar solution, respectively, using faster rate of reaction with morphological control in a single microemulsion. Despite its versatile nature, limitations were pointed out, for example, during synthesizing either ZnTe or incorporating Mn into either ZnTe or CdTe [207–210]. The parameters that influence the size and polydispersion of the synthesized NPs include—type of the solvent employed, surfactant/cosurfactants used, electrolyte, concentration of reagents and, molar ratio of water and surfactant as discussed in detail elsewhere [211].

A simple surfactant/water/oil system was found producing many types of self-assembled structures—by changing composition, one could have spheres (reverse micelles or micelles), cylinders, interconnected cylinders, and planes called lamellar phases, which could reorganize into onion-type structures [212, 213]. Numerous studies were conducted involving a variety of surfactants and system compositions (surfactant/water concentrations) in connection with investigating the growth of different NC structures [214]. However, the idea that different shapes of synthesized NCs could be due to template effect was found invalid as different shapes could be synthesized without changing the template shape, and even no template was involved in some cases at all [215, 216]. Most recently, a method was proposed for controlling the NP shapes by considering the influence of stacking faults in certain plane as observed during the synthesis of silver nanodisks with varying size characterized by HRTEM/SAED showing the presence of forbidden 1/3{422} reflections that were proposed to be promoted by the stacking faults in [110] plane [217]. From these observations, it could be concluded that defect engineering could possibly be used in influencing the shape of NPs as confirmed in case of copper system, yielding similar results [129].

Stabilized microemulsions are possible in a supercritical fluid (SCF) by using appropriate surfactant leading to smooth transition in solvent quality by pressure and temperature control required in nanoparticulate material syntheses [218, 219]. Relevantly, the advantage of using SCF-CO_2 process was demonstrated after having stabilized the microemulsion (water/P104/xylene) at higher pressure causing formations of Au NPs via KBH_4-based reduction of $HAuCl_4$, which was not feasible at ambient pressure due to inadequacy of xylene as solvent. These gold NPs were recovered by reducing the pressure to release the solvent by precipitation. In another variant, known as RESOLV (rapid expansion of a supercritical solution into a liquid solvent), a stable microemulsion of silver cations was expanded through a nozzle into solvent containing reducing agent to produce controlled morphology silver NPs [220]. Finally, water-in-SCF microemulsions were successfully used in water-in-oil microemulsions in synthesizing compounds for industrial applications.

The process involves preparing a stable dispersion using appropriate surfactant. In case of water-in-oil (w/o) microemulsions, the surfactant AOT supplemented with fluorinated cosurfactants like PFPE-PO4 (perfluoropolyether-phosphate), PFPE-NH_4 (ammonium perfluoropolyether) and F-pentanol are employed in stabilizing the dispersions in most of the liquid/supercritical alkanes applied [129, 218, 221–230].

While searching for hydrocarbon-based surfactants or polymers capable of stabilizing w/sc-CO_2 microemulsions primarily due to economic and environmental benefits, iso-steric acid was found useful in a SCF-CO_2 with <10 V/V% hexane solubilizing the reactants inside reverse micelles [231]. These microemulsions are formulated in a pressure cell to which the second reactant is added using a high-pressure syringe pump. Hence, the NP reactions that take place are similar to those occurring in normal liquid w/o microemulsions as noted in microemulsions-based synthesis of Ag NPs where NP recovery by CO_2 venting and rapid expansion method is possible [220, 227, 229, 230, 232].

3.5. Pharmacosomes (liposomes)

Besides regular inorganic/ organic NPs, there is another class of nanostructured materials involving biomolecules that have been explored well for drug/gene delivery applications, and termed as pharmacosomes derived from two terms namely—'pharmakon = drug' and 'soma = carrier' representing vesicles (drugs and carrier attached together). These are neutral particles with positive and negative charges imparting hydrophilic and hydrophobic features (involving polyphenol and phospholipids), in which the drugs are dispersed via lipid interactions (i.e., electron pair sharing, electrostatic forces, and hydrogen bonds) forming colloids, nanomicelles, vesicles and hydrogen bonded hexagonal assemblies. The carboxylic group or functional hydrogen atoms in the amino, and hydroxyl radicals of the drug molecules are converted into esters with the help of the hydroxyl moiety of the lipid forming prodrugs causing reduction in interfacial tension with larger area contacts, and improving bioavailability in addition to helping transport across the cell membrane, wall, and tissues. These prodrugs assemble into single/multiple layers, when in contact with water, forming pharmacosomes [233, 234].

The liposomes are formed by dispersing phospholipids in aqueous media followed by exposure to high shear rates using microfluidization or colloid mill in addition to mechanical dispersion involving sonication, pressure cell or membrane extrusions, freeze thawing, film hydration, microemulsion, and dried reconstituted vesicles—all initiating hydrophilic-hydrophobic interactions between phospholipids and water molecules. Liposomes are characterized by their mean particle size, zeta potential, lamellarity, encapsulation efficiency, in vitro drug release, and vesicle stability. Spherical bilayer membranes are the manifestations of the favored self-assembly features of phospholipids, which is although not limited to bilayer formations alone but also produce colloidal particles from self-aggregation of the polar lipids. Liposomes efficiently entrap even highly unstable compounds including antimicrobials, antioxidants, flavors and bioactive elements by shielding their functionality [235]. Liposomes are the latest additions to the targeted deliveries carrying hydrophilic/hydrophobic contents in nano/microsize double-layer covered spherical volumes providing effective protection from the enzymes, alkaline solutions, digestive juices, bile, and intestinal flora inside human body as well as free radicals. Liposomes not only check oxidation and degradation but also retain the double-layer barrier undamaged until the contents are delivered to the desired site.

Liposomes, discovered in the 1960s, are versatile nanocarriers owing to their compositional variability and structural properties leading to numerous applications in pharmaceutical, nutraceutical, and cosmetics sectors, wherein the herbal extracts like flavonoids, glycosides, and terpenoids are enclosed and transported from the hydrophilic to the lipophilic part of the membrane showing better bioavailability/efficacy, as noted in case of ginkgo biloba, grape seed, green tea, milk thistle, ginseng, and many other herbs already explored in therapeutic applications and dietary supplements. Liposomes are currently being used in a broad range of pharmaceutical applications showing better promises as intracellular delivery systems for antisense molecules, ribosomes, proteins/peptides, and DNA. Liposomes with enhanced drug delivery and long circulation times are currently getting clinically accepted. Liposomal drugs are known to exhibit reduced toxicities while retaining enhanced efficacy compared to the free complements [39–41]. The phospholipid liposomes were reported mimicking red blood cells by optimizing concentrations of phosphatidylserine, di-stearylphosphatidylcholine, and dipalmitoylphosphatidylcholine for a fixed concentration of lecithin and Tween® 80 using response surface methodology resulting in 112–196 nm particle size with lower efficiency encapsulation at lower levels of insulin but increasing at higher levels fulfilling the requirement for intravenous drug delivery having biodegradable and biocompatible features [236].

3.6. Phytosomal nanoparticulate materials

Phytosomes contain herbal drugs and the lipids in stoichiometric ratio in a solvent, wherein, the polar functional groups of the substrate and phosphate and ammonium groups of the polar heads of the phospholipids form the hydrogen bonds while getting attached to the phospholipid polar head and merging with the membrane. For instance, in a phosphatidylcholine and catechin complex, hydrogen bonds are established between hydroxyl groups in the phenols of the flavones and phosphate groups of the phosphatidylcholines without any change in their fatty acid chains suggesting the protected enclosure of the active components into the long aliphatic chains. These interactions form lipophilic envelope shielding the polar phospholipid as well as the constituent. The pharmacokinetic studies and the animal/human trials have confirmed the enhanced bioavailability and absorption of the lipophilic herbal extracts forming micellar constructs in water [237, 238]. Some typical examples are taken here to highlight their applications in nanomedicines.

Optimized icaritin phytosomes, prepared by solvent evaporation of icaritin in ethanol (icaritin: phospholipid = 1:3, reaction time ~1 h @50°C) showed enhanced solubility by 1.6 and 5.9 times in *n*-octyl alcohol enabling icaritin and the coprecipitate of icaritin phytosomes in PVP to dissolve in vitro better [239]. Curcumin-phytosome–loaded chitosan microspheres (Cur-PS-CMs) were reported involving curcumin-phytosomes (Cur-PSs) in chitosan microspheres via gelation (PS = 23.21 ± 6.72 μm, loading efficiency ~2.67%). In vitro curcumin release from Cur-PS-CMs was slower than that from curcumin-loaded chitosan microspheres (Cur-CMs) in pH 1.0, 4.0, 6.8, and 7.4 showing 1.67 and 1.07-fold increase in absorption of curcumin compared with Cur-PSs and Cur-CMs, respectively. The half-life of orally administered Cur-PS-CMs was longer than those of Cur-PSs and Cur-CMs by ~2 and 1.5-times, respectively, confirming oral absorption with prolonged retention time for sustained delivery of lipophilic compounds [240]. The MMC-soybean phosphatidylcholine complex-loaded PEG-lipid-PLA hybrid NPs were reported with Folate functionalization (FA-PEG-PE-PLA NPs@MMC-SPC) for targeted drug delivery and dual-controlled drug release involving hydrophobic core (PLA) loaded with MMC-SPC, an amphiphilic lipid interface layer (PE), a hydrophilic shell (PEG), and a targeting ligand (FA) on the surface, with a spherical shape, and high encapsulation efficiency (95%). In vitro cytotoxicity and hemolysis assays demonstrated the associated cyto-/hemocompatibility along with significantly prolonged blood circulation time compared to that of the free MMC with enhanced cell uptake/cytotoxicity in vitro and superior tumor accumulation and therapeutic efficacy in vivo while reducing the systemic toxicity showing a promising design of the water-soluble drug-phospholipid complex-based targeted drug delivery [241]. *Calendula officinalis* extract-based encapsulation of Au-NPs was examined with high efficiency encapsulation of chlorogenic acid and quercetin exhibiting significant antioxidant and wound-healing properties as anticipated [242]. Mitomycin C (MMC)-loaded polymer-lipid hybrid NPs were reported using single-step assembly of MMC-soybean phosphatidyhlcholine (SPC) complex and biodegradable PLA for intravenous delivery ensuring enhanced efficacy with safety in controlled release of MMC showing a significant accumulation of MMC in the nuclei with significantly higher anticancer effect compared to PLA-NP/MMC or free MMC injection in vitro *and* in vivo [243]. Curcumin phytosomes were reported (solvent evaporation based) to have free flowing powder to enhance curcumin content in different soft gel formulations employing castor oil/oleic acid, PEG 400, Miglyol 812 along with Cremophor EL and KLS P 124 surfactants revealing good stability and spherical curcumin phytosomes [244]. Self-assembled 152.5 ± 3.2 nm NPs of methotrexate (MTX)-phospholipid (PC) complex (MTX-PC NPs) were reported having drug-loading efficiency ~20.7 ± 2.4%, and sustained release behavior compared to free MTX and MTX-loaded liposomes with better promises for cancer therapy compared to traditional drug delivery systems [245]. The active lactone form of 10-hydroxycamptothecin (CPT)-soybean phosphatidylcholine (SPC) self-assembled NPs (CPT–SPC NPs) was reported showing efficient complexation between active lactone and SPC (complexation rate ~98%; 210.7 ± 6.1 nm diameter, ZP ~−24.9 ± 3.1 mV, and a high drug-loading content of 16.3 ± 0.5%) comprising of biphasic delivery (initial burst followed by sustained release). Animal imaging results indicated excellent tumor targeting in HeLa tumor-bearing nude mice [246].

The phytosomes with bioactive plant-based molecular species are poorly soluble in flavanones and terpenes. Features like biocompatibility, nontoxicity, easy to administer, reduced dosage and enhanced retention time of the liposomes and the phytosomes make potent vehicles for drug delivery as discussed [247].

Different flavones like naringin, neoeriocitrin and neohesperidin known for their antioxidant activity and phenolic content were encapsulated in phospholipid vesicles (glycerosomes, hyalurosomes and glycerol containing hyalurosomes) using a high ratio of extract/phospholipid counteracting the oxidative stress in skin cells. The glycerol containing hyalurosomes prevented the oxidative damages and death of both keratinocytes and fibroblasts by promoting their viability [248]. Berberine (BER)—a natural alternative to synthetic antidiabetic drugs, has poor gastrointestinal absorption, and low oral bioavailability limiting its clinical applications was loaded in phytosomes as berberine-phospholipid complex (P-BER) by solvent evaporation method followed by a self-assembly showing nanosize particles with negative surface charge, and excellent drug entrapment efficiency (<85%) and threefold enhanced bioavailability causing significant reductions in fasting glucose levels and improving the ability of systematic hyperlipidemia metabolism of diabetic mice [249]. The solubility and permeability study of Standardized Bacopa Extract (SBE) were reported showing improved aqueous solubility compared to the pure SBE (20-fold), or the physical mixture of SBE and the phospholipid (13-fold). Similarly, in vitro dissolution studies confirmed higher SBE release efficiency (>97%) in comparison with the pure SCE (~42%), or the physical mixture (~47%). The ex vivo studies confirmed improved permeation of SBE (>90%), compared to the pure SBE (~21%), or the physical mixture (~24%). This kind of drug-phospholipid complexation could be used for solubility enhancement of bioactive phytoconstituents [250].

3.7. Nonlayered 2D material synthesis

Besides exploring the novel features of 2D layered materials as mentioned earlier, a number of nonlayered 2D nanomaterials were also reported recently using 2D nanomaterial templates. Some of these include the examples of 2.4-nm-thick square hexagonal-close packed (hcp) gold nanosheets (hcp AuSSs) onto GO template followed by secondary growth for even thicker layers, Ag onto hcp AuSSs resulting in (100) f-oriented fcc core-shell or (110) h/(101) f-oriented hcp/fcc square nanosheets (Au@Ag); and Pt/Pd onto AuSSs inducing transformation from hcp to fcc forming core-shell Au@Pt or Au@Pd nanoplates, to name a few cases already explored. Coatings of Pt or Pd onto hcp AuSSs led to fcc Au@Pt or Au@Pd rhombic nanoplates, in which the large lattice mismatch between Pt or Pd and Au compared with Ag was believed responsible for the (101) f-oriented core-shell nanoplates. In phase transformations from hcp to fcc, the role of ligand exchange was found responsible as noted in case of transformation into (100) f-oriented fcc AuSSs by replacing oleylamine capping with thiol molecules. In another study, 0.55–0.59 nm thick and ~1 mm square freestanding a-Fe_2O_3 nanosheets were synthesized using CuO nanoplate-templates. Further, a number of 2D nanostructures including ternary/quaternary chalcogenides like $CuInS_2$, $CuIn_xGa_{1-x}S_2$, Cu_2ZnSnS_4, $Cu_{2-x}Se$ and $Cu_{1.97}S$ were prepared via the cation exchange on CuSe/CuS templates or phase transformations resulting in uniform size, shape and thickness. In addition, NiO nanosheets were reported from layered a-$Ni(OH)_2$ nanosheets through a simple annealing treatment as discussed in cited references [251–260].

Hydro/solvo-thermal syntheses were found useful in preparing a number of nonlayered 2D materials as briefly summarized here. A facile solvothermal synthesis of poly (vinylpyrrolidone) (PVP)-supported single-layer rhodium (Rh) nanosheets (0.4 nm thick with 500–600 nm edge length) was reported recently followed by a generalized method of synthesizing a number of metal oxide nanosheets, including TiO_2, ZnO, Co_3O_4, WO_3, Fe_3O_4 and MnO_2. In addition, hydro/solvothermal methods were found useful in synthesizing several nonlayer 2D nanosheets including ZnSe, ZnS, CeO_2, In_2O_3, SnO_2, Co_9Se_8 and Co_9S_8-oleylamine hybrid. The synthesis of ZnSe and ZnS nanosheets was reported by preparing lamellar organic-inorganic intermediates [(Zn_2Se_2)(n-propylamine) and (Zn_2S_2)(n-propylamine)], followed by their sonicated exfoliation to have freestanding 0.9 and ~500 nm lateral dimensions of ultrathin nanosheets followed by synthesis of CdS nanosheets (300–800 nm lateral dimension and ~4 nm thick) using diethylenetriamine (DETA) as the surfactant. In another study, atomically thin CeO_2 sheets with surface pits and ultrathin In_2O_3 porous sheets with rich oxygen vacancies were reported using hydrothermal methods followed by subsequent thermal annealing as reported in many publications [260–268].

4. Looking ahead

The novel features of 2D layered/nonlayered materials already explored in micron size samples during their preliminary studies are certainly expected to be translated into production processes with better yield, reproducibility, and process reliability once the process details are fully understood and steps optimized accordingly. The progress made in this context is possible to assess as highlighted below.

4.1. Large-scale synthesis

While examining the basic science of the nanomaterials, efforts have been made in parallel to develop the related technologies as well to support the needs of the growing industries led by innovative exploitations. In this context, it is imperative to have an idea about the market size, growth trends and support extended by the State Agencies.

With the growing trend *vis-à-vis* the market size of US$ 1.34 billion of gold NPs in 2014, it is expected to touch US$ 8 billion by 2022, primarily driven by the innovations supported by higher R&D spending, along with growing applications in healthcare and diagnostics. The emerging gold NPs market size, particularly in China and India, seems to reach US$ 2.9 billion by 2022 [269].

The emerging demands of nanoparticulate materials (NPMs) in growing industrial applications is further evidenced by the initiatives taken by the European Commission in allocating funds for the private-public partnership programs in manufacturing to meet the industrial requirements from various sectors (e.g., FP6/FP7 Projects—€240,000,000; HORIZON 2020 Program—over €1 billion) that are expected to provide scaled-up production of nanostructured materials (~100 kg/day) at lower costs [270].

A brief description of various methods that are already explored for volume production of NPMs is included here in brief to highlight the limitations that are faced in the industry despite fast rising demands from different sectors in general. Those emerging processes that have shown potentials for production after further developments of their technologies in near future are also included in the discussion.

4.2. Production technologies

NPMs (i.e., molecular sizes to 100 μm in diameter) have been produced by flame involving precursors that are sprayed either onto a heated surface or in a hot environment. Oxford University developed an electrospray process for semiconductor/metal NPs, and spray gun deposited catalysts for CNT growth that are simple and economical examples [271–275].

Alternate technique of producing metal vapors through arc discharge has been exploited equally effectively for producing metal, metal-oxides, and other compounds based NPMs in inert/oxygen/reactive gas environment. For taking care of the process reproducibility issues arising from the high temperature arcs generating high evaporation rates leading to the formation of larger size particles from vaporized metal rich carrier gas, the European project (BUONAPART-E) funded a program of upscaling and optimization of NPMs manufacture by flame pyrolysis showing preliminary achievement in form of a versatile and reliable unit capable of producing 0.1–10 g/h throughput in this context [276–278].

Another production process of metallic NPMs involving gas-phase condensation, reported way back in 1930, uses a vacuum evaporation unit attached to a separate collection chamber filled with inert/reactive gas for powder collection ensuring particle formation during rapid cooling of the arriving stream of metal atoms initiating nucleation and particle formation resulting in broad distribution of PSs as their agglomeration is quite random [275].

In a combination of ultrasonic irradiation-assisted chemical reactions followed by precipitation known as sonochemical synthesis, the molecular species are subjected to chemical reactions via energetic acoustic cavitation including formation, growth and collapse of microbubbles inside the irradiated liquid medium. This method efficiently produces the NPMs facilitating preparation of smaller PSs of different shapes at lower costs [279–281].

Milling-based production of NPMs is known since 1970 with many variants developed for preparing nanopowders of crystalline/crystalline or crystalline/amorphous, and atomic bonding-based metal/metal, metal/semiconductor, metal/ceramic, and their combinations with the advantage of low temperature working. There are two different routes of mechanical milling namely—single-phase powder milling by controlling the competing processes of fracturing and cold-welding, where particles larger than 100 nm are not cold welded, and consequently, a reduction in the average PS from 50–100 μm up to 2–20 nm is feasible. A severe plastic deformation caused by mechanical attrition at elevated temperature of 100–200°C gives rise to refinements in internal structures of the particles to produce nm size particles. The environmental sensitivity of the milling process is put to use by controlling the ambient conditions accordingly for chemical reactions to occur between the environment and the milled powders leading to a novel, cost-effective method of producing a variety of nanopowders. Mechanochemical processing (MCP) is another variant of milling used as a low temperature chemical reactor, wherein the ball mill accelerates the reaction kinetics in the powder mixture as a result of the intimate mixing and refinement of the grain structure to nm scale. For this reason, it is useful to employ a proper reactive gas environment of O_2, N_2, atmospheric air, or precursor. Oxide and nitride of Ti, Fe, V, Zr, W, Hf, Ta and Mo could, thus, be converted into NPMs in reactive milling. Similarly, in wet milling, an organic fluid is used for transforming the metal powders into nanocrystalline metal-ceramic composites comprising of individual single nanometer sized grains dispersed in a matrix. Of course, it is necessary to go for further heat treatment for ascertaining that the reaction is complete. The reduction of the process cost and the industrialization of products are achieved by using a variety of precursors for producing a large variety of NPMs in the form of oxides, carbonates, sulfates, chlorides, fluorides, hydroxides, and others. Different kinds of ball mills including tumbler mills, attrition, shaker mills, vibratory mills, planetary mills, and other variants are commercially available for mechanical attrition. A common method in all these mills is to place the material powder in a sealed container with the balls of hardened steel or tungsten carbide, while mass ratio for the ball to the powder is kept around 5:10 in case of a typical 50μm powder. Kinetic energy of balls is a function of their mass and velocity, and as a result, steel and tungsten that are high-density materials are preferred as means of milling [282–286].

Recently developed room temperature ionic liquids (RTILs) are being used as reaction media for inorganic NPMs mainly by using their preorganized structures to template porous inorganic nanomaterials and their intrinsic high charges and polarizability in affecting electrostatic and steric stabilization. For example, IL templates are used for fabricating mesoporous/super-microporous silica, and in synthesizing transition metal NPs including Ir, Au, Ag, Pt, and Pd, where the IL acts as a solvent, template, reducing agent and stabilizer. TiO_2-based NCs and microspheres, and Te- and CoPt-based nanorods, have also been fabricated using ILs [287–295].

Nanolithographic techniques involving e-beam/focused ion-beam writing, proximal probe patterning, X-ray lithography, along with dry/wet etching are found useful in fabricating a variety of nanostructured materials. By nanolithography, nanostructures and their arrays are possible to fabricate by a directed or constrained growth from one to few nm with the advantage of producing large quantities of 1D nanostructures using a wide variety of the available materials [296–300].

One of the most popular and maybe the most economical methods of nanolithography is template fabrication based suitable for growing nanowires (NWs) using electrodeposition, sol-gel or by vapor-phase followed by independently controlled NWs after removing from the templates. In this process, the ordered nanopore templates are made before filling them with the chosen materials, using one of the methods referred before. Electrodeposition of metals inside the nanopores is performed in acidic or basic baths that contain metal salts, and the metal is deposited on the cathode (working electrode), and for this reason, one end of the porous membrane is metallized with Ti and Au or Ag by sputtering or evaporation, and the anode is normally graphite or Pt with calomel or Ag/AgCl as reference electrode. The deposition conditions (DC or pulsed) affect the polycrystallinity of the nanowires.

High-temperature anneal (500–600°C) of low temperature sol-gel processed complex oxide NWs was found necessary for the required stoichiometric phase. For instance, taking out the alumina template after dipping into sol for some time, it was necessary to dry before anneal to produce the proper phase. CVD/PVD methods were also explored for growing semiconducting nanotubes, including ordered arrays of CNTs. Plasma-assisted CVD has shown adequate promises to grow aligned nanotubes of complex semiconductor or oxides in alumina templates, and in future, this technique will certainly find more applications [301].

Scanning probe microscopy (SPM)-based lithography has already been established for creating nm size patterns on metallic and semiconducting surfaces using lithography masks either via chemical route called dip-pen lithography (DPN) or SPM route based on anodic oxidation where water meniscus formation from the atmospheric moisture capillary condensation between the substrate and the tip plays an important role.

The presence of water meniscus enables the molecules to interact with the substrate for forming chemical bonds via controlled molecular transport across the region between the tip and the substrate. Another variant employs the meniscus as an electrochemical cell where the applied bias produces the metal/semiconductor nanostructures promoting nanostructures on the surface for directed assembly of nanoparticles [302–304]. Likewise, in anodic oxidation, the water meniscus forms an electrochemical cell, where moisture provides 'nanoreaction vessel' like environment for anodic oxidation after applying a negative bias to the tip with respect to the substrate, wherein the applied electric field helps in accelerating the OH^- ions to the substrate causing a self-limiting type oxidation that terminates automatically at fields below 10^7 V/cm. The geometrical features of the oxidized pattern depend on the applied bias and the radius of curvature of the tip that is micromachined for reproducible sizes, shapes and aspect ratios. By controlling all these parameters, reproducible lithography of known resolutions is performed as an advantage of this technique [305].

4.3. 2D materials production

With the emerging applications of graphene possessing novel features, systematic efforts were made in developing production methods with higher yield and improved quality of graphene and other 2D materials nanosheets. For example, the exfoliation efficiency of graphene nanosheets was improved (e.g., yield >85%; ≤3 layers; lateral size ~44 μm) using aqueous solutions of $(NH_4)_2SO_4$, Na_2SO_4, and K_2SO_4 and exhibiting hole mobility ~310 cm^2/Vs. Highly conducting films were brush coated on paper from a graphene ink for fabricating all-solid-state flexible supercapacitors delivering a high area capacitance of 11.3 mF/cm^2 [306]. A green production of graphene was reported using saccharin in aqueous solution showing that the number of graphene layers decreased with increase in the intercalation potential, while yield improved with increase in the exfoliation potential. The defect density in the exfoliated graphene layer was sensitive to the exfoliation potential as it initially increased with exfoliation potential and then eventually decreased [307]. Improved performance of exfoliation arising out melamine additive during graphite electroexfoliation was assigned to the hydrophilic force from the basal plane promoting exfoliation besides providing protection against further oxidation, leading to high-yield production of graphene of larger crystallite size. This process exhibited better performance in terms of higher uniformity (>80% in <3 layered graphene), lower oxidation density (C/O ratio of 26.17), lower defect level (I_D/I_G <0.45), and low sheet resistance of 13.5 kΩ/ϒ (95% transmittance). A graphene nanocomposite with polyvinyl butyral (PVB) exhibited an electrical conductivity of 3.3×10^{-3} S/m for the graphene-loading fraction of 0.46vol%. The continuous process for producing graphene was demonstrated, with a yield rate of 1.5 g/h [308]. The influence of a number of reducing agents (such as 2,2,6,6-tetramethylpiperidin-1-yl)oxyl (TEMPO), ascorbic acid, and sodium borohydride) was investigated to eliminate $HO^{\bullet}$ radical generated out of water electrolysis responsible for defect formation on graphene during electroexfoliation in aqueous ammonium sulfate. TEMPO-assisted exfoliation could produce large graphene sheets (~5–10 μm average), exhibiting hole mobilities <405 cm^2/Vs, very low Raman I_D/I_G ratios (<0.1), and extremely high carbon to oxygen (C/O) ratios (<25.3). High concentration graphene ink in dimethylformamide exhibited (6 mg/mL) was found useful in transparent conductive films and flexible supercapacitors [21]. Low defect concentration few-layer graphene (FLG) sheets were fabricated by a two-step electrochemical intercalation exfoliation, including a graphite foil pretreatment in sodium hydroxide solution and a subsequent further exfoliation in sulfuric acid solution. During this process, the pretreatment resulted in the expansion of the graphite foil and in turn facilitated the final exfoliation in sulfuric acid solution showing I_D/I_G of the FLG sheets as low as 0.29 while maintaining relatively high yield (>56%). In addition, the oxygen content in the FLG sheets is 8.32% with the C/O ratio of 11.02 [309]. Using *in situ* optical and electrical measurements, it was found that that solvent intercalation is the required first step and the degree of intercalation controls the thickness of the exfoliated graphene. Electrochemical decomposition of water into gas bubbles causes the expansion of graphite controlling the functionalization and lateral size of the exfoliated graphene. Both process steps proceed at different time scales and can be individually addressed through application of pulsed voltages. The potential of the presented approach was demonstrated by improving the performance of graphene-based transparent conductors by 30 times [310].

In a more recent study, graphene production was reported using HNO_3, $NaNO_3$, H_2SO_4 and H_2O_2-based exfoliation in sodium dodecylbenzene sulfonate as a surfactant [311]. Studying the influence of varying parameters like anodic bias (1–10 V), and shear field (400–74,400/s) concluded that thicker and more fragmented graphene sheets were formed at higher biases, while at potentials as low as 1 V, shear force could cause pronounced exfoliation. This process under optimum condition could produce large graphene flakes (~10 μm) with a high proportion of single, bilayer, and trilayer graphene and small I_D/I_G ratio (0.21–0.32) with only a small contribution from carbon-oxygen species [312]. Biomolecules are also attracting attentions as dispersants for 2D materials providing a number of advantages over more conventional, synthetic surfactants particularly in case of biomolecules including proteins and peptides, nucleotides and nucleic acids (RNA, DNA), polysaccharides, plant extracts and bile salts as colloidal dispersants as discussed in a recent review [313].

Graphene quantum dots (GQDs) were examined in terms of their size-dependent energy storage efficiency and optical behavior while functioning as an active material in rechargeable lithium ion batteries (LIBs). Considering three different SOC's (<05, <50 and <95%), reversible changes were noticed in the UV-VIS absorption spectra that could be explained by the mechanism of charging-discharging involving the influx/out flux of Li-ions. Some of these results could be used for understanding the energetics of Li-ion intercalation and deintercalation in multilayer graphene and related composites [314]. Single-step synthesis of halogen-functionalized graphenes (HGs) was reported using electroexfoliation of graphite in aqueous potassium halide solutions confirming the variation in the degree of halogenation between 2.32 and 0.26 atom% in fluorinated graphene (FG) and iodinated graphene (IG), respectively, which were attributed to the difference in reactivity of the halogen species generated during the exfoliation process. Among all HGs, FG has shown the superior electrocatalytic behavior for $2Br^-/Br_2$ redox reaction. The anodic (11.2 mA/cm²) and cathodic (10.7 mA/cm²) peak current densities were higher for FG than that of other halogenated graphenes. ZBB flow cell fabricated with FG as bromine electrode exhibited enhanced electrochemical performance in terms of efficiency (81% of voltaic efficiency and 72% energy efficiency) and durability up to 350 cycles [315].

Comparison of the two routes of liquid phase exfoliation (LPE) and electrochemical exfoliation (ECE) made it clear the LPE took about 13 days against 3 min for ECE process [316]. A simple and fast method of electroexfoliating graphite into graphene oxide (GO) and then its rapid reduction to graphene nanosheets (GNs) was reported using microwaves. This electroexfoliation combined with microwaves reduction offered a low-cost and efficient route to produce high-quality graphene with high yield [317].

A rapid electroexfoliation of natural Bi_2Se_3 and Bi_2Te_3 crystals in aqueous media was reported to prepare single-/few-layer nanosheets representing a simple, reagent-free, and scalable method for the fabrication of single-/few-layer nanosheets of these materials [318]. 2D TMDCs with relatively lower toxicity, higher stability in aqueous environments, and adhering well to the biological materials such as proteins are currently being considered promising for biosensing, cell imaging, diagnostics, and therapeutics. Preparation and exfoliation of 2D TMDCs showing heavily dependent features on the number of layers and lateral size were described using their liquid exfoliation from their bulk materials along with the protocols for functionalizing or modifying them [319]. Group-VI TMDCs including MoS_2 and WSe_2 being semiconductors with sizable energy band gaps offer themselves as building blocks for new generation optoelectronics particularly involving their specificity and tunability of their band gaps based on strong light-matter interactions between TMDC crystal and specific photons triggering complex phenomena like photoscattering, photoexcitation, photodestruction, photophysical modification, photochemical reaction, and photooxidation. Subsequently, photoelectric conversion devices enabled by laser excitation and the functionality extension and performance improvement in the TMDs materials via laser modification were comprehensively reviewed [320]. A green and cost-effective production process of 2D MoS_2 was reported using sonication milling (CUM) to exfoliate natural molybdenite powders to achieve few-layer MoS_2 (FL-MoS_2) nanosheets in *N*-methyl-2-pyrrolidone (NMP) with polyvinylpyrrolidone (PVP) molecules with the synergistic effect of sonication and sand milling enhancing the exfoliation efficiency, and the precursor of natural molybdenite powders minimizing the cost. The influence of various factors on exfoliation was studied by varying initial concentration of natural molybdenite powder (15–55 g/L), ultrasonic power (200–350 W), rotation speed of sand mill (1500–2250 rpm), exfoliation time (0.5–6 h), and the molar ratio of PVP unit to MoS_2 (0–1). Under the optimal condition, the yield and exfoliation rate reached as high as 21.6% (in 6 h) and 1.42 g/Lh, respectively. This process could, thus, be considered as a low-cost, green, and efficient method of producing FL-MoS_2 nanosheets from natural molybdenite powders [321].

4.4. Future perspectives

From the exceptionally faster developments taking place in the domain of 2D nanomaterials, one can easily assess about the future R&D activities in synthesizing and using still more varied combinations of ultrathin 2D nanomaterials in times to come as highlighted below.

For example, noble metals/alloys that are known as robust industrial catalyst are expected to perform better once converted into single-/few-layer forms. Similarly, the metal organic frameworks (MOFs) materials, currently found useful in gas storage, separations, and catalysis due to their tunable structures/functions, larger surface areas, and highly ordered pores, are, though, not used in electronic devices owing to their poor electrical conductivity and difficult film-forming ability, but once these shortcomings are taken care of, their device integration would open newer avenues [20].

2D TMDCs including ternary and quaternary compounds besides already known binaries like MoS_2, WS_2, $MoSe_2$, and WSe_2 are expected to offer novel applications in electronics, optoelectronics, electrocatalysis, and energy storage, even though, some of them are yet to be synthesized.

Reversible conversion from one crystal phase into the other in 2D materials is another area of potential uses offered by the phase-engineered nanomaterials. For instance, some TMDCs including MoS_2, and WS_2 exist in either of the two main crystal phases namely—2H and 1T. It is interesting to note that MoS_2 is semiconducting and metallic in 2H and 1T phase, respectively, with phase reversal caused by butyl lithium intercalation. The metallic phases of MoS_2 and WS_2 have better conductivity, whereas 2H phases exhibit enhanced electrocatalytic actions in hydrogen evolution and supercapacitors [322–325]. Similar features are anticipated in case of noble metals prepared in different crystal phases. For example, Au crystallizing in face-centered cubic (fcc) phase was also synthesized in hexagonal 2H and 4H-phases that were transformed back to fcc and thus could be used for tuning its optical, catalytic, and plasmonic properties accordingly [341].

Practical implementations of H-termination along with van der Waals epitaxy in adjusting not only the band gaps, and stability, but also other properties of germanane are currently being considered as additional scope in future. For instance, germanane is converted into germoxanene after replacing H by OH-group and attaching different ligands to polysiloxene shifting the photoluminescence features along with band gap positions. Germanane is possible to restack with other materials using different intercalation techniques like electrochemical (alkali metals insertions between few layers), organic, and similar other zero-valent metal intercalation resulting in novel properties. Highly anisotropic properties caused due to electron-hole localization in transverse direction of germanane monolayer combined with their direct band gaps make them usable in a number of electronic applications including transistors, solar cells, and cooling layers. Additionally, combined use of topotactic deintercalation with exfoliations has been found better options of synthesis, though implemented only in limited number of Zintl phase compounds. Such layered materials include polygermanane, polysilanes, siloxenes, and spinels along with others like $CaSi_2$, CaGaSi, $CaZn_2Sb_2$, and $Ca_{11}GaSb_9$ that are expected to exhibit new and better material features in the form of 2D-hydrogenated/oxygenated semiconductors with precisely tuned electronic properties. These newer families of 2D materials, when used in heterostacks comprising of layers with different physical characteristics and further modified by restacking, are expected to offer newer engineered architectures for specific applications. The process of synthesizing flat materials endowed with exceptionally high charge carrier mobility transport established at molecular scales is expected to introduce significant changes in the electron device designs employing 'atom by atom' or 'group by group' substitutional alterations in realizing p-n junctions at the desired locations in near future providing better options of touch screens, supercapacitors, batteries, fuel cells, sensors, high frequency circuits, and flexible electronics [26, 326–332].

Using 2D exfoliated nanosheets appears better suited for realizing the five types of heterointerfaces for their device applications. For instance, layer-after-layer deposition of two different nanomaterials in a vertical heterostructure is an example of type I heterointerface already demonstrated in TMDCs, h-BN, or topological insulators on graphene or in situ epitaxial growth of vertical CuS/TiS_2 type heterostructures [260, 333–336]. Similarly, type II heterointerface was reported in case of combinations of WSe_2/WS_2, $MoSe_2/MoS_2$, and $MoSe_2/WSe_2$, wherein the growth of one type of nanosheet was made to start from the edge of the other forming an in-plane 2D heterojunction [268, 337–339]. Type III is similarly realized using a vertical growth of aligned ultrathin 2D nanosheet arrays on another ultrathin 2D nanomaterial substrate to form hierarchical heterostructures. Unlike these three afore-mentioned heterointerfaces, it is also feasible using the crystal-phase concept of heterointerfaces representing another kind using the same chemical compound but with different crystal phases. Type IV heterointerface may thus be prepared using a partially converted crystal phase in a MoS_2 nanosheet by changing a part of it from its 2H phase into 1T phase resulting in an in-plane 2H-1T heterointerface. Further, type V heterointerface involving superlattices is also conceivable using binary, ternary, or multiple phase patterns in ultrathin 2D noble metal nanostructures. The phase engineered Au-nanosheets in fcc, 2H, and 4H phases, already synthesized using wet-chemical methods and validated, are good candidates for crystal-phase superlattices with the help of self-assembly or lithography. Besides binary phase heterostructures, ternary phase comprising of 2H-fcc-4H might also be explored for similar applications using ultrathin 2D Au nanostructures having promising applications in catalysis, waveguide, surface enhanced Raman spectroscopy, and many others [256–259, 340, 341].

5. Conclusions

Adding 2D materials species with the nanoparticulate inorganic, organic, and biomolecular species and invoking different aspects of molecular recognition-based self-assemblies and self-organized formations of supramolecular hierarchical complexes seems to be a feasible way of improving the smart features of a large variety of nanomaterials in addition to paving the way for introducing the intelligent features in them in due course of time. Appropriate combinations of biomolecular species endowed already with the intelligent features are going to make this transition from 'smart' to 'intelligent' materials faster and easier as they possess the basic traits required for their participation in the functioning of living organisms. Other inorganic and organic building blocks once conjugated appropriately would certainly help in accelerating various components of intelligence discussed earlier. Further, the phytosomal building blocks when combined with 2D materials would certainly provide more insight into controlling the physico-chemico-biological properties of the resultant nanomaterial species with additional assurance of their green nature as compared to other species especially in the domain of applications involving human health care.

Acknowledgements

The authors are extremely thankful for availing the opportunities of having very useful discussions with colleagues from different disciplines, graduate students, and particularly the reviewers of the research papers authored for their critical but extremely useful comments and suggestions from time to time. Kind support of the Editor extended to the authors is duly acknowledged during preparation of the manuscript.

Author details

Dinesh Kumar and Shamim Ahmad*

*Address all correspondence to: drsahmad@email.com

YMCA University of Science and Technology, Faridabad, Haryana, India

References

[1] Jha RK, Jha PK, Chaudhury K, Rana SVS, Guha SK. An emerging interface between life science and nanotechnology: Present status and prospects of reproductive healthcare aided by nano-biotechnology. Nano Reviews. 2014;**5**:22762. DOI: 10.3402/nano.v5.22762

[2] Iyer SS, Haddad YM. Intelligent materials—An overview. International Journal of Pressure Vessels and Piping. 1994;**58**:335–344

[3] Tani J, Takagi T, Qiu J. Intelligent material systems: Application of functional materials. Applied Mechanics Reviews. 1998;**51**(8):505–521

[4] Taya M. Bio-inspired design of intelligent materials. Proceedings of SPIE 5051, Smart Structures and Materials 2003: Electroactive Polymer Actuators and Devices (EAPAD), **54**; DOI: 10.1117/12.484425

[5] Saxton M. Are smart materials intelligent? The Institution of Engineering and Technology (IET) [Internet]. 2010. Available from: http://www.theiet. org/resources/inspec/support/subject-guides/asmi.cfm

[6] Ahmad S. Current status and future prospects of application specific engineered nanocurcumin compounds. Internatinal Journal of Pharmaceutical Sciences and Nanotechnology. 2016;**9**(5):3391–3451

[7] Ahmad S. Engineered nanomaterials for drug and gene deliveries—A review. Journal of Nanopharmaceutics and Drug Delivery. 2016;**3**(1):1–50

[8] Ahmad S. Band-structure-engineered materials synthesis, nanocrystals and hierarchical superstructures—Current status and future trend. International Journal of Material Science. 2016;**6**(1). DOI: 10.12783/ijmsci. 2016.0601.01

[9] Malik MA, Ramasamy K, Revaprasadu N. The recent developments in nanoparticle synthesis. Nanoscience. 2016;**3**:51–147

[10] Ahmad S. Materials by design—Prospects and challenges. Indian Journal of Engineering & Materials Sciences. 2005;**12**:299–316

[11] Ahmad S, Hashim U. Nano-herbals in human healthcare: A proposed research and development – Roadmap I, and II. ASEAN Journal on Science and Technology for Development. 2012;**29**(1):55–75

[12] Nitta SK, Numata K. Biopolymer-based nanoparticles for drug/gene delivery and tissue engineering. International Journal of Molecular Sciences. 2013;**14**(1):1629–1654

[13] Ahmad S. Organic semiconductors for device applications: Current trends and future prospects. Journal of Polymer Engineering. 2014;**34**(4):279–338

[14] Ahmad S. Device applications of band-structure-engineered nanomaterials—Current status and future trend – Review. International Journal of Nanoelectronics and Materials. 2015;**8**:129–202

[15] Dong R, Pang Y, Su Y, Zhu X. Supramolecular hydrogels: Synthesis, properties and their biomedical applications. Biomaterials Science. 2015;**3**(7):937–954

[16] Wu C-Y, Lin C-H, Yu-Chie Chen Y-C. Using glucose-bound Fe_3O_4 magnetic nanoparticles as photo-thermal agents for targeted hyperthermia of cancer cells. Journal of Nanomedicine & Nanotechnology. 2015;**6**:264. DOI: 10.4172/2157–7439.1000264

[17] Ahmad S. An affordable green energy source—Evolving through current developments of organic, dye sensitized, and perovskite solar cells. International Journal of Green Energy. 2016;**13**(9):859–906

[18] Ahmad S. Curcumins engineered drugs structure activity relationships (SARs)—A review. IOSR Journal of Pharmacy and Biological Sciences (IOSR JPBS); To be published.

[19] Rebek J Jr. Introduction to the molecular recognition and self-assembly special feature. PNAS. 2009;**106**(26):10423–10424

[20] Zhang H. Ultrathin two-dimensional nanomaterials. ACS Nano. 2015;**9**(10):9451–9469

[21] Yang G, Zhu C, DuD, Zhu J, Lin Y. Graphene-like two-dimensional layered nanomaterials: Applications in biosensors and nano-medicine. Nanoscale. 2015;**7**:14217–14231

[22] Novoselov KS, Geim AK, Morozov SV, Jiang D, Zhang Y, Dubonos SV, Grigorieva IV, Firsov AA. Electric field effect in atomically thin carbon films. Science. 2004;**306**(5696):666–669

[23] Watanabe K, Taniguchi T, Hisao Kanda H. Direct-band gap properties and evidence for ultraviolet lasing of hexagonal boron nitride single crystal. Nature Materials. 2004;**3**:404–409

[24] Novoselov KS,Geim AK, Morozov SV, Jiang D, Katsnelson MI, Grigorieva IV, Dubonos SV, Firsov AA. Two-dimensional gas of massless Dirac fermions in graphene. Nature. 2005;**438**:197–200

[25] Novoselov KS, Jiang D, Schedin F, Booth T J, Khotkevich VV, Morozov SV, Geim AK. Two-dimensional atomic crystals. PNAS. 2005;**102**(30):10451–10453

[26] Geim AK, Novoselov KS. The rise of graphene. Nature Materials. 2007;**6**:183–191

[27] Geim AK. Graphene: Status and prospects. Science. 2009;**324**:1530–1534

[28] Kong D, Randel JC, Peng H, Cha JJ, Meister S, Lai K, Chen Y, Shen ZX, Manoharan HC, Cui Y. Topological insulator nanowires and nanoribbons. Nano Letters. 2010;**10**(1):329–333

[29] Kong D, Dang W, Cha JJ, Li H, Meister S, Peng H, Liu Z, Cui Y. Few-layer nanoplates of Bi_2Se_3 and Bi_2Te_3 with highly tunable chemical potential. Nano Letters. 2010;**10**:2245–2250

[30] Liu Z, Robinson JT, Tabakman SM, Yang K, Da H. Carbon materials for drug delivery & cancer therapy. Materials Today. 2011;**14**(7–8):316–323

[31] Houssa M, Scalise E, Sankaran K, Pourtois G, Afanas'ev VV, Stesmans A. Electronic properties of hydrogenated silicene and germanene. Applied Physics Letters. 2011;**98**(22). DOI: 10.1063/1.3595682

[32] Ni Z, Liu Q, Tang K, Zheng J, Zhou J, Qin R, Gao Z, Yu D, Lu J. Tunable band gap in silicene and germanene. Nano Letters. 2012;**12**(1):113–118

[33] Radisavljevic B, Radenovic A, Brivio J, Giacometti V, Kis A. Single-layer MoS_2 transistors. Nature Nanotechnology. 2011;**6**(3):147–150

[34] Song L, Liu Z, Reddy AL, Narayanan NT, Taha-Tijerina J, Peng J, Gao G, Lou J, Vajtai R, Ajayan PM. Binary and ternary atomic layers built from carbon, boron, and nitrogen. Advanced Materials. 2012;**24**(36):4878–4895

[35] Vogt P, De Padova P, Quaresima C, Avila J, Frantzeskakis E, Asensio MC, Resta A, Ealet B, Le Lay G. Silicene: Compelling experimental evidence for graphenelike two-dimensional silicon. Physical Review Letters. 2012;**108**(15):155501

[36] Fang H, Chuang S, Chang TC, Takei K, Takahashi T, Javey A. High-performance single layered WSe_2 p-FETs with chemically doped contacts. Nano Letters. 2012;**12**:3788–3792

[37] Geim AK, Grigorieva IV. Van der Waals heterostructures. Nature. 2013;**499**:419–425

[38] Jung Y, Shen J, Cha JJ. Surface effects on electronic transport of 2D chalcogenide thin films and nanostructures. Nano Convergence. 2014;**1**(1):1–8

[39] Shehata T, Ogawara K, Higaki K, Kimura T. Prolongation of residence time of liposome by surface-modification with mixture of hydrophilic polymers. International Journal of Pharmaceutics. 2008;**359**:272–279

[40] Andreas W, Karola VU. Liposome technology for industrial purposes. Journal Drug Delivery. 2010;**2011**:9

[41] Akbarzadeh A, Rezaei-Sadabady R, Davaran S, Joo SW, Zarghami N, Hanifehpour Y, Samiei M, Kouhi M, Nejati-Koshki K. Liposome: Classification, preparation, and applications. Nanoscale Research Letters. 2013;**8**:102

[42] Misra SK, Dybowska A, Berhanu D, Luoma SN, Valsami-Jones E. The complexity of nanoparticle dissolution and its importance in nanotoxicological studies. Science of the Total Environment 2012;**438**:225–232

[43] Liedberg B, Nylander C, Lundstrom I. Surface plasmon resonance for gas detection and biosensing. Sensors and Actuators. 1983;**4**:299–304

[44] Kreibig U, Vollmer M. Optical Properties of Metal Clusters. Berlin: Springer-Verlag; 1995

[45] Stenzel O, Stendhal A, Voigtsberger K, Von Borczyskowski C. Enhancement of the photovoltaic conversion efficiency of copper phthalocyanine thin film devices by incorporation of metal clusters. Solar Energy Materials and Solar Cells. 1995;**37**:337–348

[46] Kneipp K, Wang Y, Kneipp H, Perelman LT, Itzkan I, Dasari RR, Feld MS. Single molecule detection using surface-enhanced Raman scattering (SERS). Physical Review Letters. 1997;**78**(9):1667–1670

[47] Barnes WL, Dereux A, Ebbesen TW. Surface plasmon sub-wavelength optics. Nature (London). 2003;**424**:824–830

[48] Derkacs D, Lim SH, Matheu P, Mar W, Yu ET. Improved performance of amorphous silicon solar cells via scattering from surface plasmon polaritons in nearby metallic nanoparticles. Applied Physics Letters. 2006;**89**:093103

[49] Pillai S, Catchpole KR, Trupke T, Green MA. Surface plasmon enhanced silicon solar cells. Journal of Applied Physics. 2007;**101**:093105

[50] Konda RB, Mundle R, Mustafa H, Bamiduro O, Pradhan AK, Roy UN, Cui Y, Burger A. Surface plasmon excitation via Au nanoparticles in n-CdSe/p-Si heterojunction diodes. Applied Physics Letters. 2007;**91**:191111

[51] Kim J, Lee K, Coates N, Moses D, Nguyen T, Dante M, Heeger A. Efficient tandem polymer solar cells fabricated by all-solution processing. Science. 2007;**317**:222–225

[52] Kwon M-K, Kim J-Y, Kim B-H, Park I-K, Cho C-Y, Byeon C-C, Park S-J. Surface-plasmon-enhanced light-emitting diodes. Advanced Materials. 2008;**20**:1253–1257

[53] Novotny L. Optical antennas tuned to pitch. Nature. 2008;**455**:887

[54] Yeo B-S, Stadler J, Schmid T, Zenobi R, Zhang W. Tip-enhanced Raman spectroscopy—Its status, challenges and future directions. Chemical Physics Letters. 2009;**472**:1–13

[55] Polo JA Jr, Lakhtakia A. Surface electromagnetic waves: A review. Laser and Photonics Reviews. 2011;**5**:234–246

[56] Shockley W, Queisser HJ. Detailed balance limit of efficiency of p/n junction solar cells. Journal of Applied Physics. 1961;**32**:510–519

[57] Bruchez M, Moronne M, Gin PWeiss S, Alvistos AP. Semiconductor nanocrystals as fluorescent biological labels. Science. 1998;**281**:2013–2016

[58] Coe S, Woo WK, Bawendi M, Bulovic V. Electroluminescence from single monolayers of nanocrystals in molecular organic devices. Nature. 2002;**420**:800–803

[59] Kamat PV. Quantum dot solar cells. Semiconductor nanocrystals as light harvesters. The Journal of Physical Chemistry C. 2008;**112**:18737–18753

[60] Nizamoglu S, Zengin G, Demir HV. Color-converting combinations of nanocrystal emitters for warm-white light generation with high color rendering index. Applied Physics Letters. 2008;**92**:031102

[61] Flory F, Escoubas L, Berginc G. Optical properties of nanostructured materials: A review. Journal of Nanophotonics. 2011;**5**:052502

[62] Rao JP, Geckeler KE. Polymer nanoparticles: Preparation techniques and size-control parameters. Progress in Polymer Science. 2011;**36**(7):887–913

[63] Sadat SMA, Jahan ST, Haddadi A. Effects of size and surface charge of polymeric nanoparticles on in vitro and in vivo applications. Journal of Biomaterials and Nanobiotechnology. 2016;**7**:91–108

[64] Zhao H, Lin ZY, Yildirimer L, Dhinakar A, Zhao X, Wu J. Polymer-based nanoparticles for protein delivery: Design, strategies and applications. Journal of Materials Chemistry B. 2016;**4**:4060–4071

[65] Bohrey S, Chourasiya V, Pandey A. Polymeric nanoparticles containing diazepam: Preparation, optimization, characterization, in-vitro drug release and release kinetic study. Nano Convergence. 2016;**3**:3

[66] Zhou J, Chau Y. Different oligoarginine modifications alter endocytic pathways and subcellular trafficking of polymeric nanoparticles. Biomaterial Science. 2016;**4**:1462–1472

[67] Elbaz NM, Ziko L, Siam R, Mamdouh W. Core-shell silver/polymeric nanoparticles-based combinatorial therapy against breast cancer in-vitro. Scientific Reports. 2016;**6**. Article number: 30729

[68] Ishihara K, Chen W, Liu Y, Tsukamoto Y, Inoue Y. Cyto compatible and multifunctional polymeric nanoparticles for transportation of bioactive molecules into and within cells. Science and Technology of Advanced Materials. 2016;**17**(1):300–312

[69] Salvalaio M, Rigon L, Belletti D, D'Avanzo F, Pederzoli F, Ruozi B, Marin O, Vandelli MA, Forni F, Scarpa M, Tomanin R, Tosi G. Targeted polymeric nanoparticles for brain delivery of high molecular weight molecules in lysosomal storage disorders. PLoS One. 2016;**11**(5):e0156452

[70] Lehn JM. Perspectives in supramolecular chemistry—From molecular recognition towards molecular information processing and self-organization. Angewandte Chemie International Edition. English. 1990;**29**(11):1304–1319

[71] Dickert F. Molecular imprinting in chemical sensing. TrAC Trends in Analytical Chemistry. 1999;**18**(3):192–199

[72] Rowan SJ, Cantrill SJ, Cousins GRL, Sanders JKM, Stoddart JF. Dynamic covalent chemistry. Angewandte Chemie International Edition. 2002;**41**(6):898–952

[73] Zhang S. Fabrication of novel biomaterials through molecular self-assembly. Nature Biotechnology. 2003;**21**(10):1171–1178

[74] Ariga K, Hill JP, Lee MV, Vinu A, Charvet R, Acharya S. Challenges and breakthroughs in recent research on self-assembly. Science and Technology of Advanced Materials. 2008;**9**:014109

[75] Kurth DG. Metallo-supramolecular modules as a paradigm for materials science. Science and Technology of Advanced Materials. 2008;**9**:014103

[76] Bureekaew S, Shimomura S, Kitagawa S. Chemistry and application of flexible porous coordination polymers. Science and Technology of Advanced Materials. 2008;**9**:014108

[77] Ikeda T, Stoddart JF. Electro-chromic materials using mechanically interlocked molecules. Science and Technology of Advanced Materials. 2008;**9**:014104

[78] Lee SJ, Lin W. Chiral metallo-cycles: Rational synthesis and novel applications. Accounts of Chemical Research. 2008;**41**(4):521–537

[79] Li F, Clegg JK, Lindoy LF, MacQuart RB, Meehan GV. Metallo-supramolecular self-assembly of a universal 3-ravel. Nature Communications. 2011;**2**:205

[80] Daze K. Supramolecular hosts that recognize methyllysines and disrupt the interaction between a modified histone tail and its epigenetic reader protein. Chemical Science. 2012;**3**:2695

[81] Tao AR, Habas S, Yang P. Shape control of colloidal metal nanocrystals. Small. 2008;**4**: 310–325

[82] Korte KE, Skrabalak SE, Xia Y. Rapid synthesis of silver nanowires through a $CuCl^-$ or $CuCl_2^-$ mediated polyol process. Journal of Materials Chemistry. 2008;**18**(4):437–441

[83] Wu X, Redmond PL, Liu H, Chen Y, Steigerwald M, Brus L. Photo-voltage mechanism for room light conversion of citrate stabilized silver nanocrystal seeds to large nanoprisms. Journal of the American Chemical Society. 2008;**130**(29):9500–9506

[84] Pillai ZS, Kamat PV. What factors control the size and shape of silver nanoparticles in the citrate ion reduction method? The Journal of Physical Chemistry B. 2004;**108**(3): 945–951

[85] Camden JP, Dieringer JA, Zhao J, Van Duyne RP. Controlled plasmonic nanostructures for surface-enhanced spectroscopy and sensing. Accounts of Chemical Research. 2008;**41**(12):1653–1661

[86] Dong X, Ji X, Jing J, Li M, Li J, Yang W. Synthesis of triangular silver nano prisms by stepwise reduction of sodium borohydride and tri-sodium citrate. The Journal of Physical Chemistry C. 2010;**114**(5):2070–2074

[87] Koh AL, Bao K, Khan I, Smith WE, Kothleitner G, Nordlander P, Maier SA, McComb DW. Electron energy-loss spectroscopy (EELS) of surface plasmons in single silver nanoparticles and dimers: Influence of beam damage and mapping of dark modes. ACS Nano. 2009;**3**(10):3015–3022

[88] Wiley B, Im SH, Li ZY, McLellan J, Siekkinen A, Xia Y. Maneuvering the surface plasmon resonance of silver nanostructures through shape-controlled synthesis. The Journal of Physical Chemistry B. 2006;**110**:15666–15675

[89] Wiley BJ, Chen Y, McLellan JM, Xiong Y, Li ZY, Ginger D, Xia Y. Synthesis and optical properties of silver nanobars and nanorice. Nano Letters. 2007;**7**(4):1032–1036

[90] Tao A, Sinsermsuksakul P, Yang PD. Polyhedral silver nanocrystals with distinct scattering signatures. Angewandte Chemie International Edition. 2006;**45**(28):4597–4601

[91] Xia Y, Xiong Y, Lim B, Skrabalak SE. Shape-controlled synthesis of metal nanocrystals: Simple chemistry meets complex physics? Angewandte Chemie International Edition. English. 2009;**48**(1):60–103

[92] Sun Y, Xia Y. Shape-controlled synthesis of gold and silver nanoparticles. Science. 2002;**298**(5601):2176–2179

[93] Jin R, Cao Y, Mirkin CA, Kelly KL, Schatz GC, Zheng JG. Photo-induced conversion of silver nanospheres to nano prisms. Science. 2001;**294**(5548):1901–1903

[94] Lim YP, Liu RS, She PL, Hung CF, Shih HC. Synthesis of Ag nanospheres particles in ethylene glycol by electrochemical-assisted polyol process. Chemical Physics Letters. 2006;**420**:304–308

[95] Haro-Poniatowski E, Batina N, Acosta-Garca MC, Pohl-Alfaro MA, Castillo-Ocampo P, Ricolleau C, Fort E. UV Laser Irradiation Effects on Silver Nanostructures. Radiation Effects and Defects in Solids. 2007;**162**:491-499; http://dx.doi.org/10.1080/10420150701472213

[96] Zheng X, Xu W, Corredor C, Xu S, An J, Zhao B, Lombardi JR. Laser induced growth of monodisperse silver nanoparticles with tunable surface plasmon resonance properties and a wavelength self-limiting effect. Journal of Physical Chemistry C. 2007;**111**:14962–14967

[97] Xue C, Métraux GS, Millstone JE, Mirkin CA. Mechanistic study of photomediated triangular silver nanoprism growth. Journal of the American Chemical Society. 2008;**130**(26):8337–8344

[98] Zhang J, Li S, Wu J, Schatz G, Mirkin C. Plasmon mediated synthesis of silver triangular bipyramids. Angewandte Chemie International Edition. 2009;**48**:7927–7931

[99] Kabashin AV, Delaporte P, Pereira A, Grojo D, Torres R, Sarnet T, Sentis M. Nanofabrication with pulsed lasers. Nanoscale Research Letters. 2010;**5**:454–463

[100] Qazi UY, Kajimoto S, Fukumura H. Effect of sodium dodecyl sulfate on the formation of silver nanoparticles by biphotonic reduction of silver nitrate in water. Chemistry Letters. 2014;**43**:1693–1695

[101] Sharma VK, Yngard RA, Lin Y. Silver nanoparticles: Green synthesis and their antimicrobial activities. Advances in Colloid and Interface Science. 2009;**145**:83–96

[102] Qiaoxin Z, Hao L, Xiaohui W, Xiaoliang S, Xinglong D. Fabrication and characterization of nano silver powder prepared by spray pyrolysis. Journal of Wuhan University of Technology, Materials Science Edition 2009;**24**:871–874

[103] Jana NR, Gearheart L, Murphy CJ. Wet chemical synthesis of silver nanorods and nanowires of controllable aspect ratio. Chemical Communications. 2001a;**7**:617–618. Seed mediated growth approach for shape controlled synthesis of spheroidal and rod like gold nanoparticles using a surfactant template. Advanced Materials. 2001b;**13**:1389–1393

[104] Pietrobon B, McEachran M, Kitaev V. Synthesis of size controlled faceted pentagonal silver nanorods with tunable plasmonic properties and self-assembly of these nanorods. ACS Nano. 2008;**3**:21–26

[105] Zhang Q, Li W, Moran C, Chen J, Wen LP, Xia Y. Seed mediated synthesis of Ag nano cubes with controllable edge lengths in the range of 30–200 nm and comparison of their optical properties. Journal of the American Chemical Society. 132:11372–11378

[106] Liebigs J. Ueber Versilberung und Vergoldung von Glas. Annalen der Chemie und Pharmacie. 1856;**98**:132–139

[107] Martin CR. Nanomaterials: A membrane-based synthetic approach. Science. 1994;**266**: 1961–1966

[108] Thomas A, Goettmann F, Antonietti M. Hard templates for soft materials: Creating nanostructured organic materials. Chemistry of Materials. 2008;**20**:738–755

[109] Puntes VF, Krishnan KM, Alivisatos AP. Colloidal nanocrystal shape and size control: The case of cobalt. Science. 2001;**291**:2115–2117

[110] Pastoriza-Santos I, Liz-Marzan LM. Synthesis of silver nanoprisms in DMF. Nano Letters. 2002;**2**:903–905

[111] Puntes VF, Zanchet D, Erdonmez CK, Alivisatos AP. Synthesis of HCP-Co nano disks. Journal of the American Chemical Society. 2002;**124**:12874–12880

[112] Chen S, Fan Z, Carroll DL. Silver nano disks: Synthesis, characterization, and self assembly. The Journal of Physical Chemistry B. 2002;**106**:10777–10781

[113] Chen SH, Carroll DL. Synthesis and characterization of truncated triangular silver nano plates. Nano Letters. 2002;**2**:1003–1007

[114] Maillard M, Giorgio S, Pileni MP. Silver nano disks. Advanced Materials. 2002;**14**: 1084–1086

[115] Yener DO, Sindel J, Randall CA, Adair JH. Synthesis of nanosized silver platelets in octylamine water bilayer systems. Langmuir. 2002;**18**:8692–8699

[116] Maillard M, Giorgio S, Pileni MP. Tuning the size of silver nanodisks with similar aspect ratios: Synthesis and optical properties. The Journal of Physical Chemistry B. 2003;**107**:2466–2470

[117] Zheng XW, Zhu LY, Yan AH, WangXJ, Xie Y. Controlling synthesis of silver nanowires and dendrites in mixed surfactant solutions. Journal of Colloid and Interface Science. 2003;**268**:357–361

[118] Zhang JL, Han BX, Liu MH, Liu DX, Dong ZX, Liu J, Li D, Wang J, Dong BZ, Zhao H, Rong LX. Ultrasonication induced formation of silver nanofibers in reverse micelles and small angle X ray scattering studies. The Journal of Physical Chemistry B. 2003;**107**:3679–3683

[119] Chen S, Carroll DL. Silver nano plates: Size control in two dimensions and formation mechanisms. The Journal of Physical Chemistry B. 2004;**108**:5500–5506

[120] Song Y, Yang Y, Medforth CJ, Pereira E, Singh AK, Xu H, Jiang Y, Brinker CJ, Swol FV, Shelnutt JA. Controlled synthesis of 2D and 3D dendritic platinum nanostructures. Journal of the American Chemical Society. 2004;**126**:635–645

[121] Kim JU, Cha SH, Shin K, Jho JY, Lee JC. Preparation of gold nanowires and nanosheets in bulk block copolymer phases under mild conditions. Advanced Materials. 2004;**16**:459–464

[122] Kryukov A, Stroyuk A, Zinchuk N, Korzhak A, Kuchmii S. Optical and catalytic properties of Ag2S nanoparticles. Journal of Molecular Catalysis. 2004;**221**:209–221

[123] Ah CS, Yun YJ, Park HJ, Kim WJ, Ha DH, Yun WS. Size controlled synthesis of machinable single crystalline gold nano plates. Chemistry of Materials. 2005;**17**:5558–5561

[124] Metraux GS, Mirkin CA. Rapid thermal synthesis of silver nanoprisms with chemically tailorable thickness. Advanced Materials. 2005;**17**:412–415

[125] Liu Y, Chu Y, Yang LK, Han DX, Lu ZX. A novel solution phase route for the synthesis of crystalline silver nanowires. Materials Research Bulletin. 2005;**40**:1796–1801

[126] Ni CY, Hassan PA, Kaler EW. Structural characteristics and growth of pentagonal silver nanorods prepared by a surfactant method. Langmuir. 2005;**21**:3334–3337

[127] Porel S, Singh S, Radhakrishnan TP. Polygonal gold nanoplates in a polymer matrix. Chemical Communications. 2005;**18**:2387–2389

[128] Salzemann C, Urban J, Lisiecki I, Pileni MP. Characterization and growth process of copper nanodisks. Advanced Functional Materials. 2005;**15**:1277–1284

[129] Zang DB, Song CX, Hu ZS, Zhou XD. Synthesis of silver nanoparticles with flake like shapes. Materials Letters. 2005;**59**:1760–1763

[130] Xiong Y, Washio I, Chen J, Cai H, Li ZY, Xia Y. Poly(vinyl pyrrolidone): A dual functional reductant and stabilizer for the facile synthesis of noble metal nano plates in aqueous solutions. Langmuir. 2006;**22**:8563–8570

[131] Zhang JH, Liu HY, Zhan P, Wang ZL, Ming NB. Controlling the growth and assembly of silver nanoprisms. Advanced Functional Materials. 2007;**17**:1558–1566

[132] Zhang W, Qiao X, Chen J. Synthesis of silver nanoparticles effects of concerned parameters in water/oil microemulsion. Materials Science and Engineering B. 2007;**142**:1–15

[133] Song J, Chu Y, Liu Y, Li L, Sun W. Room temperature controllable fabrication of silver nano plates reduced by aniline. Chemical Communications. 2008;**10**:1223–1225

[134] Song Y, Dorin RM, Garcia RM, Jiang YB, Wang H, Li P, Qiu Y, Swol FV, Miller JE, Shelnutt JA. Synthesis of platinum nano wheels using a bicellar template. Journal of the American Chemical Society. 2008;**130**:12602–12603

[135] Deng Z, Mansuipur M, Muscat AJ. New method to single crystal micrometer sized ultra thin silver nanosheets: Synthesis and characterization. The Journal of Physical Chemistry C. 2009;**113**:867–873

[136] Pastoriza-Santos I, Sánchez-Iglesias A, Rodríguez-González B, Liz-Marzán LM. Aerobic synthesis of Cu nano plates with intense plasmon resonances. Small. 2009;**5**:440–443

[137] Jang K, Kim HJ, Son SU. Low temperature synthesis of ultrathin rhodium nanoplates via molecular orbital symmetry interaction between rhodium precursors. Chemistry of Materials. 2010;**22**:1273–1275

[138] Garcia RM, Song Y, Dorin RM, Wang H, Moreno AM, Jiang YB, Tian YQiu Y, Medforth CJ, Coker EN, Swol FV, Miller JE, Shelnutt JA. Templated growth of platinum nano wheels using the inhomogeneous reaction environment of bicelles. Physical Chemistry Chemical Physics. 2011;**13**:4846–4852

[139] Zhang Q, Li N, Goebl J, Lu Z, Yin Y. A systematic study of the synthesis of silver nano plates: Is citrate a "magic" reagent? Journal of the American Chemical Society. 2011;**133**:18931–18939

[140] Guo H, Chen Y, Ping H, Wang L, Peng DL. One pot synthesis of hexagonal and triangular nickel copper alloy nano plates and their magnetic and catalytic properties. Journal of Materials Chemistry. 2012;**22**:8336–8344

[141] Huang X, Li Y, Chen Y, Zhou H, Duan X, Huang Y. Plasmonic and catalytic AuPd nanowheels for the efficient conversion of light into chemical energy. Angewandte Chemie International Edition. 2013;**125**:6179–6183

[142] Bagwe RP, Khilar KC. Effects of inter-micellar exchange rate on the formation of Ag nanoparticles in reverse microemulsions of AOT. Langmuir. 2000;**16**:905–910

[143] Zhang ZQ, Patel RC, Kothari R, JohnsonCP, Friberg SE, Aikens PA. Stable silver clusters and nanoparticles prepared in polyacrylate and inverse micellar solutions. The Journal of Physical Chemistry B. 2000;**104**:1176–1182

[144] Zhang WZ, Qiao XL, Chen JG, ChenQY. Self-assembly and controlled synthesis of silver nanoparticles in SDS quaternary microemulsion. Materials Letters. 2008;**62**:1689–1692

[145] Ito T, Okazaki S. Pushing the limits of lithography. Nature. 2000;**406**:1027–1031

[146] Reyntjens S, Puers R. Focused ion beam induced deposition: Fabrication of three dimensional microstructures and Young's modulus of the deposited material. Journal of Micromechanics and Microengineering. 2000;**10**:181–188

[147] Kim SE, Han YH, Lee BC, Lee JC. One pot fabrication of various silver nanostructures on substrates using electron beam irradiation. Nanotechnology. 2010;**21**: Article ID: 075302

[148] Korgel BA, Fullam S, Connolly S, Fitzmaurice D. Assembly and self-organization of silver nanocrystal superlattices: Ordered "soft spheres". The Journal of Physical Chemistry B. 1998;**102**:8379–8388

[149] Martin JE, Wilcoxon JP, Odinek J, Provencio P. Control of the inter-particle spacing in gold nanoparticle superlattices. The Journal of Physical Chemistry B. 2000;**104**:9475–9486

[150] Duan H, Wang D, Kurth DG, Möhwald H. Directing self-assembly of nanoparticles at water/oil interfaces. Angewandte Chemie International Edition. 2004;**43**:5639–5642

[151] Deng Z, Tian Y, Lee SH, Ribbe AE, Mao C. DNA encoded self-assembly of gold nanoparticles into one dimensional arrays. Angewandte Chemie International Edition. 2005;**44**:3582–3585

[152] Bigioni TP, Lin XM, Nguyen TT, Corwin EI, Witten TA, Jaeger HM. Kinetically driven self-assembly of highly ordered nanoparticle monolayers. Nature Materials. 2006;**5**:265–270

[153] Aldaye FA, Sleiman HF. Dynamic DNA templates for discrete gold nanoparticle assemblies: Control of geometry, modularity, write/erase and structural switching. Journal of the American Chemical Society. 2007;**129**:4130–4131

[154] Ofir Y, Samanta B, Rotello VM. Polymer and biopolymer mediated self-assembly of gold nanoparticles. Chemical Society Reviews. 2008;**37**:1814–1825

[155] Nykypanchuk D, Maye MM, Van der Lelie D, Gang O. DNA-guided crystallization of colloidal nanoparticles. Nature. 2008;**451**:549–552

[156] Cheng W, Campolongo MJ, Cha JJ, Tan SJ, Umbach CC, Muller DA, Luo D. Free standing nanoparticle superlattice sheets controlled by DNA. Nature Materials. 2009; 8:519–525

[157] Maye MM, Nykypanchuk D, Cuisinier M, Van der Lelie D, Gang O. Stepwise surface encoding for high-throughput assembly of nanoclusters. Nature Materials. 2009;**8**:388–391

[158] Jones MR, Osberg KD, Macfarlane RJ, Langille MR, Mirkin CA. Templated techniques for the synthesis and assembly of plasmonic nanostructures. Chemical Reviews. 2011;**111**:3736–3827

[159] Chen Y, Cheng W. DNA based plasmonic nanoarchitectures: From structural design to emerging applications. Wiley Interdisciplinary Reviews: Nanomedicine and Nanobiotechnology. 2012;**4**:587–604

[160] Elbaz J, Cecconello A, Fan Z, Govorov AO, Willner I. Powering the programmed nanostructure and function of gold nanoparticles with catenated DNA machines. Nature Communications. 2013;**4**:2000

[161] Wu Z, Dong C, LiY, Hao H, Zhang H, Lu Z, Yang B. Self-assembly of Au15 into single cluster thick sheets at the interface of two miscible high-boiling solvents. Angewandte Chemie International Edition. 2013;**125**:10136–10139

[162] Liu B, Zeng HC. Hydrothermal synthesis of ZnO nanorods in the diameter regime of 50 nm. Journal of the American Chemical Society. 2003;**125**:4430–4431

[163] Yin AX, Liu WC, Ke J, Zhu W, Gu J, Zhang YW, Yan CH. Ru nanocrystals with shape dependent surface enhanced Raman spectra and catalytic properties: Controlled synthesis and DFT calculations. Journal of the American Chemical Society. 2012;**134**: 20479–20489

[164] Yang Y, Zhong XL, Zhang Q, Blackstad LG, Fu ZW, Li ZY, Qin D. The role of etching in the formation of Ag nanoplates with straight, curved and wavy edges and comparison of their SERS properties. Small. 2014;**10**:1430–1437

[165] Kowshik M, Deshmukh N, Vogel W, UrbanJ, Kulkarni SK, Paknikar MK. Microbial synthesis of semiconductor CdS nanoparticles, their characterization, and their use in the fabrication of an ideal diode. Biotechnology and Bioengineering. 2002;**28**:583–588

[166] Sastry M, Ahmad A, Islam Khan M, Kumar R. Biosynthesis of metal nanoparticles using fungi and actinomycete. Current Science India. 2003;**85**:162–170

[167] Armendariz V, Herrera I, Peralta-Videa JR, Jose-Yacaman M, Troiani H, Santiago P, Gardea-Torresdey JR. Size controlled gold nanoparticle formation by Avena sativa biomass: Use of plants in nanobiotechnology. Journal of Nanoparticle Research. 2004;**6**:377–382

[168] Bansal V, Rautaray D, Ahmad A, Sastry M. Biosynthesis of zirconia nanoparticles using the fungus Fusarium oxysporum. Journal of Materials Chemistry. 2004;**14**:3303–3305

[169] Ankamwar B, Chaudhary M, Sastry M. Gold nanotriangles biologically synthesized using tamarind leaf extract and potential application in vapor sensing. Synthesis and Reactivity in Inorganic Metal-Organic and Nano Metal Chemistry. 2005;**35**:19–26

[170] Bansal V, Poddar P, Ahmad A, Sastry M. Room temperature biosynthesis of ferroelectric barium titanate nanoparticles. Journal of the American Chemical Society. 2006;**128**:11958–11963

[171] Gericke M, Pinches A. Biological synthesis of metal nanoparticles. Hydrometallurgy. 2006;**83**:132–140

[172] Treguer M, Cointet C, Remita H, Khatouri J, Mostafavi M, Amblard J, Belloni JJ. Dose rate effect on radiolytic synthesis of gold-silver bimetallic clusters in solution. The Journal of Physical Chemistry B. 1998;**102**:4310–4321

[173] Chen W, Cai W, Zhang L, Wang G, Zhang L. Sono-chemical processes and formation of gold nanoparticles within pores of mesoporous silica. Journal of Colloid and Interface Science. 2001;**238**:291–295

[174] Mafune F, Kohno J, Takeda Y, Kondow TJ. Dissociation and aggregation of gold nanoparticles under laser irradiation. The Journal of Physical Chemistry B. 2001; 105:9050–9056

[175] Rodríguez-Sanchez L, Blanco MC, Lopez-Quintela MA. Electrochemical synthesis of silver nanoparticles. The Journal of Physical Chemistry B. 2002;**104**:9683–9638

[176] Eustis S, Hsu HY, El-Sayed MA. Gold nanoparticle formation from photochemical reduction of Au3 by continuous excitation in colloidal solutions: A proposed molecular mechanism. The Journal of Physical Chemistry B. 2005;**109**:4811–4815

[177] Frattini A, Pellegri N, Nicastro D, de Sanctis O. Effect of amine groups in the synthesis of Ag nanoparticles using aminosilanes. Materials Chemistry and Physics. 2005;**94**:148–152

[178] Starowiicz M, Stypula B, Banas J. Electrochemical synthesis of silver nanoparticles. Electrochemistry Communications. 2006;**8**:227–230

[179] Zhang G, Wang DJ. Fabrication of heterogeneous binary arrays of nanoparticles via colloidal lithography. Journal of the American Chemical Society. 2008;**130**:5616–5617

[180] Wang L, Chen X, Zhan J, Chai Y, Yang C, Xu L, Zhuang W, Jing B. Synthesis of gold nano and micro plates in hexagonal liquid crystals. The Journal of Physical Chemistry B. 2005;**109**:3189–3194. Alexandridis, 2011

[181] You H, Yang S, Ding B, Yang H. Synthesis of colloidal metal and metal alloy nanoparticles for electrochemical energy applications. Chemical Society Reviews. 2013;**42**:2880–2904

[182] Alexandridis P. Gold nanoparticle synthesis, morphology control, and stabilization facilitated by functional polymers. Chemical Engineering & Technology. 2011; 34(1):15–28

[183] Kulkarni N, Muddapur U. Biosynthesis of metal nanoparticles: A review. Journal of Nanotechnology 2014 (2014). Article ID 510246, 8 pages; http://dx.doi.org/10.1155/2014/510246

[184] Ai J, Biazar E, Jafarpour M, Montazeri M, Majdi A, Aminifard S, Zafari M, Akbari HR, Rad HG. Nano toxicology and nanoparticle safety in biomedical designs. International Journal of Nanomedicine. 2011;**6**:1117–1127

[185] Jain N, Bhargava A, Majumdar S, Panwar J. Extracellular biosynthesis and characterization of silver nanoparticles using Aspergillus flavus NJP08: A mechanism prospective. Nanoscale. 2011;**3**:635–641

[186] Baker S, Harini BP, Rakshith D, Satish S. Marine microbes: Invisible nano factories. Journal of Pharmacy Research. 2013;**6**:383–388

[187] Makarov VV, Love AJ, Sinitsyna OV, Makarova SS, Yaminsky IV, Taliansky ME, Kalinina NO. Green nanotechnologies: Synthesis of metal nanoparticles using plants. Acta Naturae. 2014;**6**:35–44

[188] Gowramma B, Keerthi U, Mokula R, Rao DM. Biogenic silver nanoparticles production and characterization from native stain of Corynebacterium species and its antimicrobial activity. 3 Biotech. 2015;**5**:195–201

[189] Shah M, Fawcett D, Sharma S, Tripathy SK, Poinern GEJ. Green synthesis of metallic nanoparticles via biological entities. Materials. 2015;**8**:7278–7308

[190] Bollag JM, Mertz T, Otjen L. Role of microorganisms in soil bioremediation. In Bioremediation through Rhizosphere Technology; Journal of the American Chemical Society. 1994;**563**:2–10

[191] Mann S. Biomimetic Materials Chemistry. New York: VCH Publishers; 1996

[192] Stephen JR, Macnaughton SJ. Developments in terrestrial bacterial remediation of metals. Current Opinion in Biotechnology. 1999;**10**:230–233

[193] Sastry M, Ahmad A, Khan MI, Kumar R. Microbial nanoparticle production. In: Niemeyer CM, Mirkin CA, editors. Nanobiotechnology. Weinheim: Wiley-VCH; 2004

[194] Bhattacharya D, Gupta RK. Nanotechnology and potential of microorganisms. Critical Reviews in Biotechnology. 2005;**25**:199–204

[195] Kowshik M, Deshmukh N, Vogel W, Urban J, Kulkarni SK, Paknikar KM. Microbial synthesis of semiconductor CdS nanoparticles, their characterization, and their use in the fabrication of an ideal diode. Biotechnology and Bioengineering. 2002; 78:583–588

[196] Duran N, Marcato PD, Alves OL, Souza GI, Esposito E. Mechanistic aspects of biosynthesis of silver nanoparticles by several Fusarium oxysporum strains. Journal of Nanobiotechnology. 2005;**3**:1–8

[197] Mandal D, Bolander ME, Mukhopadhyay D, Sarkar G, Mukherjee P. The use of microorganisms for the formation of metal nanoparticles and their application. Applied Microbiology and Biotechnology. 2006;**69**:485–492

[198] Lengke M, Ravel B, Fleet ME, Wanger G, Gordon RA, Southam G. Mechanisms of gold bioaccumulation by filamentous cyanobacteria from gold (III) chloride complex. Environmental Science & Technology. 2006;**40**:6304–6309

[199] Mittal AK, Chisti Y, Banerjee UC. Synthesis of metallic nanoparticles using plants. Biotechnology Advances. 2013;**31**:346–356

[200] Singh R, Gautam N, Mishra A, Gupta R. Heavy metals and living systems: An overview. Indian Journal of Pharmacology. 2011;**43**:246–253

[201] Mukherjee P, Ahmad A, Mandal D, Senapati S, Sainkar SR, Khan MI, Ramani R, Parischa R, Ajayakumar PV, Alam M. Sastry M, Kumar R. Bioreduction of AuCl4 ions by the fungus, Verticillium sp. and surface trapping of the gold nanoparticles formed. Angewandte Chemie International Edition. 2001;**40**:3585–3588

[202] Iravani S. Green synthesis of metal nanoparticles using plants. Green Chemistry. 2011;**13**:2638–2650

[203] Cason JP, Miller ME, Thompson JB, Roberts CB. Solvent effects on copper nanoparticle growth behavior in AOT reverse micelle systems. The Journal of Physical Chemistry B. 2001;**105**:2297

[204] Lopez-Quintela MA, Tojo C, Blanco MC, Garcıa Rio L, Leis JR. Microemulsion dynamics and reactions in microemulsions. Current Opinion in Colloid and Interface Science. 2004;**9**:264

[205] Uskokovic V, Drofenik M. Synthesis of relatively high magnetic nanosized NiZn-ferrite in microemulsion at 450C. Surface Review and Letters. 2005;**12**:239

[206] Li L, Qing-Sheng W, Ya-Ping D, Pei-Ming W. Wurtzite ZnSe quantum dots: Synthesis, characterization and PL propertie. Materials Letter. 2005;**59**:1623

[207] Ingert D, Pileni M-P. Limitations in producing nanocrystals using reverse micelles as nanoreactors. Advanced Functional Materials 2001;**11**:136

[208] Husein M, Rodil E, Vera J. Formation of silver chloride nanoparticles in microemulsions by direct precipitation with the surfactant counterion. Langmuir. 2003;**19**:8467

[209] Husein M, Rodil E, Vera J. Formation of silver bromide precipitate of nanoparticles in a single microemulsion utilizing the surfactant counterion. Journal of Colloid and Interface Science. 2004;**273**:426

[210] He P, Shen X, Gao H. Size-controlled preparation of Cu2O octahedron nanocrystals and studies on their optical absorption. Journal of Colloid and Interface Science. 2005;**284**:510

[211] Eastoe J, Hollamby MJ, Hudson L. Recent advances in nanoparticle synthesis with reversed micelles. Advances in Colloid and Interface Science. 2006;128–130: 5–15

[212] Pileni M-P. Nanocrystals: Fabrication, organization and collective properties. CR Chimie. 2003;**6**:965

[213] Pileni M-P. The role of soft colloidal templates in controlling the size and shape of inorganic nanocrystals. Nature. 2003;**2**:145

[214] Rees GD, Evans-Gowing R, Hammond SJ, Robinson BH. Formation and morphology of calcium sulfate nanoparticles and nanowires in water-in-oil microemulsions. Langmuir. 1999;**15**:1993

[215] Filankembo A, Pileni M-P. Is the template of self-colloidal assemblies the only factor, which controls nanocrystal shapes? The Journal of Physical Chemistry B. 2000;**104**:5865

[216] Krauel K, Davies NM, Hook S, Rades T. Using different structure types of microemulsions for the preparation of poly (alkylcyanoacrylate) nanoparticles by interfacial polymerization. Journal of Control Release. 2005;**106**:76

[217] Germain V, Li J, Ingert D, Wang ZL, PileniM-P. Stacking faults in formation of silver nano disks. The Journal of Physical Chemistry B. 2003;**107**:8718

[218] Liu J, Ikushima Y, Shervani Z. Environmentally benign preparation of metal nano-particles by using water-in-CO2 microemulsions technology. Current Opinion in Colloid & Interface Science. 2003;**7**:255

[219] Eastoe J, Gold S. Self-assembly in green solvents. Physical Chemistry Chemical Physics. 2005;**7**:1352

[220] Meziani MJ, Pathak P, Beacham F, Allard LF, Sun Y-P. Nanoparticle formation in rapid expansion of water-in-supercritical carbon dioxide microemulsion into liquid solution. The Journal of Supercritical Fluids. 2005;**34**:91

[221] Ji M, Chen X, Wai CM, Fulton JL. Synthesizing and dispersing silver nanoparticles in a water-in-supercritical carbon dioxide microemulsion. Journal of the American Chemical Society. 1999;**121**:2631

[222] Cason JP, Khambaswadkar K, Roberts CB. Supercritical fluid and compressed solvent effects on metallic nanoparticle synthesis in reverse micelles. Industrial & Engineering Chemistry Research. 2000;**39**:4749

[223] Ohde H, Rodriguez JM, Ye X-R, Wai CM. Synthesizing silver halide nanoparticles in supercritical carbon dioxide utilizing a water in CO2 microemulsion. Chemical Communications. 2000; 2353–2354. DOI: 10.1039/b005924m

[224] Ohde H, Wai CM, Kim H, Kim J, Ohde M. Hydrogenation of olefins in supercritical CO2 catalyzed by palladium nanoparticles in a water-in-CO2 microemulsion. Journal of the American Chemical Society. 2002;**124**: 4540

[225] Ohde M, Ohde H, Wai CM. Catalytic hydrogenation of arenes with rhodium nanoparticles in a water-in-supercritical CO2 microemulsion. Chemical Communications. 2002;2388–2389

[226] McLeod MC, McHenry RS, Beckman EJ, Roberts CB. Synthesis and stabilization of silver metallic nanoparticles and premetallic intermediates in perfluoropolyether/CO2 reverse micelle systems. The Journal of Physical Chemistry B. 2003;**107**:2693

[227] Kitchens CL, Roberts CB. Copper nanoparticle synthesis in compressed liquid and supercritical fluid reverse micelle systems. Industrial & Engineering Chemistry Research. 2004;**43**:6070

[228] Liu J, Raveendran P, Shervani Z, Ikushima Y, Hakuta Y. Synthesis of Ag and AgI quantum dots in AOT-stabilized water-in-CO2 microemulsions. Chemistry. 2005;**11** (6):1854–1860

[229] Kitchens CL, McLeod MC, Roberts CB. Chloride ion effects on synthesis and directed assembly of copper nanoparticles in liquid and compressed alkane microemulsions. Langmuir. 2005;**21**:5166

[230] Liu J, Raveendran P, Shervani Z, Ikushima Y, Hakuta Y. Synthesis of Ag and AgI quantum dots in AOT-stabilized water-in-CO2 microemulsions. Chemistry - A European Journal. 2005;**11**:1854

[231] Bell PW, Anand M, Fan X, Enick RM, Roberts CB. Stable dispersions of silver nanoparticles in carbon dioxide with fluorine-free ligands. Langmuir. 2005;**21**:11608

[232] Lim KT, Hwang HS. Synthesis of TiO2 nanoparticles utilizing hydrated reverse micelles in CO2. Langmuir. 2004;**20**:2466

[233] Semalty A, Semalty M, Rawat BS, Singh D, Rawat MSM. Pharmacosomes: The lipid-based new drug delivery system. Expert Opinion on Drug Delivery. 2009;**6**(6): 599–612

[234] Semalty A, Semalty M, Rawat BS, Singh D, Rawat MSM. Development and evaluation of pharmacosomes of aceclofenac. Indian Journal of Pharmaceutical Sciences. 2010;72(5):576–581

[235] Laouini A, Jaafar-Maalej C, Limayem-Blouza I, Sfar S, Charcosset C, Fessi H. Preparation, characterization and applications of liposomes: State of the art. Journal of Colloid Science and Biotechnology. 2012;1(2):147–168

[236] Naeem S, Kiew LV, Chung LY, Eh Suk VR, Mahmood A, Misran MB. Optimization of phospholipid nanoparticle formulations using response surface methodology. Journal of Surfactants and Detergents. 2016;**19**(1):67–74

[237] Murray D. Phytosomes-Increase the Absorption Herbal Extract [Internet]. Available from: http://doctormurray.com/phytosomes-herbal-support-increase-the-absorption-of-herbal-extracts/

[238] Marena C, Lampertico M. Preliminary clinical development of silipide: A new complex of silybin in toxic liver disorders. Planta Medica. 1991;**57**(2):A124-A125

[239] Sheng JD, Li ZJ, Feng S, Bin JX. Preparation of icaritin phytosomes and their solid dispersions. Chinese Traditional and Herbal Drugs. 2010;**41**(9):1449–1453

[240] Zhang J, Tang Q, Xu X, Li N. Development and evaluation of a novel phytosome-loaded chitosan microsphere system for curcumin delivery. International Journal of Pharmaceutics. 2013;**448**(1):168–174

[241] Li L, Xiang D, Shigdar S, Yang W, Li Q, Lin J, Liu K, Duan W. Epithelial cell adhesion molecule aptamer functionalized PLGA-lecithin-curcumin-PEG nanoparticles for targeted drug delivery to human colorectal adenocarcinoma cells. International Journal of Nanomedicine. 2014;**9**:1083–1096

[242] Demir B, Barlas FB, Guler E, Gumus PZ, Can M, Coskunolbef MYH, Timur S. Gold nanoparticle loaded phytosomal systems: Synthesis, characterization and in vitro investigations. RSC Advances. 2014;**4**:34687–34695

[243] Yi Y, Li Y, Wu H, Jia M, Yang X, Wei H, Lin J, Wu S, Huang Y, Hou Z, Xie L. Single-step assembly of polymer-lipid hybrid nanoparticles for mitomycin C delivery. Nanoscale Research Letters. 2014;**9**:560

[244] Allam AN, Komeil IA, Abdallah OY. Phytosomal soft gel formulation: Development, optimization and physicochemical characterization. Acta Pharmaceutica. 2015;**65**: 285–297

[245] Li Y, Lin J, LiuG, Li Y, Song L, Fan Z, ZhuX, Su G, Hou Z. Self-assembly of multifunctional integrated nanoparticles loaded with a methotrexate–phospholipid complex: Combining simplicity and efficacy in both targeting and anticancer effects. RSC Advances. 2016;**6**:86717–86727

[246] Fan Z, Liu G, Li Y, Ma J, Lin J, Guo F, Hou Z, Liya Xie L. Self-assembly of the active lactone form of a camptothecin–phospholipid complex for sustained nuclear drug delivery. RSC Advances. 2016;**6**:82949–82960

[247] Guliani A, Singla R, Kumari A, Yadav SK. Liposomal and phytosomal formulations. In: Nanoscale Materials in Targeted Drug Delivery, Theragnosis and Tissue Regeneration. Yadav SK, editor. Springer Singapore: Springer Science+Business Media Singapore; 2016. ISBN: 978-981-10-0817-7

[248] Manconi M, Marongiu F, Castangia I, Manca ML, Caddeo C, Tuberoso CIG, D'hallewin G, Bacchetta G, Fadda AM. Polymer-associated liposomes for the oral delivery of grape pomace extract. Colloids and Surfaces B: Biointerfaces. 2016;**146**:910–917

[249] Yu F, Li Y, Chen Q, He Y, Wang H, Yang L, Guo S, Meng Z, Cui J, Xue M, Chen XD. Monodisperse microparticles loaded with the self-assembled berberine-phospholipid complex-based phytosomes for improving oral bioavailability and enhancing hypoglycemic efficiency. European Journal of Pharmaceutics and Biopharmaceutics. 2016;**103**:136–148

[250] Saoji SD, Dave VS, Dhore PW, Bobde YS, Mack C, Gupta D, Raut NA. The role of phospholipid as a solubility- and permeability-enhancing excipient for the improved delivery of the bioactive phytoconstituents of Bacopa monnieri. European Journal of Pharmaceutical Sciences. August 31, 2016. pii: S0928-0987(16)30345-1. DOI: 10.1016/j.ejps. 2016.08.056

[251] Huang X, Li SZ, Huang YZ, Wu SX, Zhou XZ, Li SZ, Gan CL, Boey F, Mirkin CA, Zhang H. Synthesis of Hexagonal Close Packed Gold Nanostructures. Nature Communications. 2011;**2**:292

[252] Huang XQ, Tang SH, Mu XL, Dai Y, Chen GX, Zhou ZY, Ruan FX, Yang ZL, Zheng NF. Freestanding palladium nanosheets with plasmonic and catalytic properties. Nature Nanotechnology. 2011;**6**:28–32

[253] Cheng W, He J, Yao T, Sun Z, Jiang Y, Liu Q, Jiang S, Hu F, Xie Z, He B, Yan W, Wei S. Half-unit-cell a-Fe2O3 semiconductor nanosheets with intrinsic and robust ferromagnetism. Journal of the American Chemical Society. 2014;**136**:10393–10398

[254] Wu X-J, Huang X, Liu J, Li H, Yang J, Li B, Huang W, Zhang H. Two-dimensional CuSe nanosheets with microscale lateral size: Synthesis and template-assisted phase transformation. Angewandte Chemie International Edition. 2014;**53**:5083–5087

[255] Wu X-J, Huang X, Qi X, Li H, Li B, Zhang H. Copper-based ternary and quaternary semiconductor nanoplates: Templated synthesis, characterization, and photo-electrochemical properties. Angewandte Chemie International Edition. 2014;**53**:8929–8933

[256] Fan ZX, Huang X, Tan CL, Zhang H. Thin metal nanostructures: Synthesis, properties and applications. Chemical Science. 2015;**6**:95–111

[257] Fan ZX, Huang X, Han Y, Bosman M, Wang QX, Zhu YH, Liu Q, Li B, Zeng ZY, Wu J, Shi WX, Li SZ, Gan CL, Zhang H. Surface modification-induced phase transformation of hexagonal close-packed gold square sheets. Nature Communications. 2015;**6**:6571

[258] Fan ZX, Bosman M, Huang X, Huang D, Yu Y, Ong KP, Akimov YA, Wu L, Wu J, Liu Q, Png CE, Gan CL, Yang PD, Zhang H. Stabilization of 4H hexagonal phase in gold nanoribbons. Nature Communications. 2015;**6**:7684

[259] Fan Z, Zhu Y, Huang X, Han Y, Wang Q, Liu Q, Huang Y, Gan CL, Zhang H. Synthesis of ultrathin fcc Au@Pt and Au@Pd core-shell nanoplates from hcp Au square sheets. Angewandte Chemie International Edition. 2015;**127**:5764–5768

[260] Tan C, Hua Zhang H. Wet-chemical synthesis and applications of non-layer structured two-dimensional nanomaterials. Nature Communications. 2015;**6**:7873

[261] Sun Y, Sun Z, Gao S, Cheng H, Liu Q, Piao J, Yao T, Wu C, Hu S, Wei S, Xie Y. Fabrication of flexible and freestanding zinc chalcogenide single layers. Nature Communications. 2012;**3**:1057

[262] Zhang X, Zhang J, Zhao J, Pan B, Kong M, Chen J, Xie Y. Half-metallic ferromagnetism in synthetic Co9Se8 nanosheets with atomic thickness. Journal of the American Chemical Society. 2012;**134**:11908–11911

[263] Sun Y, Liu Q, Gao S, Cheng H, Lei F, Sun Z, Jiang Y, Su H, Wei S, Xie Y. Pits confined in ultrathin cerium (IV) oxide for studying catalytic centers in carbon monoxide oxidation. Nature Communications. 2013;**4**:2899

[264] Sun Y, Lei F, Gao S, Pan B, Zhou J, Xie Y. Atomically thin tin dioxide sheets for efficient catalytic oxidation of CO. Angewandte Chemie International Edition. 2013;**52**:10569–10572

[265] Xu Y, Zhao WW, Xu R, Shi YM, Zhang B. Synthesis of ultrathin CdS nanosheets as efficient visible-light-driven water splitting photo-catalysts for hydrogen evolution. Chemcal Communications. 2013;**49**:9803–9805

[266] Duan X, Wang C, Shaw JC, Cheng R, Chen Y, Li H, Wu X, Tang Y, Zhang Q, Pan A, Jiang J, Yu R, Huang Y, Duan X. Lateral epitaxial growth of two-dimensional layered semiconductor heterojunctions. Nature Nanotechnology. 2014;**9**:1024–1030

[267] Lei F, Sun Y, Liu K, Gao S, Liang L, Pan B, Xie Y. Oxygen vacancies confined in ultrathin indium oxide porous sheets for promoted visible-light water splitting. Journal of the American Chemical Society. 2014;**136**:6826–6829

[268] Sun Z, Liao T, Dou Y, Hwang SM, Park MS, Jiang L, Kim JH, Dou SX. Generalized self-assembly of scalable two-dimensional transition metal oxide nanosheets. Nature Communications. 2014;**5**:3813

[269] Industry Analysis. Available from: https://www.gminsights.com/industry-analysis/gold-nanoparticles-market

[270] CORDIS; European Commission, Community Research and Development Information. Available from: http://cordis.europa.eu/

[271] Gurav A, Kodas T, Pluym T, Xiong Y. Aerosol processing of materials. Aerosol Science and Technology. 1993;**19**(4):411–452

[272] Kodas TT, Hampden-Smith MJ. Aerosol Processing of Materials. 6th ed. New York: Wiley-VCH; ISBN: 978-0-471-24669-5

[273] Kammler HK, Mueller R, Senn O, Pratsinis SE. Synthesis of silica-carbon particles in a turbulent H2-air flame aerosol reactor. AIChE Journal. 2001;**47**(7):1533–1543

[274] Luther W. editor. Industrial Application of Nanomaterials-Chances and Risks. Dusseldorf: Future Technologies Division; 2004

[275] Gohier A, Kim KH, Norman E, Gorintin L, Bondavalli P, Cojocaru CS. Spray-gun deposition of catalyst for large area and versatile synthesis of carbon nanotubes. Applied Surface Sciences. 2012;**258**:6024

[276] Swihart MT. Vapor-phase synthesis of nanoparticles. Current Opinion in Colloid & Interface Science. 2003;**8**:127

[277] Skandan G, Singhal A. In: Gogotsi Y, editor. Nanomaterials Handbook. Florida: Taylor & Francis Group; 2006

[278] BUONAPART-E. Available from: www.buonapart-e.eu

[279] Mason TJ. Sono-chemistry. Cambridge: Royal Society of Chemistry; 1990

[280] Zhou SM, Feng YS, Zhang LD. Sonochemical synthesis of large-scale single crystal CdS nanorods. Materials Letters. 2003;**57**:19

[281] Esmaeili-Zare M, Salavati-Niasari M. Proceedings of the International Conference Nanomaterials: Applications and Properties. Ukraine; 2013 (NAP, 2013), pp. 1–3

[282] Calka A. Formation of titanium and zirconium nitrides by mechanical alloying. Applied Physical Letters. 1991;**59**:13

[283] Calka A, Williams JS. Synthesis of nitrides by mechanical alloying. Materials Science Forum. 1992;**88–90**:787–794

[284] El-Eskandarany MS, Sumiyama K, Aoki K, Suzuki K. Reactive ball mill for solid state synthesis of metal nitrides powders. Materials Science Forum. 1992;**88–90**:801–808

[285] Fect H. In: Edelstein AS, Cammaratra RC, editors. Nanomaterials: Synthesis, Properties and Applications. Washington, DC: Taylor & Francis; 1996

[286] Kuhn HA, Lee PW, Erturk T. A Fracture Criterion for Cold Forming. Powder Metallurgy. 2nd ed. USA: American Society of Metals;1998

[287] Dupont J, Fonseca G, Fichtner P, Teixeira S. Transition-metal nanoparticles in imidazolium ionic liquids: Recycable catalysts for biphasic hydrogenation reactions. Journal of the American Chemical Society. 2002;**124**:4228

[288] Zhou Y, Antonietti MJ. Synthesis of very small TiO_2 nanocrystals in a room temperature ionic liquid and their self-assembly toward mesoporous spherical aggregates. Journal of the American Chemical Society. 2003;**125**:49

[289] Nakashima T, Kimizuka N. Interfacial synthesis of hollow TiO2 microspheres in ionic liquids. Journal of the American Chemical Society. 2003;**125**:6386

[290] Zhu YJ, Wang WW, Qi RJ, Hu XL. Microwave-assisted synthesis of single-crystalline tellurium nanorods and nanowires in ionic liquids. Angewandte Chemie International Edition. 2004;**43**:1410

[291] Zhou Y, Schattka JH, Antonietti M. Room-temperature ionic liquids as template to monolithic mesoporous silica with wormlike pores via a sol-gel nano-casting technique. Nano Letters. 2004;**4**(3):477–481

[292] Zhou Y, Antonietti M. A Series of highly ordered, super-micro-porous, lamellar silicas prepared by nano casting with ionic liquids. Chemical Materials. 2004;**16**(3):544–550

[293] Itoh H, Naka K, Chujo Y. Synthesis of gold nanoparticles modified with ionic liquid based on the imidazolium cation. Journal of the American Chemical Society. 2004;**126**:3026

[294] Zhou Y. Recent advances in ionic liquids for synthesis of inorganic nanomaterials. Current Nanoscience. 2005;**1**:35–42

[295] Wang Y, Yang H. Synthesis of CoPt nanorods in ionic liquids. Journal of the American Chemical Society. 2005;**127**:5316

[296] Dunn PN. X-rays future—a cloudy picture. Solid State Technology. 1994;**37**:49

[297] Dagata JA. Device fabrication by scanned probe oxidation. Science. 1995;**270**:5242

[298] Cerrina F, Marrian C. A path to nanolithography. MRS Bulletin. 1996;**21**(12):1996:56–62

[299] Hong SH, Zhu J, Mirkin CA. Multiple ink nanolithography: Toward a multiple-pen nano-plotter. Science. 1999;**286**:5439

[300] Heiz U, Schneider WD. Cluster at surfaces. Critical Reviews in Solid State and Materials Sciences. 2001;**26**(4):251–290

[301] Zha J, Roggendorf H. In: Brinker CJ, Sherer GW, editors. Sol-Gel Science. New York: Academic Press; 1990.

[302] Piner RD, Mirkin CA. Effect of water on lateral force microscopy in air. Langmuir. 1997;**13**:6864

[303] Li Y, Maynor BW, Liu J. Electrochemical AFM "Dip-Pen" nanolithography. Journal of the American Chemical Society. 2001;**123**:2105

[304] Demers LM, Park SJ, Taton TA, Li Z, Mirkin CA. Orthogonal assembly of nanoparticle building blocks on dip-pen nano lithographically generated templates of DNA. Angewandte Chemie International Edition. English. 2001;**40**(16):3071–3073

[305] Carbrera N, Mott NF. Theory of the oxidation of metals. Reports on Progress in Physics. 1949;**12**:163

[306] Parvez K, Wu Z-S, Li R, Liu X, Graf R, Feng X, Müllen K. Exfoliation of graphite into graphene in aqueous solutions of inorganic salts. Journal of the American Chemical Society. 2014;**136**(16):6083–6091

[307] Punith Kumar MK, Shanthini S, Srivastava C. Electrochemical exfoliation of graphite for producing graphene using saccharin. RSC Advances. 2015;**5**:53865–53869

[308] Chen C-H, Yang S-W, Chuang M-C, Woon W-Y, and Su C-Y. Towards the continuous production of high crystallinity graphene via electrochemical exfoliation with molecular in situ encapsulation. Nanoscale. 2015;**7**:15362–15373

[309] Huang X, Li S, Qi Z, Zhang W, Ye W, Fang Y. Low defect concentration few-layer graphene using a two-step electrochemical exfoliation. Nanotechnology, 2015;**26**. Number 105602

[310] Hofmann M, Chiang W-Y, Nguyn T D, Hsieh Y-P. Controlling the properties of graphene produced by electrochemical exfoliation. Nanotechnology. 2015;**26**:335607 (6pp)

[311] Hamra AAB, Lim HN, Chee WK, Huang NM. Electro-exfoliating graphene from graphite for direct fabrication of supercapacitor. Applied Surface Sciences. 2016; **360**:213–223

[312] Shinde DB, Brenker J, EastonCD, Tabor RF, Neild A, Majumder M. Shear assisted electrochemical exfoliation of graphite to graphene. Langmuir. 2016;**32**(14):3552–3559

[313] Paredes JI, Munuera JM, Villar-Rodil S, Guardia L, Ayán-Varela M, Pagán A, Aznar-Cervantes SD, Cenis JL, Martínez-Alonso A, Tascón JMD. Impact of covalent functionalization on the aqueous processability, catalytic activity, and biocompatibility of chemically exfoliated MoS2 nanosheets. ACS Applied Materials and Interfaces. 2016;**8**(41):27974–27986

[314] Kundu A, Nandi S, Das P, Nandi AK. Facile and green approach to prepare fluorescent carbon dots: Emergent nanomaterial for cell imaging and detection of vitamin B2.Journal of Colloid and Interface Science. 2016;**468**:276–283

[315] Munaiah Y, Ragupathy P, Vijayamohanan K. Pillai VK. Single-step synthesis of halogenated graphene through electrochemical exfoliation and its utilization as electrodes for Zn-Br redox flow battery. Journal of the Electrochemical Society. 2016;**163**(14):A2899-A2910

[316] Li-Shang L, Westwood A, Brydson R. Graphene Synthesis via Electrochemical Exfoliation of Graphite Nano Platelets in Aqueous Sulphuric Acid. Available from: http://eprints.whiterose.ac.uk/102051/3/Westwood GRAPHENE% 20SYNTHESIS%20 VIA%20ELECTROCHEMICAL%20.pdf

[317] Wazir AH, Kundi IW. Synthesis of graphene nano sheets by the rapid reduction of electrochemically exfoliated graphene oxide induced by microwaves. Journal of the Chemical Society of Pakistan. 2016;**38**(01):11–16

[318] Ambrosi A, Sofer Z, Luxa J, Pumera M. Exfoliation of layered topological insulators Bi2Se3 and Bi2Te3 via electrochemistry. ACS Nano. 2016;**10**(12):11442–11448

[319] Nguyen TP, Sohn W, Oh JH, Jang HW, Kim SY. Size-dependent properties of two-dimensional MoS2 and WS2. The Journal of Physical Chemistry C. 2016;**120**(18):10078–10085

[320] Lu J, Liu H, Tok ES, Sow C-H. Interactions between lasers and two-dimensional transition metal dichalcogenides. Chemical Society Reviews. 2016;**45**:2494–2515

[321] Dong H, Chen D, Wang K, Zhang R. High-yield preparation and electrochemical properties of few-layer MoS2 nanosheets by exfoliating natural molybdenite powders directly via a coupled ultrasonication milling process. Nanoscale Research Letters. 2016;**11**:409

[322] Voiry D, Mohite A, Chhowalla M. Phase engineering of transition metal dichalcogenides. Chemical Society Reviews. 2015;**44**:2702–2712

[323] Voiry D, Yamaguchi H, Li JW, Silva R, Alves DCB, Fujita T, Chen MW, Asefa T, Shenoy VB, Eda G, Chhowalla M. Enhanced catalytic activity in strained chemically exfoliated WS2 nanosheets for hydrogen evolution. Nature Materials. 2013;**12**:850–855

[324] Kappera R, Voiry D, Yalcin SE, BranchB, Gupta G, Mohite AD, Chhowalla M. Phase-engineered low-resistance contacts for ultrathin MoS2 transistors. Nature Materials. 2014;**13**:1128–1134

[325] Acerce M, Voiry D, Chhowalla M. Metallic 1T phase MoS2 nanosheets as super-capacitor electrode materials. Nature Nanotechnology. 2015;**10**:313–318

[326] Kauzlarich SM. Chemistry, structure, and bonding of Zintl phases and ions. New York: Wiley-VCH; 1996

[327] Vogg G, Brandt MS, Stutzmann M. Polygermyne; A prototype system for layered germanium polymers. Advanced Materials. 2000;**12**:1278–1281

[328] Hajnal Z, Vogg G, Meyer LJ-P, Szücs B, Brandt MS, Frauenheim T. Band structure and optical properties of germanium sheet polymers. Physical Review B. 2001;**64**:033311

[329] Brandt MS, Vogg G, Stutzmann M. Silicon-and germanium-based sheet polymers and Zintl phases. In: Jutzi P, Schubert U, editors. Silicon Chemistry. New York: Wiley-VCH; 2003

[330] Kauzlarich, S M, Shawna R. Browna, SR, and G. Jeffrey Snyder, GJ. Zintl phases for thermoelectric devices. Dalton Trans., 2007, 2099–2107 © The Royal Society of Chemistry 2007

[331] Koski KJ, Cha JJ, Reed BW, Wessells CD, Kong D, Cui Y. High density chemical intercalation of zero-valent copper into Bi2Se3 nanoribbons. Journal of the American Chemical Society. 2012;**134**:7584–7587

[332] Koski KJ, Cui Y. The new skinny in two-dimensional nanomaterials. ACS Nano. 2013;**7**(5):3739–3743

[333] Dang WH, Peng HL, Li H, Wang P, Liu ZF. Epitaxial heterostructures of ultrathin topological insulator nanoplate and graphene. Nano Letters. 2010;**10**:2870–2876

[334] Shi YM, Zhou W, Lu AY, Fang WJ, Lee YH, Hsu AL, Kim SM, Kim KK, Yang HY, Li L-J, et al. van der Waals epitaxy of MoS2 layers using graphene as growth templates. Nano Letters. 2012;**12**:2784–2791

[335] Lin M, Wu D, Zhou Y, Huang W, Jiang W, Zheng WS, Zhao SL, Jin CH, Guo YF, Peng HL, et al. Controlled growth of atomically thin In2Se3 flakes by van der Waals epitaxy. Journal of the American Chemical Society. 2013;**135**:13274–13277

[336] Lin YC, Lu N, Perea-Lopez N, Li J, Lin Z, Peng X, Lee CH, Sun C, Calderin L, Browning PN, et al. Direct synthesis of van der Waals solids. ACS Nano. 2014;**8**:3715–3723

[337] Huang CM, Wu SF, Sanchez AM, Peters JJP, BeanlandR, Ross JS, Rivera P, Yao W, Cobden DH, Xu XD. Lateral heterojunctions within monolayer MoSe2-WSe2 semiconductors. Nature Materials. 2014;**13**:1096–1101

[338] Gong YJ, Lin JH, Wang XL, Shi G, Lei SD, Lin Z, Zou XL, Ye GL, Vajtai R, Yakobson BI, et al. Vertical and in-plane heterostructures from WS2/MoS2 monolayers. Nature Materials. 2014;**13**:1135–1142

[339] Li M-Y, Shi Y, Cheng C-C, Lu L-S, Lin Y-C, Tang H-L, Tsai M-L, Chu C-W, Wei K-H, He J-H, et al. Epitaxial growth of a monolayer WSe2/MoS2 lateral p-n junction with an atomically sharp interface. Science. 2015;**349**:524–528

[340] Huang X, Li H, Li S, Wu S, Boey F, Ma J, Zhang H. Synthesis of gold square-like plates from ultrathin gold square sheets: The evolution of structure phase and shape. Angewandte Chemie International Edition. 2011;**50**:12245–12248

[341] Huang X, Li S, Huang Y, Wu S, Zhou X, Li S, Gan CL, Boey F, Mirkin CA, Zhang H. Synthesis of hexagonal close-packed gold nanostructures. Nature Communications. 2011;**2**:292

Molybdenum Disulfide-Based Photocatalysis:Bulk-to-Single Layer Structure and Related Photomechansim for Environmental Applications

Surya Veerendra Prabhakar Vattikuti and
Chan Byon

Additional information is available at the end of the chapter

http://dx.doi.org/10.5772/67825

Abstract

Bulk-to-single layer molybdenum disulfide (MoS_2) is widely used as a robust candidate for photodegradation of organic pollutants, hydrogen production, and CO_2 reduction. This material features active edge sites and narrow band gap features, which are useful for generating reactive species in aqueous suspensions. However, the high-charge carrier recombination, photocorrosion, unstable sulfide state, and formation of Mo-S-O links during photocatalytic reactions limit its applicability. Thus, research has focused on improving the performance of MoS_2 by tailoring its bulk-to-single layer structure and combining it with other semiconductor materials to improve the photocatalytic performance. Different strategies have been successfully applied to enhance the photocatalytic activity of MoS_2, including tailoring of the surface morphology, formation of heterojunctions with other semiconductors, doping, and modification with excess sulfur or carbon nanostructures. This review describes the influence of starting precursors, sulfur sources, and synthetic methods to obtain heterostructured morphologies and study their impact on the photocatalytic efficiency. Finally, the relevance of crystal facets and defects in photocatalysis is outlined. Future applications of MoS_2 with tailoring and tuning physicochemical properties are highlighted.

Keywords: layered materials, molybdenum disulfide, photocatalyst, pollutants, nanomaterials

1. Introduction

The environment continues to become more polluted due to industrialization. However, traditional chemical methods that deal with environmental pollution have been unable to meet the requirements of saving energy and environmental protection. Environmental problems induced by toxic and organic pollutants that are hard to degrade (such as halides, dioxins, pesticides, and dyes) are important issues for human well-being and development. The sun is an abundant source of energy and sustains life on earth; and photocatalysis has been studied extensively for waste water recycling in various industries to remove organic pollutants using photocatalysts. The use of photocatalysts for waste water treatment is promising for meeting increasing water recycling demands without compromising the quality of our environment.

Although photocatalysis is successful in laboratory studies, there are some technological problems that hamper the extensive commercial applicability of this technique. Optimum utilization and commercial viability of photocatalysis could possibly be achieved by replacing the expensive and technically complex artificial light sources with low cost and renewable energy from sunlight as a natural excitation source. In recent decades, photocatalysts with high activities based on transition metal dichalcogenides (TMDs) have been used extensively for environmental applications such as air purification, water disinfection, hazardous waste remediation, and water purification. TMDs materials such as MQ_2 (M = Mo, Nb, Re, V, W and Q = S, Se) have gained much attention due to their unique properties for a wide range of applications, since the nanoscale form of these inorganic materials was discovered.

Tenne et al. [1] discovered spherical fullerene-like nanoparticles of molybdenum disulfide (MoS_2) and tungsten disulphide (WS_2) nanotubes in 1992. Since then, research on these materials and their tribological properties has intensified. One-dimensional (1D) or two-dimensional (2D) structures of TMDs materials have remarkable properties, such as chemical inertness, anisotropy, photocorrosion resistance, electronic properties, tribological properties, and photocatalytic behavior [2–8]. They also have good catalytic properties and resistance to sulfur poisoning [9]. MoS_2 and WS_2 are the most prominent family of TMDs materials and are most commonly used in layered structured forms. They have a layered, close-packed hexagonal crystal structure confined in vertically stacked monolayers that bond together by weak van der Waals forces. However, the structural and morphological features of these materials widely depend on the synthesis strategies. The starting materials, surfactants, sulfur sources, and solvents play crucial roles in the structural and morphological features.

The different morphologies of MoS_2 and WS_2 include nanospheres, few-layered nanosheets, nanofibers, nanotubes, and nanorods. These forms have attracted extensive interest owing to their intriguing physical properties and prospects for applications in nanoelectronics, electrochemistry, catalysis, and lubrication. However, the performance characteristic of these materials depends on the particle size, shape, and structure. These materials are synthesized by various approaches, including chemical vapor deposition, photothermal, sonochemical, solvothermal, hydrothermal, and two-step electrochemical synthesis methods.

2. Synthetic methods

2.1. Synthesis of MoS_2 nanomaterials

In the reaction stage, the size and shape control during the synthesis of MoS_2 nanomaterials are crucial for obtaining well-defined materials with specific properties. The morphology and

http://dx.doi.org/10.5772/67825

size of MoS_2 play a significant role in catalysis, sensors, and other applications. Shape control is also important for applications in photochemistry and fuel cell catalysis. Other factors include monodispersity, avoiding agglomeration, and surface functionalization.

The wide variety of synthetic methods for obtaining MoS_2 nanomaterials can be divided into three main groups: (i) mechanical methods (e.g., grinding, ultrasonic cracking, or milling), (ii) liquid phase methods (e.g., sol-gel, hydrothermal, or wet chemical methods), and (iii) gas phase techniques (e.g., chemical vapor deposition or laser ablation deposition). Liquid phase techniques are used most often due to their simplicity, low cost, and the wide variety of different sizes, shapes, and surface functionalities that can be obtained. Also, the size and agglomeration are effectively controlled by functionalization of the nanomaterial surface with surfactants. In some cases, these surfactants are used to control the shape and to promote growth in a specific direction by selective binding to some crystalline faces. **Figure 1** shows the SEM and TEM images of MoS_2 nanosheets synthesized by solvothermal approach [6]. These MoS_2 nanosheets were obtained as a few lamellar layers using thiourea as a sulfur source.

The combination of size- and shape-dependent physical properties along with their simple fabrication and processing techniques make MoS_2 nanomaterials, a promising candidate for a wide range of applications. The properties of the individual particles and their mutual interactions determine important features of nanomaterial systems. For example, optical properties are highly dependent on the size, shape, and crystallinity of the MoS_2 nanomaterials. However, controlled synthesis with a narrow size distribution and uniform shape remains an important issue in photocatalytic applications.

2.2. Synthesis of MoS_2 thin films

Chemical vapor deposition (CVD) is one of the most popular methods for fabricating thin films of few-layered MoS_2 nanosheets [10, 11]. In addition, impurity-assisted methods are

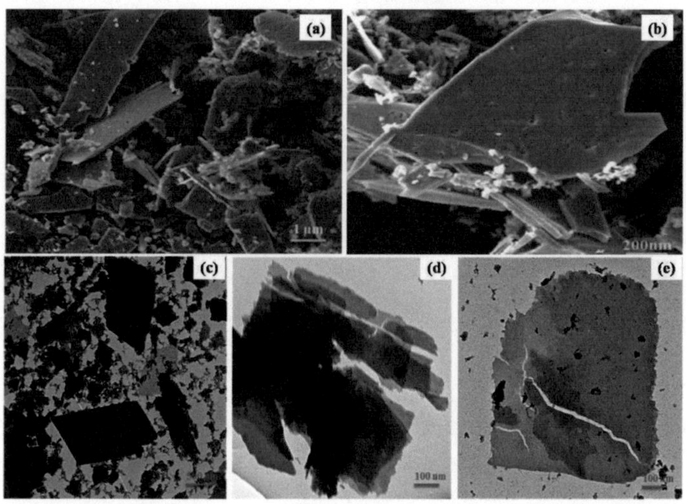

Figure 1. SEM (a,b) and TEM (c-e) images of solvothermally- synthesized MoS_2 nanosheets [6].

presently gaining much attention for increasing the grain size and decreasing the growth temperature. The growth conditions of single-layer MoS_2 in a CVD system depend on the nature of the substrate and the surface treatments used. Kinetic effects and on-off stoichiometric growth conditions can be used to produce different shapes in MoS_2 nanosheets, including star and dendrite shapes. The controllability and reproducibility of shape control are still being improved. In photocatalyst applications, the crystallinity and morphology strongly affect the device performance. Heterostructured MoS_2-based layered composite structures have recently been a focus in fundamental semiconductor technology.

The simplest way to develop a monolayer MoS_2 analog to graphene is the Scotch-tape exfoliation method from bulk MoS_2 along the direction of van der Waals interaction. Repeated exfoliation of bulk MoS_2 decreases the number of MoS_2 layers, eventually producing few- to single-layer MoS_2. However, this method is difficult for large-scale production due to the poor size controllability. To synthesize MoS_2 thin films on a large scale, three important techniques are mostly applied by researchers: thermal vapor sulfurization (TVS), dip coating, and CVD. CVD is the most prominent technique and is used with gasified Mo- and S-containing species that react and deposit on the surface of a substrate. The major difference between the CVD and TVS approaches is the state of Mo source: a solid Mo source is used in TVS, whereas a gasified Mo source is used in a CVD system. CVD systems can be classified according to energy source as (i) hot-walled thermal CVD, (ii) plasma CVD, and (iii) metal organic CVD, among others.

Hot-walled thermal CVD systems are widely used to synthesize MoS_2 nanosheets. Both sources of Mo and sulfur are gasified and transported to the surface of a substrate, adsorbed, and decomposed into reactive atoms, which results in the formation of covalent bonds and the growth of MoS_2 nanosheets. Byproducts and unreacted species are removed by the carrier gas. MoO_3, $Mo(CO)_6$, and $MoCl_5$ are widely used as Mo sources, while elemental sulfur and H_2S are used as sulfur sources, and N_2 or Ar is used as a carrier gas. $Mo(CO)_6$ has recently attracted attention and was successfully applied for large-scale production of MoS_2 thin film.

CVD systems can also be classified according to the design and features into four categories: (i) single-zone, (ii) two-zone, (iii) three-zone, and (iv) two-flow CVD systems. Two-flow CVD systems are the most sophisticated and are designed with three-zone CVD systems and a one-zone CVD system. The main features of a two flow-system are (i) independent control of the temperatures of the Mo and sulfur sources and the flow rates of each carrier gas for both sources, and (ii) the Mo source can be completely adjusted or stopped during when ramping and lowering the temperatures. Therefore, the growth rate of thin film can be controlled easily with a two-flow CVD system compared to conventional ones. However, careful attention is needed to obtain high-quality MoS_2 nanosheets.

Generally, the growth rate depends on various parameters, including the nature of the source materials, the temperature ranges of the sources and substrates, the system pressure, the vacuum levels, the type of substrates, and the type of carrier gas. However, unified conditions for the synthesis of MoS_2 nanosheets have not been developed due to the lack of a common practice and a unique CVD system design. The temperature ranges of the Mo and sulfur sources are 500–800°C and 130–300°C, respectively. The reaction temperature range of CVD systems is 650–1000°C, and the range of low rate of the carrier gas is 1–800 sccm. The duration ranges from 30 s to 60 min. Some researchers synthesize MoS_2 nanosheets under reduced pressure or at atmospheric pressure.

Generally, the different shapes of synthesized MoS_2 nanosheets reflect the crystal structure. Hexagonal nanosheets consist of S-zig-zag and Mo-zig-zag termination sides and form triangle shapes that can be grown under different Mo and S source ratios. If either a small excess of the Mo or sulfur source is used, the shape becomes a truncated triangle instead of a perfect triangle. Initially, the MoS_2 nanosheets grow as a hexagonal shape, which changes to a triangle over time. When the Mo and sulfur sources reach a critical point, the MoS_2 nanosheets become star-shaped. For instance, a large excess amount of the sulfur source can facilitate the growth of only Mo-zig-zag termination sides in hexagonal MoS_2 nanosheets with suppression of S-zig-zag termination side growth, which results in the formation of a star shape.

Dendritic MoS_2 nanosheets form in the case of large flow rates of Mo and sulfur sources. These large flow rates create a thinner boundary layer, which results in the formation of dendritic shapes. Nonoptimal growth conditions cause round-shaped MoS_2 nanosheets with poor crystallinity. Surface treatment of a suitable substrate is very important for obtaining highly crystalline MoS_2 nanosheets. A hydrophilic substrate can be used to obtain few-layer MoS_2 nanosheets, whereas monolayer MoS_2 nanosheets can be obtained by using a superhydrophobic substrate. However, the exact mechanism of forming a monolayer on a superhydrophobic substrate is still not clear.

Park et al. [12] fabricated a thin film transistor (TFT) based on CVD-grown single-layer MoS_2, and the photoresponsive current and voltage characteristics of the TFT were measured with varying intensities of incident light. The photocurrent and mobility increased with increasing light intensity due to the contribution of photoinduced charge carriers from the valance band and trap states of the single-layered MoS_2. An exfoliated single-layer MoS_2-based TFT by Lin et al. exhibited higher mobility than the one based on CVD-grown MoS_2 [13]. However, the main advantage of CVD is the relatively large area of samples with homogenous qualities [14]. The CVD method has been used to grow MoS_2 directly on different dielectric substrates [15]. MoS_2 with photocatalytic properties have become an interesting candidate for the photodegradation of organic dye, hydrogen evolution, and CO_2 reduction, with the advantages of chemical and photostability.

3. Photocatalytic properties of MoS_2

The photocatalysis process involves the conversion of solar energy into chemical energy. The main goal for researchers is maximum utilization of the solar energy to enable the practical use of photocatalysts. However, photocatalytic ability has still been limited due to the fast recombination effect of electron-hole pairs and an insufficient absorption coefficient. Therefore, enhancing the efficiency of photocatalysts under visible light still remains a challenge for practical applications.

Many semiconductor oxides, sulfides, and nitrides have been used as photocatalysts for various applications. However, most sulfides and nitrides have lower band gaps, which limit usage due to stability issues of these materials in an aqueous medium. This issue is an important one to solve to promote the use of these materials. The band gap of the oxides is higher than that of sulfides, and the absorption edge is only in the UV region. This limits the usage in the solar spectrum, although the material has good stability in an aqueous medium. The main characteristics of an ideal photocatalyst are maximum absorption in the visible solar spectrum, favorable band edges for promoting reactions, environmental friendliness, low cost, good stability, and reusability [16].

The band gap value of MoS_2 (1.9 eV) photoelectrodes and their considerably lower valence band edge than the water oxidation potential are favorable for water splitting through photo-electro-chemical techniques. The theoretical photocurrent density of chemically exfoliated MoS_2 is ~17.6 mA/cm^2 at 0.0 V vs. RHE under solar irradiation [17], and the solar energy conversion efficiency is ~18.7% for an ideal PEC cell. However, the photocatalytic activity of MoS_2 is limited by the factors mentioned thus far, which lead to low efficiencies and require larger potential to promote the photo-assisted water oxidation process [18]. Many researchers have tried to overcome these drawbacks of MoS_2 by decreasing the recombination rate through forming a composite or heterogeneous structure, as well as enhancing the conductivity by doping with metal and promoting the charge carrier transferability [19–22].

Apart from water-splitting applications, the photocatalytic performance of MoS_2 can be used for the degradation of organic compounds in waste water treatment applications. Vattikuti et al. [23] reported the mechanism of the degradation of rhodamine B (RhB) dye. The photosensitization of the RhB dye first takes place when charge transfer occurs from the valence band of the dye to the conduction band of the photocatalyst. This is followed by the initial photocatalytic reaction, where MoS_2 generates electron-hole pairs under photoirradiation. The electrons transfer from the valence band of MoS_2 to the conduction band and settle down holes within the valence band. The photoinduced electrons of MoS_2 produce intermediate superoxide radicals (O_2^-) by responding with chemisorbed oxygen on the photocatalyst surface and oxygen in the aqueous solution. The O_2^- radicals react with dissociated water (H^+) to form $\cdot HO_2$ and H_2O_2. In addition, hydroxyl groups (OH^-) are formed on the catalyst surface by the reaction with photoinduced holes (h^+) by absorbed water (OH_2). Thus, these generated radicals along with intermediate species react with RhB dye and degrade it into nontoxic organic compounds as follows:

$$RhB + h\upsilon \rightarrow RhB^{\cdot} \quad (1)$$

$$RhB^{\cdot} + MoS_2 \rightarrow RhB^{\cdot} + MoS_2\,(e_{CB}^{-}) \quad (2)$$

$$MoS_2 + h\upsilon \rightarrow MoS_2\,(e_{CB}^{-} + h_{VB}^{+}) \quad (3)$$

$$H_2O \rightarrow H^+ + OH^- \quad (4)$$

$$e_{CB}^{-} + O_2 \rightarrow O_2^{\bullet -} \quad (5)$$

$$O_2^{\bullet -} + H^+_{aq} \rightarrow HO_2^{\cdot} \quad (6)$$

$$HO_2^{\cdot} + HO_2^{\cdot} \leftrightarrow H_2O_2 + O_2 \quad (7)$$

$$h_{VB}^{+} + OH^-_{aq} \rightarrow OH^{-} \quad (8)$$

$$RhB^{\cdot}/RhB^{\cdot +} + (O_2^{\bullet -}, HO_2^{\cdot}, H_2O_2, OH^{-}) \rightarrow \text{Degradation products....} \quad (9)$$

http://dx.doi.org/10.5772/67825

MoS_2 is also a good photocatalyst for photocatalytic oxidative desulfurization [24]. Sulfur compounds in fuel convert into SOx, which causes air pollution and acid rain. Generally, the hydrodesulfurization (HDS) process is widely employed commercially to remove sulfur species at high temperature (350°C) and pressure (7 MPa). Recently, photooxidative desulfurization has been popular because it is economical and has high efficiency [25, 26], which can provide cleaner and more efficient removal of sulfur species from petroleum fuel oils [27]. Lia et al. reported that CeO_2/MoS_2 and attapulgite showed excellent electron transfer within the composite and favor the desulfurization process under solar irradiation [24]. MoS_2-assisted nanocomposite systems have led to a new era in research and show promise as a high-activity and low-cost photocatalyst for applications such as deep desulfurization. Thurston et al. [28] and Wilcoxon et al. [29] reported on MoS_2 nanoparticles with diameters at 3–4.5 nm as a catalyst for the degradation of phenol, 4-chlorophenol, and pentachlorophenol under visible light irradiation.

4. Literature reviews on MoS_2 photocatalytic mechanism

We emphasize three different forms of MoS_2 that have been studied. Ongoing research on MoS_2 nanoparticles as a photocatalyst is addressed first, followed by studies associated with MoS_2 composites. This section concludes with a discussion on thin-coated MoS_2.

4.1. Unary MoS_2 photocatalyst

A high aspect ratio plays a key role in the photocatalytic activity of materials, and researchers have concentrated on reducing the size of the photocatalyst and improving the photocatalytic activities of these materials by making nanoscale MoS_2. Many approaches have been used to synthesize nanocrystalline MoS_2 with different morphologies, including ultrasonic cracking [30], hydrothermal methods [31–33], chemical synthesis [34], combustion methods [35–38], wet chemical methods, and coprecipitation methods [23, 38–40].

Vattikuti et al. synthesized MoS_2 multiwall nanotubes (MWNTs) by a wet chemical method assisted by H_2O_2 solvent as a growth promotor [41]. The photocatalytic performance of MoS_2 MWNTs was applied to the degradation of RhB. The MoS_2 MWNTs exhibited excellent photocatalytic performance compared to pure MoS_2. The higher photocatalytic activity of MoS_2 MWNTs was ascribed to the large number of active sites with a high specific surface area. The performance of the optimal amount of 0.5 wt% MoS_2 MWNTs was attributed to the higher transfer of electrons and holes during the photoreaction, which effectively suppressed the recombination of the electron-hole pairs and enhanced the degradation efficiency.

Zhou et al. [42] hydrothermally synthesized porous MoS_2 without any sacrificial template using sodium molybdate and thioacetamide as Mo and S sources. Porous MoS_2 showed 89.2% degradation efficiency of MB under 150 min of visible light irradiation. MB photodegradation in the presence of porous MoS_2 was obtained with a pseudo-first-order kinetic reaction rate of 0.01484 min^{-1}. Polycrystalline porous MoS_2 shows attractive photocatalytic activities that are ascribed to the active edge sites. Sheng et al. [43] synthesized flower-like MoS_2 spheres via the hydrothermal method and studied the effects of excess sulfur source on the flower-like MoS_2 structure. To obtain the flower-like MoS_2 spheres, MoO_3, and potassium thiocyanate (KSCN) were used as Mo and S sources with different S/Mo ratios. The optimal S/Mo ratio of 2.75 resulted in the highest degradation rate of MB with a degradation rate of 0.03833 min^{-1} under 90 min of visible light irradiation. The increased photocatalytic performance was ascribed to the increased exposed area of the [43] facets with the optimal S/Mo ratio in the hydrothermal synthesis environment. The sheet thickness of the MoS_2 spheres increased with the S/Mo ratio and enhanced the photocatalytic activity.

Liu et al. [44] produced MoS_2 nanosheets by a hydrothermal method with H_2SiO_3 (silicic acid) hydrogel containing ammonium molybdate hydrate and thiourea precursors. MoS_2 nanosheets were obtained by removing the H_2SiO_3. These MoS_2 nanosheets have a high specific surface area (S_{BET}) of 37.8 m^2g^{-1} and present notable absorption of MO under visible light rather than ultraviolet light in 70 min of irradiation. Different shapes of MoS_2 nanosheets were obtained by varying the concentration of silicic acid with MoS_2 molar ratios of 2.5 and 0.8, such as leaf-shaped and flower-shaped MoS_2 nanosheets. These provide steric hindrance for MoS_2 nanosheet growth. The amount of hydroxyl radicals was highest at pH 2 and decreased when increasing to pH 9. The OH group plays a major role in MO photodegradation in the catalytic system. The reaction time, initial concentration, catalyst dosage, and local structures are also key factors that affect the photocatalytic performance of the materials.

4.2. Binary MoS_2 photocatalysts

This section reviews the effect of heterogeneous structures or composite forms of MoS_2 on the photocatalytic properties. Combining MoS_2 with metals or nonmetals and semiconductor materials is a common practice for enhancing photocatalytic performance by facilitating and promoting efficient charge transfer at the interfaces. Similar attempts have been made for other classes of materials to improve photocatalytic activity, including Fe_2O_3, TiO_2, and ZnO.

Thurston et al. [28] demonstrated that MoS_2 nanoparticles with diameter of 8–10 nm could not photodegrade phenol under visible light due to poor light absorption. Hence, they sensitized TiO_2 nanoparticles with MoS_2 nanoparticles, which enabled photodegradation under visible light irradiation. This composite structure showed a blue shift in absorbance due to quantum confinement of the charge carriers [28, 45]. We recently reported improved photocatalytic performance of MoS_2 nanosheets decorated with mesoporous SnO_2 nanospheres by a facile two-step method [46]. We also observed the photocatalytic effect in the degradation of RhB with less than 50 min of UV light irradiation. The supported mesoporous SnO_2 nanoparticles significantly suppressed the recombination of electron-hole pairs compared to pure MoS_2 photocatalyst material. The improved photocatalytic performance of the MoS_2/SnO_2 composite was explained by two mechanisms: (i) the absorption ability of the MoS_2 nanosheets with active edges and (ii) enhanced electron transfer from SnO_2 to the MoS_2 nanosheets. This heterostructured composite facilitated effective electron transfer

http://dx.doi.org/10.5772/67825

from the CB of SnO_2 to the MoS_2 nanosheets and suppressed the recombination effect. Therefore, the SnO_2-decorated MoS_2 nanocomposite showed better photocatalytic performance than pure MoS_2. Photocorrosion is the main reason for the lower photocatalytic activity of the pure MoS_2.

Pourabbas et al. [47] synthesized a hybrid MoS_2/TiO_2 composite using a modified hydrothermal method. The changes from the normal hydrothermal method included using sodium lauryl sulfate as a surface-active agent with 1-octanol as a cosurfactant and varying reaction temperature. The hybrid composite was used as a photocatalyst for the photo-oxidative removal of phenol. The composite showed enhanced photocatalytic performance in the phenol degradation under both UV (70 min) and visible light (24 min) compared to pure TiO_2 and MoS_2. The complete mineralization of phenol during the photo-oxidation reaction in 145 min of UV irradiation was indicated by HPLC chromatograms. Zhou et al. [48] and Bai et al. [49] did similar work and evaluated the photocatalytic performance of the MoS_2/TiO_2 composite for photodegradation of MB under visible light irradiation.

MoS_2-coated TiO_2 nanobelt composites showed excellent photocatalyst properties for RhB degradation under 33 min of visible light irradiation. The matched energies of the TiO_2@ MoS_2 composite are favorable for the charge transfer and suppress the recombination of electron-hole pairs. The photocatalytic hydrogen production was also enhanced. Liu et al. [50] synthesized a composite of TiO_2 nanobelts decorated with MoS_2 nanoparticles using a two-step hydrothermal method. The photocatalytic degradation of the TiO_2/MoS_2 composite was evaluated with RhB under 90 min of visible light irradiation. The sample with 40 wt% MoS_2 nanoparticles decorated on TiO_2 nanobelts showed the best photocatalytic performance, which was attributed to the prevented recombination of photoinduced electron-hole pairs. This sample showed a high photocatalytic reaction rate constant that is about 4.78 times that of pure TiO_2.

Cao et al. [51] synthesized MoS_2/TiO_2 hybrid composites by a two-step hydrothermal route. The MoS_2/TiO_2 hybrid composite showed excellent photocatalytic performance in the degradation of RhB in 100 min of visible-light irradiation in comparison to pure forms. The improvement in photocatalytic activity of the composite was mainly ascribed to the properly matching CB and VB energy levels and the enhanced separation efficiency of photoinduced electron-hole pairs at interfacial contacts of the composite. Wang et al. [52, 53] reported the *in situ* deposition of Ag_3PO_4 on graphene-like MoS_2 nanosheets via a wet chemical route. The goal was to improve the photocatalytic performance for the degradation of RhB in 20 min of visible light irradiation (>400 nm). The improved photocatalytic performance of the heterostructure of Ag_3PO_4/MoS_2 composite is ascribed to the efficient separation of photoinduced electron-hole pairs within the photocatalyst.

Ding et al. [54] synthesized a MoS_2-GO hydrogel composite using a hydrothermal method for MB degradation under 60 min of solar light irradiation. This composite showed enhanced photocatalytic performance in the degradation of MB with a maximum degradation rate of 99% for 60 min under solar light irradiation. The improvement was attributed to the increased light absorption and suppressed recombination effect of semiconductor photocatalysis.

Zhang et al. [55] synthesized MoS_2/rGO photocatalyst for the fluorescence detection of glutathione in a ·OH radical elimination system based on the reducing ability of glutathione under visible light irradiation. The MoS_2/rGO composite efficiently generated ·OH radicals and reduced ·OH radicals by the absorption of glutathione under visible light, which is reflected by a reduction of the fluorescence intensity due to the elimination of ·OH radicals. This kind of photocatalyst can be effectively implemented for the identification of glutathione in commercial drugs and human serum.

Wang et al. [56] synthesized $MoS_2/Bi_2O_2CO_3$ composites for RhB photodegradation under 150 min of visible light irradiation by a simple hydrothermal method. The effect of photocatalyst concentration on the photocatalytic efficiency was observed. This composite has more active sites of MoS_2 on $Bi_2O_2CO_3$, which promoted the photocatalytic performance by absorbing and decomposing more RhB pollutant than pure $Bi_2O_2CO_3$. The remarkable enhancement in the photocatalytic activity could be ascribed to the synergistic effect between the MoS_2 and $Bi_2O_2CO_3$ in the heterostructured composite. Li et al. [57] reported $MoS_2/BiVo_4$ hetero-nanoflower composites as an excellent photocatalyst for MB degradation with less than 120 min of sunlight irradiation.

Li et al. [58] successfully synthesized a 2D heterojunction photocatalyst of g-C_3N_4 coupled with MoS_2 nanosheets using a simple impregnation and calcination method. The g-C_3N_4/MoS_2 composite promoted the charge transfer and improved the separation efficiency of photo-induced electron–hole pairs in RhB and MO degradation under 180 min of visible light irradiation. Jo et al. [59] synthesized MoS_2 nanosheets loaded with ZnO-gC_3N_4 ternary photocatalyst for MB photodegradation under 60 min of UV-visible light irradiation. The ternary nanocomposite significantly improved the lifetime of charge carriers and facilitated effective migration and charge separation at the interface.

Zhang et al. [60] synthesized a ternary composite system of TiO_2/MoS_2@zeolite using a facile ultrasonic-hydrothermal synthesis method with $TiCl_4$ as a Ti source and zeolite as a carrier. The photocatalytic performance was investigated for MO degradation for 60 min under xenon long-arc lamps as a visible light source. The photoinduced electrons and holes are collected in the CB of MoS_2 and the VB of TiO_2. The more negative bottom CB energy of MoS_2 and more positive top CB energy of TiO_2 allow the photoinduced electrons in the CB of MoS_2 to reduce the absorbed O_2 into $\cdot O_2^-$. ·OH can be produced easily in the VB of TiO_2, and the $\cdot O_2^-$ and ·OH active species lead to MO degradation. **Figure 2** shows the possible photocatalytic mechanism of both unary and binary photocatalysts.

Hu et al. [62] synthesized MoS_2/Kaolin composites by calcining a MoS_3/kaolin precursor in H_2 under strong acidic conditions. The composite had a specific surface area of 16 m^2g^{-1} and showed a positive photocatalytic effect on MO degradation under 105 min of visible light irradiation. This performance was attributed to the good absorption capacity in the visible light region. The photocatalyst has remarkable stability and can be regenerated and reused via filtration. The deactivating photocatalyst could be reactivated even after photocatalytic reaction at 450°C for 30 min under H_2. The photocatalytic performance of exfoliated MoS_2 was also investigated, and the relationship between the morphology of nano-MoS_2 and the photocatalytic properties was discussed [63]. The photocatalytic performance of this and

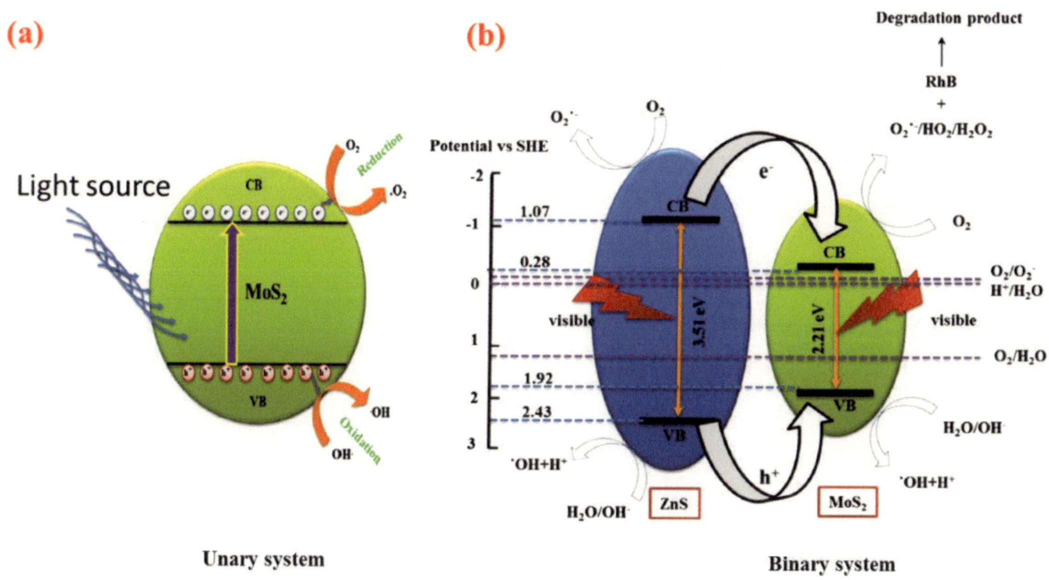

Figure 2. Schematic diagram of photocatalytic mechanism of (a) unary [8] and (b) binary photocatalyst [61].

other heterostructured composites are influenced by the quantity of photocatalyst, initial concentration of pollutant or dye, pH, irradiation time, type of light source, and degradation temperature.

5. Summary

Layered MoS_2 materials have attracted continuously increasing interest and demand, and preparation techniques have been successfully developed. We have a provided a detailed overview of the photocatalytic performance of MoS_2 nanomaterials, three different types of MoS_2 photocatalyst systems were distinguished according to their structural components(single component, heterostructured, and doped MoS_2). There is great interest in preparing various MoS_2 photocatalyst systems by novel strategies, as well as hierarchical MoS_2 structures with special functionalities. Therefore, there is ongoing effort to develop new MoS_2 materials with novel structures and their applications.

The importance of MoS_2 photocatalysts has been highlighted for the degradation of pollutant from contaminated waste water through solar light irradiation. There have been a number of advances in this field, including the development of materials with lower band gap, low cost, and increased stability and reusability. These developments make MoS_2 photocatalyst a promising candidate for further practical advances in the future. However, degradation rates are still generally low, the materials are somewhat unstable over repeated usage, and there is great variability in the reported reduction rate and efficiencies of these systems. It is of great importance that reduction rates be reproduced from one lab to another, and repeatability and reusability are currently some of the significant deficiencies in the field. In future, scientists should focus on material design and the realization of practical applications.

Acknowledgements

This work was conducted under the framework of the National Research Foundation of Korea (NRF) and funded by the Ministry of Science, ICT, and Future Planning (2014 R1A2A2A01007081).

Author details

Surya Veerendra Prabhakar Vattikuti[1]* and Chan Byon[2]

*Address all correspondence to: vsvprabu@gmail.com

1 School of Mechanical Engineering, Yeungnam University, Gyeongsan-si, Gyeongsangbuk-do, Republic of Korea

2 School of Mechanical and Nuclear Engineering, Ulsan National Institute of Science and Technology, Ulsan, Republic of Korea

References

[1] Tenne R, Margulis L, Genut M, Hodes G. Polyhedral and cylindrical structures of tungsten disulphide. Nature. 1992;**360** (6403):444–446.

[2] Youngblood WJ, Lee SA, Maeda K, Mallouk ThE. Visible light water splitting using dyesensitized oxide semiconductors. Accounts of Chemical Research. 2009;**42**(12): 1966–1973.

[3] Rao CNR, Nath M. Inorganic nanotubes. Dalton Transactions. 2003;**1**:1.

[4] Splendiani A, Sun L, Zhang Y, Li T, Kim J, Chim CY, Galli G, Wang F. Emerging photoluminescence in monolayer MoS_2. Nano Letters. 2010;**10**(4):1271.

[5] Visic B, Dominko R, Klanjsek Gunde R, Hauptman N, Skapin SD, Remskar M. Optical properties of exfoliated MoS_2 coaxial nanotubes – Analogues of graphene. Nanoscale Research Letters. 2011;**6**:593.

[6] Prabhakar Vattikuti SV, Byon C, Venkata Reddy C, Venkatesh B, Shim J. Synthesis and structural characterization of MoS_2 nanospheres and nanosheets using solvothermal method. Journal of Materials Science. 2015;**50**:5024–5038.

[7] Prabhakar Vattikuti SV, Byon C. Synthesis and characterization of molybdenum disulfide nanoflowers and nanosheets: Nanotribology. Journal of Nanomaterials. 2015;**2015**, Article ID 710462:1–11.

[8] Prabhakar Vattikuti SV, Byon C, Venkata Reddy Ch. Synthesis of MoS_2 multi-wall nanotubes using wet chemical method with H_2O_2 as growth promoter. Superlattices and Microstructures. 2015;**85**:124–132.

[9] Lukowski MA, Daniel AS, Meng F, Forticaux A, Li L, Jin S. Enhanced hydrogen evolution catalysis from chemically exfoliated metallic MoS_2 nanosheets. Journal of American Chemical Society. 2013;**135**(28):10274–10277.

[10] Han SA, Bhatia R, Kim SW. Synthesis, properties and potential applications of two-dimensional transition metal dichalcogenides. Nano Convergence. 2015;**2**:17.

[11] Li X, Zhu H. Two-dimensional MoS_2: Properties, preparation, and applications. Journal of Materiomics. March 2015;**1**(1):33–44.

[12] Park HJ, Kim MS, Kim J, Joo J. Photo-responsive transistors of CVD grown single-layer MoS_2 and its nanoscale optical characteristics. Current Applied Physics. 2016;**16**:1320–1325.

[13] Lin MW, Kravchenko II, Fowlkes J, Li X, Puretzky AA, Rouleau CM, Geohegan DB, Xiao K. Thickness-dependent charge transport in few-layer MoS2 field-effect transistors. Nanotechnology. 2016;**27**:165203.

[14] Yin Z, Li H, Li H, Jiang L, Shi Y, Sun Y, Lu G, Zhang Q, Chen X, Zhang H. Single-Layer MoS2 Phototransistors. ACS Nano. 2012;**6**:74.

[15] Chen J, Guo Y, Jiang L, Xu Z, Huang L, Xue Y, Geng D, Wu B, Hu W, Yu G, Liu Y. Chemical Vapor Deposition of High-Quality Large-Sized MoS2 Crystals on Silicon Dioxide Substrates. Advanced Materials. 2014;**26**:1348.

[16] Liu G, Niu P, Yin LC, Cheng HM. α-Sulfur crystals as a visible-light-active photocatalyst. Journal of American Chemical Society. 2012;**134**:9070–9073.

[17] Chen Sh, Thind SS, Chen A. Nanostructured materials for water splitting – State of the art and future needs: A mini-review. Electrochemistry Communications. 2016;**63**:10–17.

[18] Yu H, Zhou W, Li G, Jin R. Some strategies in designing highly efficient photocatalysts for degradation of organic pollutants in water. ACS Symposium Series. 2014;**1186**:139–160 ISBN13: 9780841230187eISBN: 9780841230194.

[19] Barber J, Tran PD. From natural to artificial photosynthesis. 2013. Jornal of the royal society interface, 2013; **10**:20120984.

[20] Gao M, Liang J, Zheng Y, Xu Y, Jiang J, Gao Q, Li J, Yu SH. An efficient molybdenum disulfide/cobalt diselenide hybrid catalyst for electrochemical hydrogen generation. Nature Communications. 2015;**6**:5982–5987.

[21] Mohamed RM, McKinney DL, Sigmund WM. Enhanced nanocatalysts. Materials Science and Engineering R 2012;**73**:1–13.

[22] Bai S, Jiang W, Li Zh, Xiong Y. Surface and interface engineering in photocatalysis. ChemNanoMat. 2015;**1**:223–239.

[23] Vattikuti SVP, Byon Ch, Reddy ChV. Synthesis of MoS_2 multi-wall nanotubes using wet chemical method with H_2O_2 as growth promoter. Superlattices and Microstructures. 2015;**85**:124–132.

[24] Lia X, Zhang Z, Yao Ch, Lu X, Zhao XX, Ni Ch. Attapulgite-CeO_2/MoS_2 ternary nanocomposite for photocatalytic oxidative desulfurization. Applied Surface Science. 2016;**364**:589–596.

[25] Yang D, Yang S, Jiang ZY, Yu SN, Zhang JL, Pan FS, Cao XZ, Wang BY, Yang J. Polydimethyl siloxane–graphene nanosheets hybrid membranes with enhanced pervaporative desulfurization performance. Journal of Membrane Science. 2015;**487**:152–161.

[26] Wang R, Wan JB, Li YH, Sun HW. An improvement of MCM-41 supported phosphoric acid catalyst for alkylation desulfurization of fluid catalytic cracking gasoline. Fuel. 2015;**143**:504–511.

[27] Zhu WS, Xu YH, Li HM, Dai BL, Xu H, Wang C, Chao YH, Liu H. Photocatalytic oxidative desulfurization of dibenzothiophene catalyzed by amorphous TiO_2 in ionic liquid. Korean Journal of Chemical Engineering. 2014;**31**:211–217.

[28] Thurston TR, Wilcoxon JP. Photooxidation of organic chemicals catalyzed by nanoscale MoS_2. The Journal of Physical Chemistry B. 1999;**103**:11–17.

[29] Wilcoxon JP. Catalytic photooxidation of pentachlorophenol using semiconductor nanoclusters. The Journal of Physical Chemistry B. 2000;**104**:7334–7343.

[30] Zheng X, Zhu L, Yan A, Bai Ch, Xie Y. Ultrasound-assisted cracking process to prepare MoS_2 nanorods. Ultrasonics Sonochemistry. 2004;**11**:83–88.

[31] Pu FL, Chi Ch, Zakari S, Liew T, Yarmo MA, Huang NM. Preparation of transition metal sulide nanoparticles via hydrothermal route. Sains Malaysiana. 2010;**39**(2):243–248.

[32] Lin H, Chen X, Li H, Yang M, Qi Y. Hydrothermal synthesis and characterization of MoS_2 nanorods. Materials Letters. 2010;**64**:1748–1750.

[33] Chen X, Li H, Wang Sh, Yang M, Qi Y. Biomolecule-assisted hydrothermal synthesis of molybdenum disulfide microspheres with nanorods. Materials Letters. 2012;**66**:22–24.

[34] Tian Y, Zhao J, Fu W, Liu Y, Zhu Y, Wang Z. A facile route to synthesis of MoS_2 nanorods. Materials Letters. 2005;**59**:3452–3455.

[35] Mukasyan AS, Manukyan Kh. Combustion/micropyretic synthesis of atomically thin two-dimensional materials for energy applications. Current Opinion in Chemical Engineering. 2015;**7**:16–22.

[36] Rao CNR, Thomas PJ, Kulkarni GU. Nanocrystals: Synthesis, Properties and Applications. 2007. 95, Springer Berlin Heidelberg. ISBN 978-3-540-68752-8.

[37] Gonzalez-Cortes SL, Xiao T, Rodulfo-Baechler SMA, Green MLH. Impact of the urea–matrix combustion method on the HDS performance of Ni-MoS_2/-Al_2O_3 catalysts. Journal of Molecular Catalysis A: Chemical. 2005;**240**:214–225.

[38] Gonzalez-Cortes SL, Xiao T, Lin T, Green MLH. Influence of double promotion on HDS catalysts prepared by urea-matrix combustion synthesis. Applied Catalysis A: General. 2006;**302**:264–273.

[39] Hu KH, Wang YR, Hu XG, Wo HZ. Preparation and characterization of ball-like MoS_2 nanoparticles. Materials Science and Technology. 2007;**23**:242–246.

[40] Yu H, Liu Y, Brock SL. Synthesis of discrete and dispersible MoS_2 nanocrystals. Inorganic Chemistry. 2008;**47**:1428–1434.

[41] París JRS, Montes V, Boutonnet M, Järås S. Higher alcohol synthesis over nickel-modified alkali-doped molybdenum sulfide catalysts prepared by conventional coprecipitation and coprecipitation in microemulsions. Catalysis Today. 2015;**258**:294–303.

[42] Zhou Zh, Lin Y, Zhang P, Ashalley E, Shafa M, Li H, Wu J, Wang Zh. Hydrothermal fabrication of porous MoS_2 and its visible light photocatalytic properties. Materials Letters. 2014;**131**:122–124.

[43] Sheng B, Liu J, Li Z, Wang M, Zhu K, Qiu J, Wang J. Effects of excess sulfur source on the formation and photocatalytic properties of flower-likeMoS_2 spheres by hydrothermal synthesis. Materials Letters. 2015;**144**:153–156.

[44] Liu W, Hu Q, Mo F, Hu J, Feng Y, Tang H, Yea Sh, Miao H. Photo-catalytic degradation of methyl orange under visible light by MoS_2 nanosheets produced by H_2SiO_3 exfoliation. Journal of Molecular Catalysis A: Chemical. 2014;**395**:322–328.

[45] Wilcoxon JP, Samara GA. Strong quantum-size effects in a layered semiconductor: MoS_2 nanoclusters. Physical Review B. 1995;**51**:7299–7302.

[46] Vattikuti SVP, Byon Ch, Reddy ChV, Ravikumar RVSSN. Improved photocatalytic activity of MoS_2 nanosheets decorated with SnO_2 nanoparticles. RSC Advances. 2015;**5**:86675–86684.

[47] Pourabbas B, Jamshidi B. Preparation of MoS_2 nanoparticles by a modified hydrothermal method and the photo-catalytic activity of MoS_2/TiO_2 hybrids in photo-oxidation of phenol. Chemical Engineering Journal. 2008;**138**:55–62.

[48] Zhou H, Gu T, Yang D, Jiang Zh, Zeng J. Photocatalytic degradation of methylene blue on MoS_2/TiO_2 nanocomposite. Advanced Materials Research. 2011;**197–198**:996–999.

[49] Bai S, Wang L, Chen X, Du J, Xiong Y. Chemically exfoliated metallic MoS_2 nanosheets: A promising supporting co-catalyst for enhancing photocatalytic performance of TiO_2 nanocrystals. Nano Research. 2015;**8**(1):175–183.

[50] Liu H, Lv T, Zhu Ch, Su X, Zhu Zh. Efficient synthesis of MoS_2 nanoparticles modified TiO_2 nanobelts with enhanced visible-light-driven photocatalytic activity. Journal of Molecular Catalysis A: Chemical 2015;**396**:136–142.

[51] Cao L, Wang R, Wang D, Li X, Jia H. MoS_2-hybridized TiO_2 nanosheets with exposed {001} facets to enhance the visible-light photocatalytic activity. Materials Letters 2015;**160**:286–290.

[52] Wang P, Shi P, Hong Y, Zhou X, Yao W. Facile deposition of Ag_3PO_4 on graphene-like MoS_2 nanosheets for highly efficient photocatalysis. Materials Research Bulletin 2015;**62**:24–29.

[53] Wang L, Chai Y, Ren J, Ding J, Liu Q, Dai W. Ag_3PO_4 nanoparticles loaded on 3D flower-like spherical MoS_2: A highly efficient hierarchical heterojunction photocatalyst. Dalton Transactions 2015;**44**:14625–14634.

[54] Ding BY, Zhou Y, Nie W, Chen P. MoS_2–GO nanocomposites synthesized via a hydrothermal hydrogelmethod for solar light photocatalytic degradation of methylene blue. Applied Surface Science 2015;**357**:1606–1612.

[55] Zhang N, Ma W, Han DX, Wang L, Wu T, Niu L. The fluorescence detection of glutathione by · OH radicals' elimination with catalyst of MoS_2/rGO under full spectrum visible light irradiation. Talanta 2015;**144**:551–558.

[56] Wanga Q, Yun G, Bai Y, An N, Lian J, Huang H, Su B. Photodegradation of rhodamine B with $MoS_2/Bi_2O_2CO_3$ composites under UV light irradiation. Applied Surface Science 2014;**313**:537–544.

[57] Li H, Yu K, Lei X, Guo B, Fu H, Zhu Z. Hydrothermal synthesis of novel $MoS_2/BiVO_4$ hetero-nanoflowers with enhanced photocatalytic activity and a mechanism investigation. The Journal of Physical Chemistry C 2015;**119**:22681–22689.

[58] Li J, Liu E, Ma Y, Huc X, Wan J, Sun L, Fan J. Synthesis of MoS_2/g-C_3N_4 nanosheets as 2D heterojunction photocatalysts with enhanced visible light activity. Applied Surface Science 2016;**364**:694–702.

[59] Jo W, Lee YJ, Selvam NCS. Synthesis of MoS_2 nanosheets loaded ZnO-g-C_3N_4 nanocomposites for enhanced photocatalytic applications. Chemical Engineering Journal 2016;**289**:306–318.

[60] Zhang W, Xiao X, Zheng L, Wan C. Fabrication of TiO_2/MoS_2@zeolite photocatalyst and its photocatalytic activity for degradation of methyl orange under visible light. Applied Surface Science 2015;**358**:468–478.

[61] Vattikuti SVP, Byon Ch, Reddy ChV. ZrO_2/MoS_2 heterojunction photocatalysts for efficient photocatalytic degradation of methyl orange. Electronic Materials Letters. 2016;**12**(6):812–823.

[62] Hu KH, Liu Zh, Huang F, Hu XG, Han ChL. Synthesis and photocatalytic properties of nano-MoS_2/kaolin composite. Chemical Engineering Journal 2010;**162**:836–843.

[63] Hu KH, Hu XG. Formation, exfoliation and restacking of MoS_2 nanostructures. Materials Science and Technology 2009;**25**:407–414.

Chapter 10

Advance in Tribology Study of Polyelectrolyte Multilayers

Yanbao Guo and Deguo Wang

Additional information is available at the end of the chapter

http://dx.doi.org/10.5772/67571

Abstract

This review introduced the preparation and structural characterization of polyelectrolyte multilayers in recent years and also summarized the tribology research progress of the polyelectrolyte multilayers, including tribological properties, surface adhesion characteristics, and wear resistance properties. Statistics analysis indicated that nanoparticles-doped polyelectrolyte multilayers present better friction and wear performance than pristine polyelectrolyte multilayers. Furthermore, the in situ growth method resulted in improved structural order of nanoparticles composite molecular deposition film. In situ nanoparticles not only reduced the molecular deposition film surface adhesion force and friction force but also significantly improved the life of wear resistance. That was due to the nanoparticles that possessed a good load-carrying capacity and reduced the mobility of the polymer-chain segments, which can undergo reversible shear deformation. Based on this, further research direction of in situ nanoparticles molecular deposition film was proposed.

Keywords: tribology, polyelectrolyte multilayers, nanoparticles, friction, antiwear

1. Introduction

With the development of micro-/nanotechnology, the integration of miniaturized mechanical components with microelectronic components has spawned a new technology; it is well known as microelectromechanical systems (MEMS)/nanoelectromechanical systems (NEMS) [1]. Due to the large surface-area-to-volume ratios of surface and bulk micromachine, micromechanism brings the role of surface and interfacial forces into the foreground [2]. Thus, the surface adhesion and friction greatly affect the safety and service life of MEMS or micromechanisms. So the need of developing nanolubricating technique is becoming increasingly urgent. Thin organic films as a more effective modification for antifriction treatments involve the application of a molecular film to the micromachine surface [3]. In the recent two decades, with the development of MEMS, magnetic storage, and recording systems, Langmuir-Blodgett (LB) films and self-assembled (SA) films were applied to reduce the frictional force between the contacted surfaces of different materials. However, both LB films and SA films were limited in use because of their own shortages: instability and the cost of the apparatus for the LB film, and the difficulty of multilayers assembling for the SA film [4].

Decher et al. [5–8] introduced a method for preparing multilayers by the consecutive deposition of oppositely charged polyelectrolytes from dilute aqueous solution by way of intermolecular electrostatic forces onto charged substrates. Polyelectrolyte multilayers (PEMs) through this molecular deposition process have been intensively investigated in recent years. The popularity of this molecular deposition procedure is due to its simplicity, versatility, and systematical control over the structure and the thickness of the resulting films. Moreover, the materials used in molecular deposition studies can be macromolecules [6], small organic molecules [7] or inorganic compounds [8], biomacromolecules such as proteins [9], DNA [10], or even colloids [11]. Previous work has reported that polyelectrolyte multilayers can decrease the adhesive force on a surface [12], so as to modify the friction surface and reduce the friction force [13, 14]. This shows that the polyelectrolyte multilayers are ideally suited to combat the tribological challenges in MEMS. Up to now, the friction and wear behaviors of polyelectrolyte multilayers were studied popularly. Pavoor et al. has reported that the preparation of a poly(acrylic acid) (PAA) and poly(allylamine hydrochloride) (PAH) polyelectrolyte multilayers using layer-by-layer method, and the capacity for these multilayers induced wear reduction at large scales under dry condition [15]. The mechanical property of polyelectrolyte multilayers was characterized by using a nanoindentation tester. The results indicated that the elastic modulus and hardness of the films were independent of the contact depth over a range of penetration where artifacts arising from the free surface and the underlying glass substrate were insignificant [16]. Furthermore, nanoparticles can reduce friction and enhance the lubrication performance of the lubricant [17]. Nanoparticles were also applied to prepared composite polyelectrolyte multilayers, such as TiO_2 [18], graphite oxide (GO) [18], graphene oxide [19], SiO_2 [20], Ag [21], and so on. The tribological investigation of nanoparticles composite polyelectrolyte multilayers shows that the nanoparticles in the polymer films could enhance the antiwear life of polyelectrolyte multilayers. Then, it indicated that these composite films can improve the tribological performance greatly.

In this review, the correlation between the parameters and the performance was calculated. We defined chemical composition parameter as the molecular structure and ionization behavior of polyelectrolyte. As a result, it substantially facilitates our fundamental understanding in the mechanism of lubrication and antiwear performance that enables us to design molecular film lubricants with well tribological properties.

2. Effects of chemical composition of polyelectrolyte

Since the invention of layer-by-layer method, many polyelectrolytes are used for preparing multilayers. It mainly includes polycations and polyanions. These groups dissociate in aqueous solutions. Polyelectrolyte thus has properties of both electrolytes as salts and polymer as high-molecular-weight compounds. Like salts, their solutions are electrically conductive. Like polymers, their solutions are often viscous. Charged molecular chains are commonly present in soft-matter systems. Due to its ionization properties, different polyelectrolytes have been utilized in the formation of ultrathin materials known as polyelectrolyte multilayers. During layer-by-layer deposition, a suitable growth-charged substrate is dipped back and forth between dilute baths of positively and negatively charged polyelectrolyte solutions. As shown in the schematic diagram (**Figure 1**), during each dip process an amount of polyelectrolyte molecular is adsorbed on the substrate and the surface charge is reversed.

The statistic analysis was conducted based on the experimental results collected from 42 papers that were related to tribological study of polyelectrolyte multilayers, as listed in **Table 1**. Based on **Table 1**, we can divide reported polyelectrolytes into different types based on their characteristic (strong or weak polyelectrolyte, terminal group). It can be found that the main polyelectrolytes used in references were PAA, PAH, poly(diallyl dimethylammonium chloride) (PDDA), and poly(4-styrenesulfonic acid) sodium salt (PSS). Then, we principally review these polyelectrolyte multilayers including the nanoparticles composite films.

The factors affecting the micro- or nanofrictional behavior of polyelectrolyte multilayers include the characteristics of molecule assembled, surface morphology, and mechanical properties of the films.

The characteristics of the outermost layer molecules forming the films have essential effects on the frictional properties of the polyelectrolyte multilayers. **Table 1** lists the different structures adsorbed on the substrate according to the last adsorbed layer polyelectrolyte. **Figure 2** shows the friction coefficient comparison based on different outermost layer polyelectrolyte. According to the results, the polyelectrolyte multilayer-modified substrate shows a lower friction coefficient than that for the bare substrate (glass or silicon). It demonstrated the lubricating

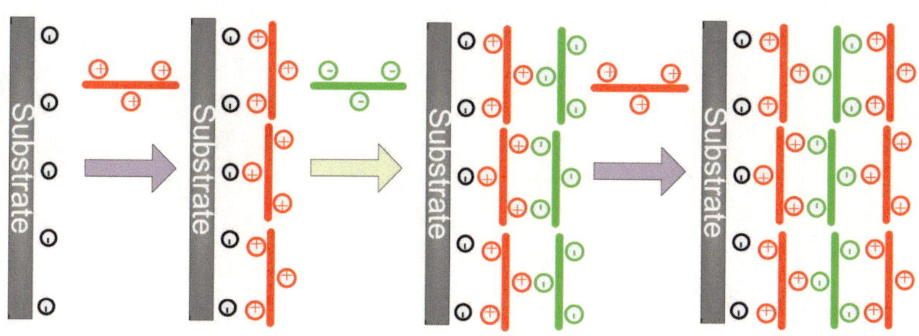

Figure 1. The schematic diagram of deposition process of polyelectrolyte multilayers (redrawn from Refs. [5–8]).

Materials	Abbreviation of molecular	Cationic or anionic (C or A)	References
Poly(acrylic acid)	PAA	A	[15, 16, 18, 20–22, 24–26, 34, 35, 38, 39, 41–43, 45, 47, 49, 51, 55, 57]
Poly(allylamine hydrochloride)	PAH	C	[15, 16, 18, 20, 21, 24, 25, 33, 36, 38–40, 44, 45, 52, 54–57]
Poly(diallyl dimethylammonium chloride)	PDDA	C	[22, 23, 26–28, 30–32, 34, 35, 43, 46–51, 53]
Polyethylenimine	PEI	C	[25, 26, 33, 39, 40, 42, 44, 52]
Poly(4-styrenesulfonic acid) sodium salt	PSS	A	[23, 25, 27, 28, 31–33, 36, 40, 44, 46, 48, 50, 52–54]
Poly(L-lysine)	PLL	C	[29]
Hyaluronic acid	HA	A	[29, 37, 44]
Natural biopolymers chitosan	CHI	C	[37]
Deoxyribonucleic acid	DNA	A	[33]
Diazoresin	DR	C	[41]

Table 1. Summary of polyelectrolytes used for tribology.

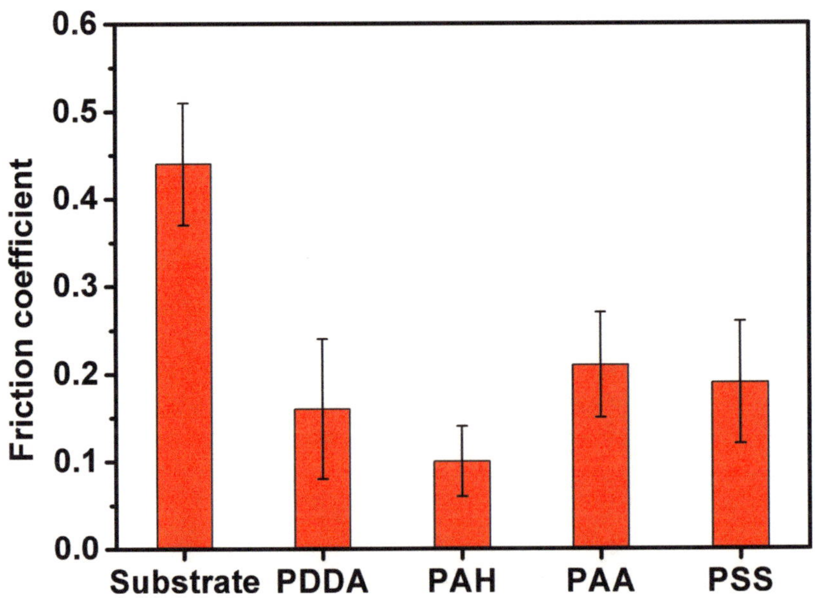

Figure 2. Friction coefficient comparison based on the outermost layer polyelectrolyte.

http://dx.doi.org/10.5772/67571

and antifriction properties of polyelectrolyte multilayers. The polyelectrolyte deposited on the substrate can reduce the friction force, which is mainly influenced by the surface layer. In this case, PAH layer was believed to be the lowest friction coefficient. From the macrofriction tests, the polycationic layer is a better element for improving tribological performance.

Pavoor et al. [15, 57] investigated the tribological behavior of PAH/PAA film on UHMWPE. To examine the behavior of PEM structure in the presence of biological lubricant solution, PAH/PAA film with PAA as the last adsorbed polyelectrolyte is used. The lubricant solution was prepared by diluting the bovine calf serum to a protein concentration of 23 g/L, containing 20 mMol of the sodium salt of ethylenediaminetetraacetic acid (EDTA) and 0.2% by weight of sodium azide. Moreover, a film with PAH as the last adsorbed layer was also investigated. For film with different periods of time immersing in the lubricant solution, the average thickness after immersion was observed by using a profilometer (as shown in **Figure 3**). It shows that a slight decrease in thickness was observed initially for the PAA-topped layer films, but without an obvious effect on the thickness for the PAH adsorbed last. The corresponding friction tests show that the average friction coefficient slightly increased after PEMs deposited on UHMWPE surface. But the use of the PAH/PAA film reduced ultra-high molecular weight polyethylene (UHMWPE) wear by up to 33% compared with the uncoated sample. Furthermore, the capacity for PEM-induced wear reduction was confirmed at larger scales of tests in the dry state a pin on disk tester [57].

It is well known that the mechanical properties influence the tribological behavior of lubricant. Pavoor et al. [16] investigated the mechanical characteristics of PAH/PAA PEMs using nanoindentation tester. The elastic modulus and hardness of the films were demonstrably

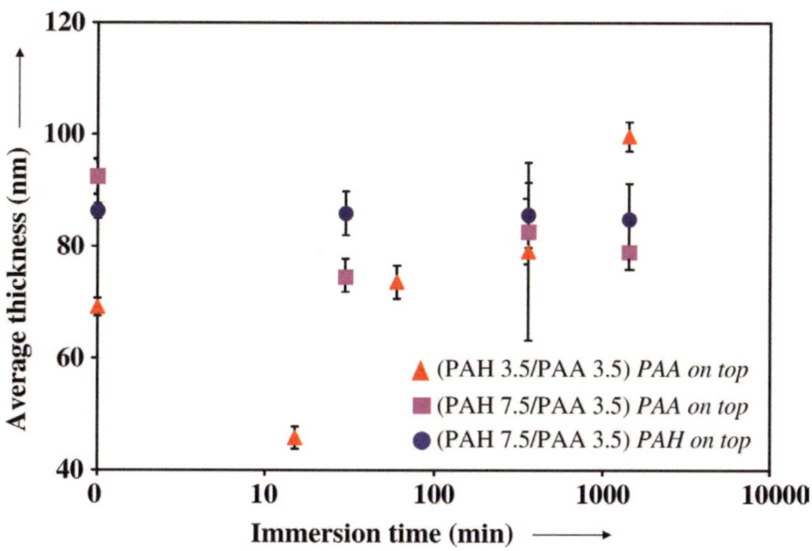

Figure 3. Average PEM thickness (on glass substrates) after immersion in bovine calf serum–containing lubricant for different times (adapted from Ref. [15]).

free of the influence by the underlying supporting material. The mechanical properties of these PEMs can be altered significantly by varying the pH values of the PAH and PAA assembly solutions. Furthermore, in the dry state, the PEM behaviors are virtually unaffected by whether the linkages among the functional groups of the parent polyelectrolytes are ionic or covalent in nature. The essence of mechanical strength of PEMs lies in the linkage density in the film. The modulus values of these interpenetrated polyelectrolyte structures in the swollen state (in water) are about two orders of magnitude lower than the corresponding values in dry conditions. Then, in the swollen state, chemical cross-links do augment the modulus values due to the smaller amount of water uptake in cross-linked PEMs. Gao et al. [58] composed capsules of PSS/PDDA using layer-by-layer deposition on melamine formaldehyde colloidal templates. The results indicated that the elasticity modulus of the PSS/PDDA multilayer as obtained by the osmotic pressure method is 136 MPa, which is considerably smaller than that of the PSS/PAH PEMs. The apparent difference between the PSS/PAH capsules and PSS/PDDA capsules with regard to their stability and elasticity is explained as the result of the different chemical nature of PAH and PDDA. It is due to an overall weaker interaction between PDDA and PSS compared with the PAH and PSS interaction.

3. Effect of nanoparticle doped into PEMs

In recent years, nanoparticles composite ultrathin films have increasingly attracted a wide attention. Due to the assemble characteristic of PEMs, many nanoparticles as formation materials were introduced for preparing composite PEMs. **Table 2** lists the different nanoparticles used to prepare nanoparticles-doped PEMs. For example, Cassagneau et al. synthesized nanoparticles or nanoplates and used these nanoparticles to prepare composite multilayers [59, 60]. Feng et al. [56] prepared surface-charged graphite oxide (GO) solution and formed PAH/GO multilayers by layer-by-layer method. Furthermore, they synthesized PAA-coated

Nanoparticles/nanoplates	Abbreviation	References
Titanium dioxide	TiO_2	[18, 56, 59]
Graphite oxide	GO	[18, 60]
Silicon dioxide	SiO_2	[20]
Copper	Cu	[23, 32, 48, 53]
Copper sulfide	CuS	[22, 35, 43, 49, 51]
Zinc sulfide	ZnS	[47, 49, 66]
Copper hydroxide	$Cu(OH)_2$	[50, 69]
Rare earth	RE	[36]
Silver	Ag	[21, 30, 34, 46, 54, 61, 65, 68]
Gold	Au (or complex)	[34, 46, 52, 55, 64, 67]

Table 2. Summary of different nanoparticles used to prepare nanoparticle-doped PEMs.

TiO_2 nanoparticles and prepared PAH/PAA(TiO_2) multilayers by using the same method [18]. Nanocomposite (PAH-PSS) containing Au or Ag film was fabricated on a silicon substrate using spin-assisted layer-by-layer self-assembly technique [52, 54] where the fabrication process is shown in **Figure 4**. Xiao et al. [30] synthesized sodium citrate-protected silver nanoparticles and prepared PDDA/Ag composite films. AFM images indicated that the additional nanoparticles were formed on the substrate. With the number of deposited cycles and increasing deposition time, the particles show an obvious tendency to aggregate, and large particle clusters are observed in the sample.

The above-mentioned method as commonly used step is that the prepared nanoparticles or nanoplatelets (surface modified) were processed using surface modification firstly, and then the substrates were coated by nanoparticles or nanoplatelets and polymers layer by layer using molecular deposition method. However, it is difficult to prepare nanoparticles composite MD films using the above-mentioned steps, since it is difficult to control the size of nanoparticles, preventing the reunion of nanoparticles and making the nanoparticles dispersed in the film uniformity. Recently, polyelectrolyte multilayers have emerged as a useful vessel for novel nanomaterial synthesis [61–64]. The studies show that various nanomaterials with desirable shape and composition can be synthesized by loading the reactants into the PEMs interior and then performing appropriate reactions such as reduction, sulfuration, and so on. This reaction can be called as in situ synthesis. Logar et al. [65, 66] has synthesized the Ag and ZnS nanoparticles in situ in PAH/PAA PEMs. Gold nanoparticles could be synthesized in the PEMs [67]. This indicated that the in situ nucleation and growth of nanoparticles appeared in PEM films. This technique can be used for synthesizing nanoparticles in the polymer structure domain as nanoreactor that can bind metal cations from an aqueous solution. Then, the postbinding chemistries include reduction, hydroxide, sulfidation, and growth nanoparticles from the cationic precursors [49, 50, 68]. For instance, the schematic diagram of the preparation process of in situ CuS/ZnS nanoparticles hybrid PEMs is shown in **Figure 5**.

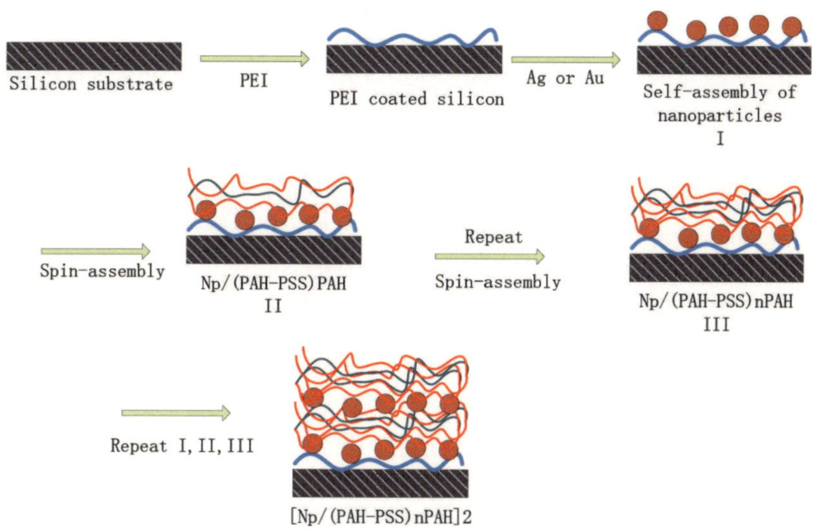

Figure 4. Schematic of fabrication of silver nanoparticle-polyelectrolyte multilayers (redrawn from Refs. [52] and [54]).

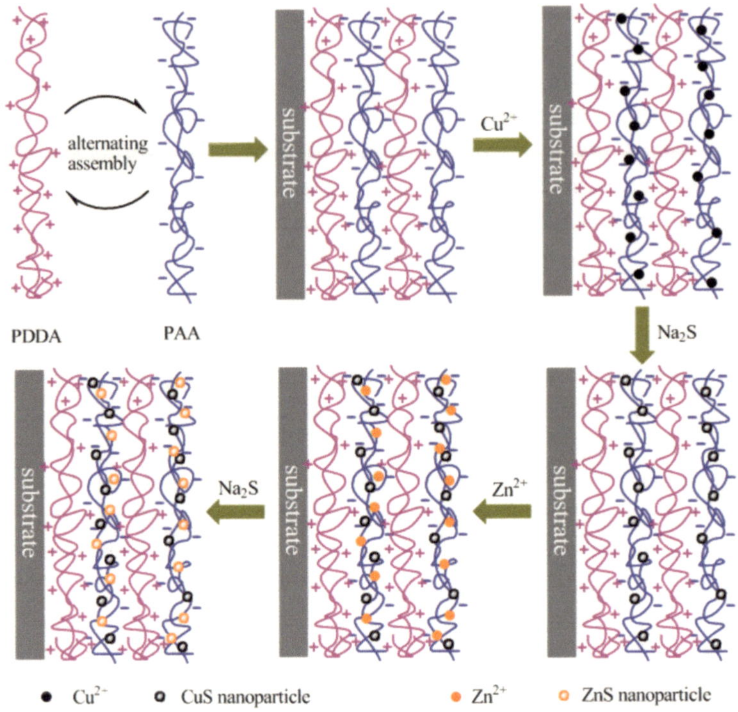

Figure 5. The schematic diagram of preparation of in situ CuS/ZnS nanoparticles hybrid polyelectrolyte multilayer thin film (adapted from Ref. [49]).

Moreover, some polyelectrolyte solution mixed with metal ions can form polymer complexes. The study shows that the in situ gold and silver nanoparticles in polyelectrolyte multilayer film can be prepared by alternate immersion of a substrate in PDDA-$AuCl^{4-}$ complexes solution and PAA-Ag^{+} complexes solution followed by the reduction of the metal cations (Au^{3+}, Ag^{+}) through immersion of $NaBH_4$ solution [34]. The AFM, SEM, TEM, XPS, and UV-vis spectrum measurements demonstrated the Au and Ag nanoparticles distributed in the polyelectrolyte multilayers uniformly.

The tribological properties of nanoparticles-doped composite PEMs were investigated by microtribometer [21–23, 52–56]. The statistic results of friction and wear behaviors are shown in **Figure 6**. It can be seen that the friction coefficient of nanoparticles-doped composite PEMs has a slight decrease compared with pure PEMs (contrast **Figure 2**). However, the antiwear life has an obvious improvement with nanoparticles hybrid into PEM films. The antiwear life (sliding time or reciprocation cycles) was determined by the time or cycles when the friction coefficient increased sharply. The results indicated that the PEMs filled with nanoparticles showed better tribological performance than the pristine PEMs. This is quite evident that nanoparticles within the PEMs can enhance antiwear life of PEMs. From the tribometer tests, this is due to the nanoparticles possessing good load-carrying capacity and decreasing the mobility of the polymer-chain segments which can undergo reversible shear deformation, and the higher shear force must be overcome during relative motion, which results in the increased friction coefficient and shorter antiwear life.

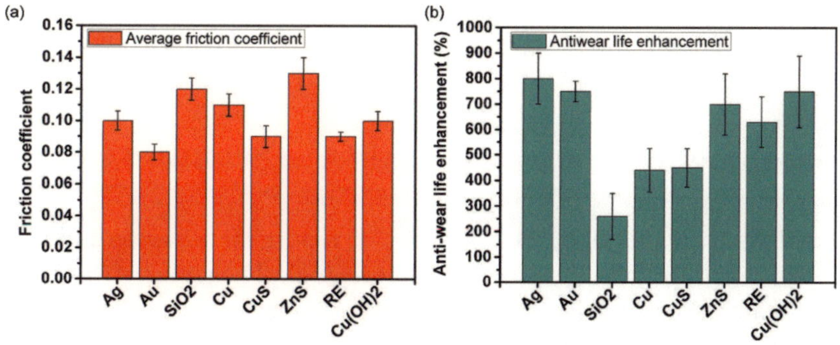

Figure 6. Tribological performance comparison based on different nanoparticle-doped PEMs: (a) average friction coefficient and (b) antiwear enhancement.

More recently, Yang and Guo researched the tribological properties of PEMs filled with in situ nanoparticles. Some conclusions were presented. The in situ nanoparticles exhibited higher durability because of the inorganic nanoparticles enhancing the load-carrying capacity. The main wear mechanism was adhesive wear, and the in situ nanoparticles mitigated plastic deformation and slowed down adhesive wear [69]. AFM has been used extensively to measure adhesive force between surfaces at nanoscale. The surface adhesive force is between the AFM tip and the film surface by force-curves mode under ambient condition. AFM tests showed that the adhesive forces between the probe and the sample surface decreased, indicating that the surface interactions between the probe and the sample are reduced by compositing in situ nanoparticles [22, 49]. The adhesive force is mainly dependent on the surface interaction between the probe and the sample surfaces. It is well known that, when the surface is hydrophilic, they would easily form meniscus by adsorbed water molecules in air, thus they had higher adhesive force. From the ultrapure water-contact angle measurement, the nanoparticles-filled PEMs film was more hydrophobic than the pristine PEM film and show a lower surface adhesion.

Surface wettability of a solid surface can be controlled by two factors: one is the change and design of geometrical structure (such as surface roughness) in micro- or nanosurface structure [70–72]; the other is surface modification with chemical component (surface-free energy) [73, 74]. Therefore, hydrophobic surfaces can be obtained by using low surface energy materials such as fluoroalkylsilane [75, 76] and wax [77, 78]. In the meantime, the hydrophobicity of the surface can also be increased by enhancing surface roughness [79]. The previous study shows that the surface wetting and adhesion can be changed by using layer-by-layer method [80]. These surface properties have an effect on the tribological behavior of PEMs. A 4.5-bilayer PDDA/PSS PEM on quartz or silicon plate was prepared by spin-assisted layer-by-layer assembly technique. Then, PDDA/PSS doped with Cu nanoparticles was built by immersing PDDA/PSS PEMs in Cu^{2+} and $NaBH_4$ solutions to allow nucleation of Cu nanoparticles [23]. The relationship between water wettability and tribological properties of PDDA/PSS doped with Cu nanoparticles films showed that the adhesion and tribological behavior were closely related to their wettability. That is, the PEMs with stronger hydrophobicity have a lower surface energy, which show a lower friction and longer antiwear life. Guo et al. [36] fabricated lanthanum hybrid PEMs using the layer-by-layer and self-assembly methods. The (PAH/PSS)/RE film has a larger water-contact angle and lower surface energy than PAH/PSS PEMs. The microtribology study showed that the (PAH/PSS)/RE film with a very low friction coefficient of about 0.09 and a longer antiwear life was obtained than that for the pure PAH/PSS film.

4. Nanotribology of PEMs

The PEMs are desired for the application of MEMS to reduce the adhesion and friction. AFM is a mighty instrument for investigating the nanotribological behavior of PEMs. Wang et al. [4] have investigated the nanotribological of different molecular deposition films by AFM. The test results showed that the process of the anionning and the depositing of a monolayer molecular film on an Au substrate and the process of decorating an alkyl-terminal to molecular deposition film surfaces were all capable of lowering the frictional force and improving the nanolubrication property. Moreover, the film decorated with alkyl chains had lower frictional forces than the undecorated films. Accordingly, it provided a significant thought to seek for a new nanolubrication film. Zhang et al. [18] investigated the surface roughness, hardness, and nanofriction force of PAH/GO and PAH/PAA(TiO_2) PEMs. Both of them were heated to change the film forming dynamic force from electrostatic force to covalent bond so as to increase the bonding strength of the films. The surface roughness increased, and the friction force was significantly decreased after heating for PAH/GO PEMs. However, the surface roughness and nanofriction force of PAH/PAA(TiO_2) PEMs both decreased slightly after heating. The AFM images of the five-layer PAH/PAA(TiO_2) before and after heating are shown in **Figure 7**. It was found that these films had a much smaller friction force than their substrates and the friction force was dependent on the morphology and/or hardness of the films.

The nanofriction coefficient of the PDDA/GO was found to decrease with increasing load. However, the decrease in nanofriction coefficient reduced noticeably and maintained if the load was beyond a specific value [81]. The effect of load on nanofrictional properties of different Ag nanoparticles-doped PEMs, whose number of bilayers was one to five, respectively, was investigated by Guo et al. [21]. The nanofriction increased linearly for different bilayer composite PEMs with the increasing load. With the bilayer number (thickness) increasing, the fitting friction coefficient decreased slightly from 0.06 (one and two bilayers) to 0.05 (three to five bilayers).

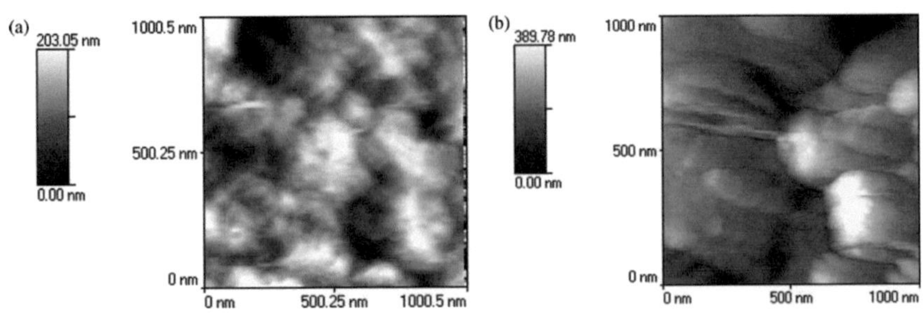

Figure 7. AFM images of five-layer PAH/PAA(TiO_2) before (a) and after (b) heating (adapted from Ref. [18]).

The effects of scanning speed of AFM tip on the nanotribological behavior of PEMs were investigated [30]. All the substrates deposited with PDDA/Ag film were found with a lower nanofriction force than the clean substrate at different scanning speeds with a normal load of 5 nA. And with the increase of the scan rate, the increasing amplitude of the friction force of all the films was apparently less than that of the clean substrate. It might be concluded that the films decrease the adhesion force of the substrates. Also, the friction force of the trilayer film was the smallest. This was corresponding with the surface roughness. In these experiments, the surface roughness of trilayer film was the smallest among the films.

Guo et al. [43] prepared PDDA/PAA and containing CuS PEMs on quartz and glass substrates. It was found that the CuS nanoparticles were homogeneously distributed throughout the whole film. And these films had a much smaller friction force than their substrates and higher antiwear life than pristine PDDA/PAA PEMs. **Figure 8** shows the surface wearing capacity of the pristine and CuS nanoparticles-doped PEMs. From the wear tests, it can be found that the wearing capacity increases with the times of reciprocating scan with AFM tip. The pristine PEM's antiwear capacity is lower than that of the CuS nanoparticles-doped PEMs. From the curves, the wearing capacity of pristine PEMs was about 90 scanning times. But the CuS nanoparticles-doped PEMs have not been destroyed after 100 scanning times. Moreover, they also prepared the in situ Au nanoparticles hybrid PEMs [55]. The nanotribological investigation showed that the PAH/PAA PEMs with in situ Au nanoparticles have a lower surface adhesion and friction force than the pure PAH/PAA PEMs. It is due to the nanoparticles in situ synthesized in polyelectrolyte multilayers making the surface morphology change and load-carrying capacity to increase. Under a normal load, the pure PAH/PAA multilayer deformation against a probe tip was larger than the PAH-Au/PAA film. From the friction force tests, it can be found that the PEMs with in situ nanoparticles exert good load capacity. Therefore, a slight and gradual increase of friction forces could be observed. Furthermore, with the increasing of normal load, the deformation of pure PAH/PAA PEMs enhanced, which caused the friction force to increase.

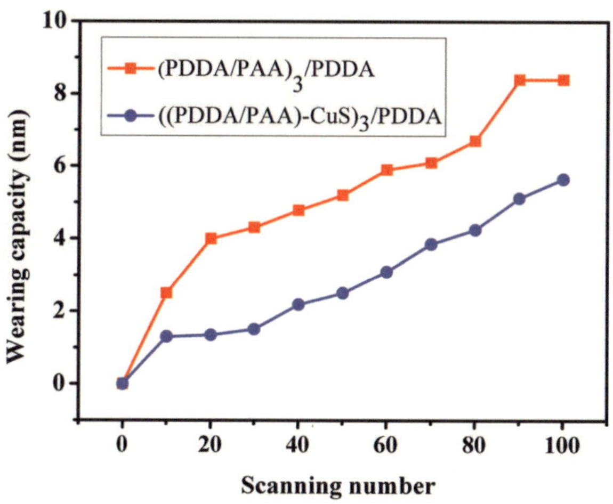

Figure 8. The surface-wearing capacity of pristine and CuS nanoparticle-doped PEMs (adapted from Ref. [43]).

5. Conclusions

Very promising prospects for the tribological application of PEM films have been put forward by recent research. The investigation of PEMs has been extended due to their modification of substrate and reduction of surface friction performance. However, some challenges still exist in applying PEM films into practical application. Thus, for preparing more varieties of PEM films that have better tribological performance and promote the use of PEMs in MEMS or NEMS, some key problems in the research of PEMs should be addressed, such as optimizing conditions of preparation, improvement of the antiwear performance, friction and antiwear mechanism, as well as clarifying the parameters influencing the tribological properties of PEM films. Some new nanolubricant materials could be introduced into PEMs to improve the tribological behavior of PEMs, thereby enhancing the bonding strength between layers and substrates to further improve the antiwear life. Furthermore, more attention needs to be paid to research the antifriction, antiwear, and repair mechanism of in situ nanoparticles-doped PEMs, as well as the tribological investigation in different conditions, such as atmosphere, liquid medium, and so on in order to assess the application prospect of these PEM films.

Acknowledgements

This research is supported by the Beijing Natural Science Foundation (No. 3162024), the National Natural Science Foundation of China (No. 51305459), Tribology Science Foundation of the State Key laboratory of Tribology (No. SKLTKF14A08), and the Science Foundation of China University of Petroleum, Beijing (Nos. 2462017BJB06, C201602).

Author details

Yanbao Guo[1,2]* and Deguo Wang[1,2]

*Address all correspondence to: gyb@cup.edu.cn

1 College of Mechanical and Transportation Engineering, China University of Petroleum, Beijing, China

2 Beijing Key Laboratory of Process Fluid Filtration and Separation, Beijing, China

References

[1] Maboudian R., Ashurst W.R., Carraro C. Tribological challenges in micromechanical systems. Tribology Letters. 2002;**12**(2):95–100.

[2] Maboudian R. Surface processes in MEMS technology. Surface Science Reports. 1998;**30**: 207–269.

[3] Bhushan B., Kasai T., Kulik G., Barbieri L., Hoffmann P. AFM study of perfluoroalkylsilane and alkylsilane self-assembled monolayers for anti-stiction in MEMS/NEMS. Ultramicroscopy. 2005;**105**:176–188.

[4] Wang Q.B., Gao M.L., Zhang S.W. Tribological application of molecular deposition (MD) films. Materials Science and Engineering C. 1999;**10**:127–1129.

[5] Decher G. Multilayer thin films. New York, NY: Wiley-VCH; 2002.

[6] Decher G. Fuzzy nanoassemblies: toward layered polymeric multicomposites. Science. 1997;**277**:1232–1237.

[7] Decher G., Hong J.D., Schmitt J. Buildup of ultrathin multilayer films by a self-assembly process: I. Consecutively alternating adsorption of anionic and cationic polyelectrolytes on charged surface. Thin Solid Films. 1992;**210–211**:831–835.

[8] Decher G. Layered nanoarchitectures via directed assembly of anionic and cationic molecules. Comprehensive Supramolecular Chemistry. 1996;**9**:507–528.

[9] Lvov Y.M., Lu Z., Schenkman J.B. Zu X., Rusling J.F. Direct electrochemistry of myoglobin and cytochrome $P450_{cam}$ in alternate layer-by-layer films with DNA and other polyions. Journal of the American Chemical Society. 1998;**120**:4073–4080.

[10] Taton T.A., Mucis R.C., Mirkin C.A., Letsinger R.L. The DNA-mediated formation of supramolecular mono-and multilayered nanoparticle structures. Journal of the American Chemical Society. 2000;**122**:6305–6306.

[11] Schmitt J., Decher G., Fressick W.J., Brandow S.L., Geer R.E., Shashidhar R., et al. Metal nanoparticle/polymer superlattice films: fabrication and control of layer structure. Advanced Materials. 1997;**9**:61–65.

[12] Wang D.G., Zhang S.W., Gao M.L. Nano-scratch study of molecular deposition films on silicon wafers using nano-indentation. Tribotest. 2003;**9**:231–237.

[13] Wang Q.B., Gao M.L., Zhang S.W. Nanofriction properties of molecular deposition films. Science in China, Series B. 2000;**43**(2):138–142.

[14] Wang Q.B., Gao M.L., Zhang S.W. Study on nanofriction properties of molecular deposition (MD) films on the Au substrate by atomic force microscopy. Progress in Natural Science. 1999;**9**:938–941.

[15] Pavoor P.V., Gearing B.P., Muratoglu O., Cohen R.E., Bellare A. Wear reduction of orthopaedic bearing surfaces using polyelectrolyte multilayer nanocoatings. Biomaterials. 2006;**27**:1527–1533.

[16] Pavoor P.V., Bellare A., Strom A., Yang D., Cohen R.E. Mechanical characterization of polyelectrolyte multilayers using quasi-static nanoindentation. Macromolecules. 2004; **37**:4865–4871.

[17] Dai W., Kheireddin B., Gao H., Liang H. Roles of nanoparticles in oil lubrication. Tribology International. 2016;**102**:88–98.

[18] Zhang S.W., Feng D.P., Wang D.G. Tribological behaviors of composite molecular deposition films. Tribology International. 2005;**38**:959–965.

[19] Humood M., Qin S., Song Y., Polychronopoulou K., Zhang Y., Grunlan J.C., et al. The influence of graphene reduction and polymer crosslinking on improving the interfacial properties of multilayer thin films. ACS Applied materials and Interfaces. 2017;9(1):1107–1118.

[20] Yang G., Ma H., Yu L., et al. Preparation and characterization of layer-by-layer self-assembled polyelectrolyte multilayer films doped with surface-capped SiO_2 nanoparticles. Journal of Colloid and Interface Science. 2009;**333**:776–781.

[21] Guo Y., Wang D., Liu S. Tribological behavior of in situ Ag nanoparticles/polyelectrolyte composite molecular deposition films. Applied Surface Science. 2010;**256**:1714–1719.

[22] Guo Y.B., Wang D.G., Zhang S.W. Adhesion and friction of nanoparticles/polyelectrolyte multilayer films by AFM and micro-tribometer. Tribology International. 2011;**44**:906–915.

[23] Yang G., Chen X., Zhang Z., Song S., Yu L., Zhang P. Adhesion and tribological property of Cu nanoparticles-doped hydrophobic polyelectrolyte multilayers. Lubrication Science. 2012;**24**:313–323.

[24] Johansson E., Blomberg E., Lingström R., Wågberg L. Adhesive interaction between polyelectrolyte multilayers of polyallylamine hydrochloride and polyacrylic acid studied using atomic force microscopy and surface force apparatus. Langmuir. 2009;**25**: 2887–2894.

[25] Jiang C., Markutsya S., Tsukruk V.T. Compliant, robust, and truly nanoscale free-standing multilayer films fabricated using spin-assisted layer-by-layer assembly. Advanced Materials. 2004;**16**:157–161.

[26] Wang B., Liu L., Chen K., Chen L., Feng J., Gao C. Compression-inhibited pore formation of polyelectrolyte multilayers containing weak polyanions: a scanning force microscopy study. ChemPhysChem. 2006;**7**:590–596.

[27] Wang L., Lin Y., Su Z. Counterion exchange at the surface of polyelectrolyte multilayer film for wettability modulation. Soft Matter. 2009;**5**:2072–2078.

[28] Ma H., Yang G., Yu L., Zhang P. Effect of counterparts on the friction and wear behavior of polyelectrolyte multilayers. Tribology International. 2009;**42**:869–874.

[29] Francius G., Hemmerle J., Ohayon J., Schaaf P., Voegel J-C., Picart C., et al. Effect of crosslinking on the elasticity of polyelectrolyte multilayer films measured by colloidal probe AFM. Microscopy Research and Technique. 2006;**69**:84–92.

[30] Xiao Y., Wang D., Zhang S., Guo Y., Gao M. Fabrication and nano-tribological behaviors of PDDA/Ag NPs composite molecular deposition films. In: Luo J., Meng Y., Shao T., Zhao Q., editors. Advanced Tribology-Proceedings of CIST2008 & ITS-IFToMM2008; 24th-27th September 2008; Beijing. Springer-Verlag, Berlin Heidelberg; 2009. p. 499–502.

[31] Yang J., Song H., Chen B., Tang H., Li C. Fabrication of superoleophobic surfaces with controllable liquid adhesion from polyelectrolyte multilayer film. RSC Advances. 2014;**4**:14227.

[32] Yang G., Yu L., Chen X., Zhang P. Hydrophobic surfaces of spin-assisted layer-by-layer assembled polyelectrolyte multilayers doped with copper nanoparticles and modified by fluoroalkylsilane. Applied surface Science. 2009;**255**:4097–4101.

[33] Gong H., Garcia-Turiel J., Vasilev K., Vinogradova O.L. Interaction and adhesion properties of polyelectrolyte multilayers. Langmuir. 2005;**21**:7545–7550.

[34] Guo Y.B., Wang D.G., Liu S.H., Zhang S.W. Fabrication and tribological properties of polyelectrolyte multilayers containing in situ gold and silver nanoparticles. Colloids and Surfaces A: Physicochemical and Engineering Aspects. 2013;**417**:1–9.

[35] Guo Y., Wang D., Liu S., Zhang S. Investigation of tribological behaviors of polymer molecular deposition films by tribometer based on interferometer. Tribology Letters. 2011;**41**:515–524.

[36] Guo Y.B., Wang D.G., Liu S.H., Zhang S.W. Investigation of tribological behavior of lanthanum hybrid polyelectrolyte molecular deposition films. Surface and Coatings Technology. 2011;**205**:2924–2930.

[37] Bongaerts J.H.H., Cooper-White J.J., Stokes J.R. Low biofouling chitosan-hyaluronic acid multilayers with ultra-low friction coefficients. Biomacromolecules. 2009;**10**: 1287–1294.

[38] Dai X., Zhang Y., Guan Y., Yang S., Xu J. Mechanical properties of polyelectrolyte multilayer self-assembled films. Thin solid films. 2005;**474**:159–164.

[39] Hemmerlé J., Roucoules V., Fleith G., Nardin M., Ball V., Lavalle P., et al. Mechanically responsive films of variable hydrophobicity made of polyelectrolyte multilayers. Langmuir. 2005;**21**:10328–10331.

[40] Picart C., Sengupta K., Schilling J., Maurstad G., Ladam G., Bausch A.R., et al. Microinterferometric study of the structure, interfacial potential, and viscoelastic properties of polyelectrolyte multilayer films on a planar substrate. Journal of Physical and Chemical B. 2004;**108**:7196–7205.

[41] Cao T., Wei F., Yang Y., Huang L., Zhao X., Cao W. Microtribologic properties of a covalently attached nanostructured self-assembly film fabricated from fullerene carboxylic acid and diazoresin. Langmuir. 2002;**18**:5186–5189.

[42] Humood M., Chowdhury S., Song Y., Tzeng P., Grunlan J.C., Polycarpou A.A. Nanomechanical behavior of high gas barrier multilayer thin films. ACS Applied Materials and Interfaces. 2016;**8**:11128–11138.

[43] Guo Y., Wang D., Zhang S. Nanotribological behaviors of in situ nanoparticles doped molecular deposition films. Chinese Science Bulletin. 2009;**54**:4571–4576.

[44] Zeng Y., Klitzing R. Oscillatory forces of nanoparticle suspensions confined between rough surfaces modified with polyelectrolytes via the layer-by-layer technique. Langmuir. 2012;**28**:6313–6321.

[45] Pavoor P.V., Gearing B.P., Gorga R.E., Bellare A., Cohen R.E. Engineering the friction-and-wear behavior of polyelectrolyte multilayer nanoassemblies through block copolymer surface capping, metallic nanoparticles, and multiwall carbon nanotubes. Journal of Applied Polymer Science. 2004;**92**:439–448.

[46] Guo Y., Wang D., Zhang S., Xiao Y. Tribological properties of hybrid $Ag_4SiW_{12}O_{40}$ nanoparticles multilayer films. In: Luo J., Meng Y., Shao T., Zhao Q., editors. Advanced Tribology-Proceedings of CIST2008 & ITS-IFToMM2008; 24th-27th September 2008; Beijing. Springer-Verlag, Berlin Heidelberg; 2009.p. 250–254.

[47] Yang G., Ma H., Wu Z., Zhang P. Tribological behavior of ZnS-filled polyelectrolyte multilayers. Wear. 2007;**262**:471–476.

[48] Yang G., Chen X., Ma K., Yu L., Zhang P. The tribological behavior of spin-assisted layer-by-layer assembled Cu nanoparticles-doped hydrophobic polyelectrolyte multilayers modified with fluoroalkylsilane. Surface and Coatings Technology. 2011;**205**: 3365–3371.

[49] Guo Y.B., Wang D.G., Liu S.H., Zhang S.W. Synthesis and tribological properties of CuS/ZnS nanoparticles doped polyelectrolyte multilayers. Surface Engineering. 2013;**29**: 17–22.

[50] Yang G., Wu Z., Zhang P. Study on the tribological behaviors of polyelectrolyte multilayers containing copper hydroxide nanoparticles. Tribology Letters. 2007;**25**:55–60.

[51] Guo Y.B., Wang D.G., Liu S.H., Zhang S.W. Shear of molecular deposition films on glass substrates determined by tribometer. Science China Technological Sciences. 2011;**54**:1005–1010.

[52] Feng Z., Yan F. Preparation and tribological studies of nanocomposite films fabricated using spin-assisted layer-by-layer assembly. Surface and Coatings Technology. 2008;**202**:3290–3297.

[53] Yang G., Geng Z., Ma H., Wu Z., Zhang P. Preparation and tribological behavior of Cu-nanoparticle polyelectrolyte multilayers obtained by spin-assisted layer-by-layer assembly. Thin Solid Films. 2009;**517**:1778–1783.

[54] Feng Z., Yan F. Preparation and resistivity study of silver nanocomposite films using spin-assisted layer-by-layer assembly. Surface and Interface Analysis. 2008;**40**:1523–1528.

[55] Guo Y., Yang L., Wang D., Liu S., Zhang S. Preparation and nanotribological properties of polyelectrolyte multilayers with in situ Au nanoparticles. Progress in Organic Coatings. 2015;**88**:164–171.

[56] Feng D., Wang D., Zhang S. Preparation and nanotribological behavior of PAH/graphite oxide molecular deposition film. Science in China, Series E. 2004;**47**:385–390.

[57] Pavoor P.V., Gearing B.P., Bellare A., Cohen R.E. Tribological characteristics of polyelectrolyte multilayers. Wear. 2004;**256**:1196–1207.

[58] Gao C., Leporatti S., Moya S., Donath E., Möhwald H. Stability and mechanical properties of polyelectrolyte capsules obtained by stepwise assembly of poly(styrenesulfonate sodium salt) and poly(diallyldimethyl ammonium) chloride onto melamine resin particles. Langmuir. 2001;**17**:3491–3495.

[59] Cassagneau T., Fendler J.H., Mallouk T.E. Optical and electrical characterizations of ultrathin films self-assembled from 11-Aminoundecanoic acid capped TiO_2 nanoparticles and polyallylamine hydrochloride. Langmuir. 2000;**16**:241–246.

[60] Cassagneau T., Guérin F., Fendler J.H. Preparation and characterization of ultrathin films layer-by-layer self-assembled from graphite oxide nanoplatelets and polymers. Langmuir. 2000;**16**:7318–7324.

[61] Wang T.C., Rubner M.F., Cohen R.E. Polyelectrolyte multilayer nanoreactors for preparing silver nanoparticle composites: controlling metal concentration and nanoparticle size. Langmuir. 2002;**18**:3370–3375.

[62] Shi X., Shen M., Möhwald H. Polyelectrolyte multilayer nanoreactors toward the synthesis of diverse nanostructured materials. Progress in Polymer Science. 2004;**29**:987–1019.

[63] Joly S., Kane R., Radzilowski L., Wang T., Wu A., Cohen R.E., et al. Multilayer nanoreactors for metallic and semiconducting particles. Langmuir. 2000;**16**:1354–1359.

[64] Chia K.K., Cohen R.E., Rubner M.F. Amine-rich polyelectrolyte multilayer nanoreactors for in situ gold nanoparticle synthesis. Chemistry of Materials. 2008;**20**:6756–6763.

[65] Logar M., Jančar B., Suvorov D., Kostanjšek R. In situ synthesis of Ag nanoparticles in polyelectrolyte multilayers. Nanotechnology. 2007;**18**:325601.

[66] Logar M., Jančar B., Rečnik A., Suvorov D. Controlled synthesis of pure and doped ZnS nanoparticles in weak polyion assemblies: growth characteristics and fluorescence properties. Nanotechnology. 2009;**22**:085705.

[67] Chen H., Dong S. A method to construct polyelectrolyte multilayers film containing gold nanoparticles. Talanta. 2007;**71**:1752–1756.

[68] Zhou J., Xu S., Xu W., Zhao B., Ozaki Y. In situ nucleation and growth of silver nanoparticles in membrane materials: a controllable roughened SERS substrate with high reproducibility. Journal of Raman Spectroscopy. 2008;**40**:31–37.

[69] Yang G.B., Zhao Z., Wu Z.S., et al. Investigation of friction and wear behavior of $Cu(OH)_2$-doped polyelectrolyte multilayers. Tribology. 2006;26(4): 300–305. (In Chinese.)

[70] Cheng Y.T., Rodak D.E., Wong C.A., Hayden C.A. Effects of micro-and nano-structures on the self-cleaning behaviour of lotus leaves. Nanotechnology. 2006;**17**:1359.

[71] Ranella A., Barberoglou M., Bakogianni S., Fotakis C., Stratakis E. Tuning cell adhesion by controlling the roughness and wettability of 3D micro/nano silicon structures. Acta Biomaterialia. 2010;**6**:2711–2720.

[72] Jung Y.C., Bhushan B. Contact angle, adhesion and friction properties of micro-and nanopatterned polymers for superhydrophobicity. Nanotechnology. 2006;**17**:4970.

[73] Rupp F., Scheideler L., Olshanska N., De Wild M., Wieland M., Geis-Gerstorfer J. Enhancing surface free energy and hydrophilicity through chemical modification of microstructured titanium implant surfaces. Journal of Biomedical Materials Research Part A. 2006;**76**:323–334.

[74] Goddard J.M., Hotchkiss J.H. Polymer surface modification for the attachment of bioactive compounds. Progress in Polymer Science. 2007;**32**:698–725.

[75] Sugimura H., Ushiyama K., Hozumi A., Takai O. Micropatterning of alkyl-and fluoroalkylsilane self-assembled monolayers using vacuum ultraviolet light. Langmuir. 2000;**16**:885–888.

[76] Ishizaki T., Masuda Y., Sakamoto M. Corrosion resistance and durability of superhydrophobic surface formed on magnesium alloy coated with nanostructured cerium oxide film and fluoroalkylsilane molecules in corrosive NaCl aqueous solution. Langmuir. 2011;**27**:4780–4788.

[77] Koch K., Hartmann K.D., Schreiber L., Barthlott W., Neinhuis C. Influences of air humidity during the cultivation of plants on wax chemical composition, morphology and leaf surface wettability. Environmental and Experimental Botany. 2006;**56**:1–9.

[78] Guo Y.B., Yang L., Wang D.G. Preparation and hydrophobic behaviours of polystyrene composite coating. Surface Engineering. 2016;**32**:95–101.

[79] Bhushan B., Jung Y.C. Wetting study of patterned surfaces for superhydrophobicity. Ultramicroscopy. 2007;**107**:1033–1041.

[80] Yoo D., Shiratori S.S., Rubner M.F. Controlling bilayer composition and surface wettability of sequentially adsorbed multilayers of weak polyelectrolytes. Macromolecules. 1998;**30**:4309–4318.

[81] Feng D.P., Wang D.G., Zhang S.W. The preparation of graphite oxide molecular deposition films and its study by atomic force microscopy. Tribology. 2003;**23**(1):1–4. (In Chinese.)

Chapter 11

Thermal Radiative Wavelength Selectivity of Nanostructured Layered Media

Yi Zheng

Additional information is available at the end of the chapter

http://dx.doi.org/10.5772/67395

Abstract

Thermal radiative transport yields unique thermal characteristics of microscopic thin films—wavelength selectivity. This chapter focuses on a methodology about adjusting the wavelength selectivity of thin films embedded with nanoparticles in the far-field and near-field regimes. For nanostructured layered media doped with nanoparticles, Maxwell-Garnett-Mie theory is applied to determine the effective dielectric function for the calculation of radiative thermal transport. The thermal radiative wavelength selectivity can be affected by volume fraction and/or the size of the embedded nanoparticles in thin films. To characterize wavelength selectivity and optical property of nanostructured materials, both real and imaginary parts of effective refractive index need to be analyzed. It has been shown that the nanoparticles made of polar or metallic materials have different influence on thermal radiative wavelength selectivity of microscopic thin films.

Keywords: thermal radiation, wavelength-selective, far field, near field, nanostructured layered media

1. Introduction

Most naturally occurring materials exhibit a broad range of emission spectrum. However, thermal and optical properties of nanomaterials and nanostructures are significantly different than that of bulk materials. They are the basis of development of selective thermal emitters and absorbers that are crucial for the development of solar cells and thermophotovoltaics (TPVs) [1]. Wavelength-selective emitters also have a wide range of potential applications such as biosensors, chemical sensors [2, 3], thermal cooling, and thermal detectors [4]. It has been shown that one-dimensional (1-D) metallo-dielectric periodic structures display great selective emission properties in the infrared and visible regions [5]. Multilayered structures of thin films (1-D photonic crystals) of polar materials can also be used to develop selective emitters [6]. The property of multilayered structures to exhibit wavelength selectivity can be explained by the presence of surface phonon polaritons (SPhPs) for polar materials and surface plasmon polaritons (SPPs) for metals [5, 7]. It has also been demonstrated that two-dimensional (2-D) or three-dimensional (3-D) photonic crystals can be used to develop selective emitters [8, 9]. However, thin-film-layered structures are easy to design and fabricate. Calculation of their emission spectra is also relatively simple.

Wavelength selectivity of 1-D photonic crystals is a far-field phenomenon. When the distance between two objects is of the order of the dominant thermal wavelength, the radiative heat transfer is enhanced to many orders of magnitude due to the coupling of surface waves and is referred to as near-field thermal radiation [10]. If the materials support SPhPs or SPPs, the near-field radiative flux can be shown to be inversely proportional to the square of the distance. The enhancement of heat transfer does not take place at all wavelengths but only at specific wavelengths [10]. This wavelength selectivity in the near field is exhibited by thin films as well as bulk materials. Wavelength selectivity in the near-field limit is due to the coupling of SPPs or SPhPs across the two surfaces [10].

While many articles dedicated to the design and fabrication of selective emitters can be found in the literature, the use of nanoparticles specifically for the application of selective emitters is relatively sparse [11]. Optical properties of materials doped with nanoparticles have been investigated before [12, 13]. Experimental and analytical study of thermal coatings doped with nanoparticles such as Gonome et al. [14, 15] can also be found in literature. However, emissive properties of nanoparticles embedded thin films have not been studied in detail to the best of our knowledge. This chapter presents the multilayered structures, which contain nanoparticles (NPs) doped into the thin films, which are suitable for any of the potential applications in both the far field and the near field. In this chapter, we investigate a methodology that can be used to develop selective thermal emitters for a desired wavelength band. Ideally, one may want to develop a selective emitter for specific wavelength band as per the requirements. As the emission spectrum displays peaks at the wavelengths, which are characteristics of the refractive index of the material, changing the thickness of the film allows control over only a narrow spectral band. We propose to dope the top layer (thin film) with nanoparticles to change the dielectric properties of the material. The usual Maxwell-Garnett equation for effective medium approximation is often employed for such an analysis disregarding the sizes of doped materials [16]. Here, we apply the Maxwell-Garnett-Mie formulation [17] for effective medium approximation to calculate the dielectric function of a composite that consists of a host material embedded with nanoparticles of various sizes and volume fractions, and extend the same approach to calculate radiative heat transfer for thin films doped with nanoparticles. Thin-film structure with nanoparticles would be easy to fabricate as submicron thin films embedded with nanoparticles have been fabricated before [18, 19]. We aim to study the effect on the wavelength selectivity of thin films due to combination of surface polaritons of the films and the nanoparticles and their effects in the near-field radiative heat transfer and spectral heat flux. We consider hypothetical cases of thin film embedded with nanoparticles although the fabrication of these particular NP-embedded films discussed here is relatively unknown. We choose SiC and polystyrene (PS) as the host materials (for thin films). SiC is chosen as a host because it has high permittivity in the infrared region and supports SPhP. Polystyrene is chosen because it does not support either SPPs or SPhPs. BN, which supports SPhP, and Au, which supports SPPs, are picked for the material of inclusion (NPs).

The structure of this chapter is as follows. In Section 2, we present the theoretical background and analytical expressions used for the calculation of emissivity of thin-film structures, calculation of heat transfer between closely placed half spaces, and the application of Maxwell-Garnett-Mie theory. In Section 3, we discuss the results of our calculations obtained using the formulations outlined in Section 2. Subsequently, the ideas and conclusions of this chapter are narrated in Section 4.

2. Theoretical fundamentals

Consider a structure having N-layer media having (N - 1) interfaces. By solving the boundary conditions at the interfaces, one can obtain the expression for the generalized reflection coefficient at the interface between region i and region $i + 1$ and is given by [20]

$$\tilde{R}^{(\mu)}_{i,i+1} = \frac{R^{(\mu)}_{i,i+1} + \tilde{R}^{(\mu)}_{i+1,i+2} e^{2jk_{i+1,z}(d_{i+1}-d_i)}}{1 + R^{(\mu)}_{i,i+1}\tilde{R}^{(\mu)}_{i+1,i+2} e^{2jk_{i+1,z}(d_{i+1}-d_i)}} \tag{1}$$

where $j = \sqrt{-1}$, $R^{(\mu)}_{i,i+1}$ is the Fresnel reflection coefficient at the interface between the layer i and i + 1, and $\tilde{R}^{(\mu)}_{i+1,i+2}$ is the generalized reflection coefficient at the interface between the layer i + 1 and i + 2, μ = s (or p) refers to transverse electric (or magnetic) polarization, z = - d_i is the location of the ith interface. $k_{i,z} = \sqrt{\varepsilon_i(\omega)\omega^2/c^2 - k_\rho^2}$ is the normal z-component of the wave vector in medium i wherein $\varepsilon_i(\omega)$ is the relative permittivity of the medium i as a function of angular frequency ω, c is the speed of light in vacuum, and k_ρ is the magnitude of the in-plane wave vector. With $\tilde{R}^{(\mu)}_{N,N+1}$ = 0, the above equation provides a recursive relation to calculate the reflection coefficients $\tilde{R}^{(\mu)}_{i,i+1}$ in all regions. The generalized transmission coefficient for the layered slab is given by

$$\tilde{T}^{(\mu)}_{1,N} = \prod_{i=1}^{N-1} e^{2jk_{iz}(d_i-d_{i-1})} S^{(\mu)}_{i,i+1} \tag{2}$$

where $S^{(\mu)}_{i,i+1} = T^{(\mu)}_{i,i+1}/\left(1 - R^{(\mu)}_{i,i+1}\tilde{R}_{i+1,i+2}(\mu)e^{2jk_{i+1,z}(d_{i+1}-d_i)}\right)$. Alternatively, the generalized reflection and transmission coefficients can also be calculated using transfer matrix approach [21]. The hemispherical emissivity is given by the expression [6]

$$e(\omega) = \frac{c^2}{\omega^2} \int_0^{\omega/c} dk_\rho k_\rho \sum_{\mu=s,p} \left(1 - \left|\tilde{R}_{h1}^{(\mu)}\right|^2 - \left|\tilde{T}_{h1}^{(\mu)}\right|^2\right) \tag{3}$$

where $\tilde{R}_{h1}^{(\mu)}$ and $\tilde{T}_{h1}^{(\mu)}$ are the polarized effective reflection and transmission coefficients, respectively, as shown in **Figure 1(a)**.

Far-field radiative heat transfer is described by Planck's law of thermal radiation [22]. This description, however, does not account for the presence of evanescent (surface) waves that dominate only near the boundaries. The expression of radiative transfer between closely spaced objects can be derived using dyadic Green's function formalism [23], and is given by

$$q_{1\to 2}(T_1, T_2, L) = \int_0^\infty \frac{d\omega}{2\pi}[\theta(\omega, T_1) - \theta(\omega, T_2)]T_{1\to 2}(\omega, L) \tag{4}$$

where $\theta(\omega, T) = (\hbar\omega/2)coth(\hbar\omega/2k_BT)$ is the energy of harmonic oscillator at frequency ω and temperature T, h is the reduced Planck constant, and k_B is the Boltzmann constant. The function $T_{(1\to 2)}$ (ω) corresponds to the spectral transmissivity in radiative transfer between media 1 and 2 separated by distance L and is expressed as [23]

$$\begin{aligned} T_{1\to 2}(\omega) = & \int_0^{\omega/c} \frac{dk_\rho k_\rho}{2\pi} \sum_{\mu=s,p} \frac{\left(1 - \left|\tilde{R}_{h1}^{(\mu)}\right|\right)^2 \left(1 - \left|\tilde{R}_{h2}^{(\mu)}\right|\right)^2}{\left|1 - \tilde{R}_{h1}^{(\mu)} \tilde{R}_{h2}^{(\mu)} e^{2jk_{hz}L}\right|^2} \\ & + \int_{\omega/c}^{\infty} \frac{dk_\rho k_\rho}{2\pi} \sum_{\mu=s,p} \frac{4\text{Im}\left(\tilde{R}_{h1}^{(\mu)}\right)\text{Im}\left(\tilde{R}_{h2}^{(\mu)}\right)e^{-2|k_{hz}|L}}{\left|1 - \tilde{R}_{h1}^{(\mu)} \tilde{R}_{h2}^{(\mu)} e^{2jk_{hz}L}\right|^2} \end{aligned} \tag{5}$$

where $\tilde{R}_{h1}^{(\mu)}$ and $\tilde{R}_{h2}^{(\mu)}$ are polarized k_{hz} effective reflection coefficients of the two half spaces (calculated in the absence of other half space) and is the z-component of wave vector in

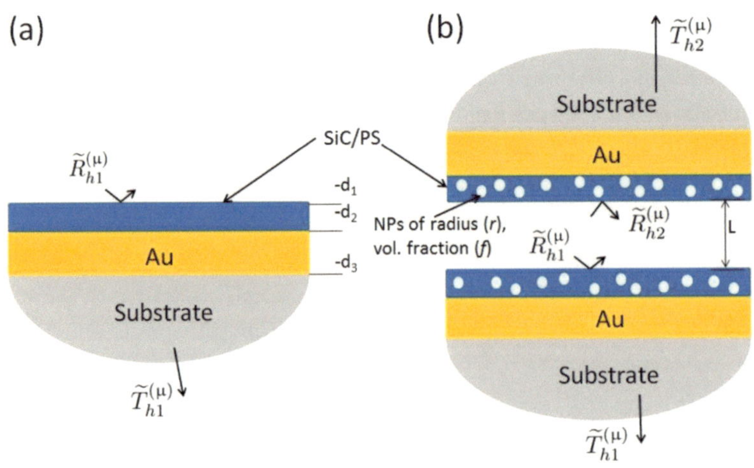

Figure 1. Configurations of nanoparticle-embedded thin films. (a) Thin film of SiC or polystyrene (PS) on Au thin film of thickness 1 μm, and (b) two half spaces with multilayers with vacuum in between. The top is SiC or PS film embedded with nanoparticles of varying radius and volume fraction on Au thin film.

http://dx.doi.org/10.5772/67395

vacuum. The first term in Eq. (5) corresponds to propagating waves, whereas the second term describes the thermal transport due to evanescent waves, and its contribution is significant only for small values of gap L.

Clausius-Mossotti equation for the effective dielectric function, of the composite medium containing nanoparticle inclusions in a host material, is given by [24, 25]

$$\varepsilon_{eff} = \varepsilon_m \left(\frac{r^3 + 2\alpha_r f}{r^3 - \alpha_r f} \right) \tag{6}$$

where ε_m is the dielectric function of the matrix, α_r is the electric dipole polarizability, and r and f are the radius and volume fraction of nanoparticles, respectively. The size-dependent extension of Maxwell-Garnett formula can be obtained by deriving an expression for electric dipole polarizability using Mie theory [17]

$$\alpha_r = \frac{3jc^3}{2\omega^3 \varepsilon_m^{3/2}} a_{1,r} \tag{7}$$

where $a_{1,r}$ is the first electric Mie coefficient given by

$$a_{1,r} = \frac{\sqrt{\varepsilon_{\rm np}}\psi_1(x_{\rm np})\psi_1{}'(x_m) - \sqrt{\varepsilon_m}\psi_1(x_m)\psi_1{}'(x_{\rm np})}{\sqrt{\varepsilon_{\rm np}}\psi_1(x_{\rm np})\xi_1{}'(x_m) - \sqrt{\varepsilon_m}\xi_1(x_m)\psi_1{}'(x_{\rm np})} \tag{8}$$

where ψ_1 and ξ_1 are Riccati-Bessel functions of the first order given by $\psi_1\ (x) = x\ j_1\ (x)$ and $\xi_1\ (x) = xh_1^{(1)}(x)$, where j_1 and $h_1^{(1)}$ are the first-order spherical Bessel functions and spherical Hankel functions of the first kind, respectively. Here, "′" indicates the first derivative. $x_m = \omega r\sqrt{\varepsilon_m}/c$ and $x_{np} = \omega r\sqrt{\varepsilon_{np}}/c$ are the size parameters of the matrix and the nanoparticles, respectively; $\varepsilon_{\rm np}$ being the dielectric function of nanoparticles.

Effective medium approximation method is applicable when average distance between inclusions is much smaller than the wavelength of interest. If the dielectric inclusions of radius r can be imagined to be arranged in simple cubic lattice of lattice constant a, the condition for validity of the effective medium approximation is $\lambda_h \gg a > 2r$. Where λ_h is the wavelength in the host material [26]. We emphasize that since we use the approximation for thin films doped with nanoparticles, its use may not be correct when the particle size is comparable to the thickness of the films. Also, as discussed by Liu et al. [27], it can be argued that the use of effective medium theory (EMT) is questionable at nanoscale distances. Although such might be the case for the near-field calculations presented here, its detailed analysis is beyond the scope of this work and is left for future work. Despite the limitations of Maxwell-Garnett-Mie theory sand its application in the near-field regime, the results obtained should provide general trends and give considerable insight into optical properties of artificial materials. Further investigations by direct numerical simulation may be necessary to confirm the validity of EMT [28]. Moreover, these results will be constructive when judging the validity of the EMT in the near field by direct numerical calculations. We would like to keep these points open for speculation.

3. Results

The dielectric function is related to real (n) and imaginary (κ) parts of refractive index as $\sqrt{\varepsilon} = n + j\kappa$. For SiC and BN, the dielectric function has the form as [29, 30]

$$\varepsilon(\omega) = \varepsilon_\infty \left(\frac{\omega^2 - \omega_{LO}^2 + j\omega\gamma}{\omega^2 - \omega_{TO}^2 + j\omega\gamma} \right) \tag{9}$$

where ω_{TO} and ω_{LO} are transverse and longitudinal optical phonon frequencies, respectively, and γ is the damping constant. For SiC, the constants ε_∞, ω_{TO}, ω_{LO}, and γ are equal to 6.7, 9.83 × 10^{-2} eV, 0.12 eV, and 5.90 × 10^{-2} eV, respectively. The values of ε_∞, ω_{TO}, ω_{LO}, and γ for BN are 4.46, 0.1309 eV, 0.1616 eV, and 6.55 × 10^{-2} eV, respectively. Data for the bulk gold (Au) are taken from Johnson and Christy [31]. **Figure 2(a)** considers the case of SiC film doped with NPs of BN. SiC film of 0.4 μm is on the top of Au film of 1 μm deposited on a substrate. The effect of change in NPs volume fraction (*f*) is studied. The volume fraction of BN nanoparticles is changed from 0% to 30% while maintaining the radius of 25 nm. Thin film of pure SiC exhibits emission peaks at $\lambda_n^{SiC} \approx$ 10.33 μm and $\lambda_n^{SiC} \approx$ 13 μm. λ_n is the wavelength at which the real part of the refractive index becomes zero (zero-index material) [6, 32]. λ_κ is the wavelength at which the real part of refractive index (n) is large while the imaginary part of refractive index (κ) is small [6]. These peaks are attributed to the presence of SPhPs, and the characteristic wavelengths of the dielectric function of SiC. The appearance of new peaks upon 5% inclusion of BN nanoparticles has been observed at $\lambda \approx$ 8.5 μm and $\lambda \approx$ 11.5 μm. When the volume fraction of NPs is increased further, each of these peaks splits into two giving rise to a total of six peaks. Locations of these peaks do not correspond to the characteristic wavelengths of BN ($\lambda_n^{BN} \approx$ 7.6 μm and $\lambda_\kappa^{BN} \approx$ 9.8 μm). In addition, there exists a small shift in the emission peak at λ_κ^{SiC}. This suggests an interaction between SiC matrix and BN NPs. Consider the case with 30% inclusion of BN. **Figure 3(a)** shows that the mixture has two additional locations where the refractive index is zero ($\lambda_{n1}^{mixture} \approx$ 8.5 μm and $\lambda_{n2}^{mixture} \approx$ 11.7 μm). Moreover, at two more points n is large while imaginary part of refractive index is small, namely $\lambda_{k1}^{mixture} \approx$ 9 μm and $\lambda_{k2}^{mixture} \approx$ 11.1 μm.

These wavelengths correspond to the additional peaks. While the additional peaks are at the location of the characteristics of the refractive index of the mixture, it is interesting to note that peaks at ≈10.33 μm and ≈12.98 μm have no or little shift even at large volume fraction of 30%, because they are characteristic wavelengths of the host. Inclusion of BN leads to new SPhP leading to new peaks. **Figure 2(b)** shows the effect of Au nanoparticles in SiC thin film. When particle size is small, especially when the size is comparable to the mean free path of the free electrons, confinement effects become significant [25, 33]. The optical properties of metallic nanoparticles have shown size dependence [34]. We utilize a size-dependent dielectric function for Au nanoparticles of radius *r* that takes care of electron scattering, which is given by [35]

$$\varepsilon(\omega, r) = \varepsilon_b(\omega) + \frac{\omega_p^2}{\omega^2 + j\omega\gamma_0} - \frac{\omega_p^2}{\omega^2 + j\omega(\gamma_0 + Av_f/r)} \tag{10}$$

where ε_b, ω_p, v_f, and γ_0 are the bulk dielectric function, the plasma frequency, the Fermi velocity of free electrons, and the electron damping, respectively. The values of ε_b are taken

http://dx.doi.org/10.5772/67395

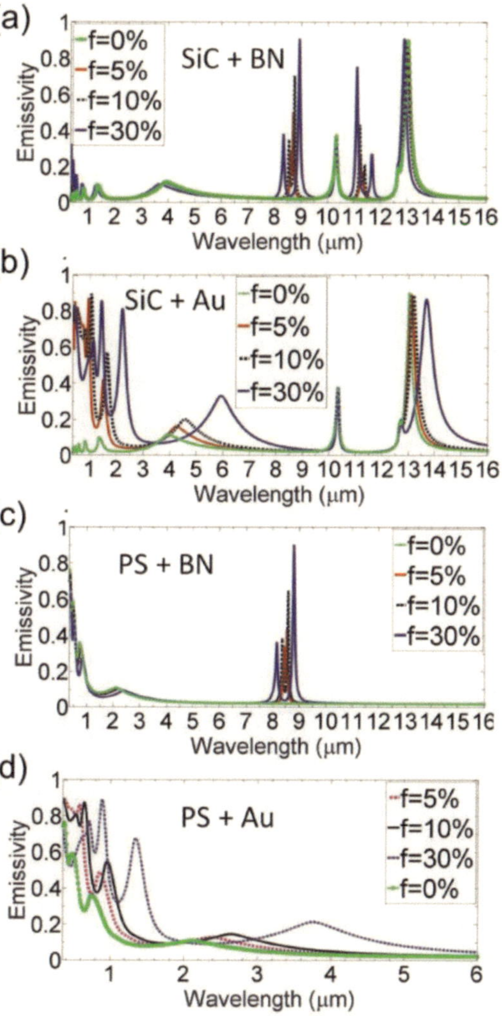

Figure 2. Emissivity spectra for SiC or PS thin film of thickness 0.4 μm embedded with BN or Au nanoparticles of radius 25 nm and various volume fractions. (a) BN nanoparticle-embedded SiC thin film, (b) Au nanoparticle-embedded SiC thin film, (c) BN nanoparticle-embedded PS thin film, and (d) Au nanoparticle-embedded PS thin film.

from Johnson and Christy [31]. The parameters ω_p, v_f, and γ_0 are taken to be 9.06 eV, 0.077 eV, and 1.4×10^6 m/s, respectively. The proportionality constant A that depends on the electron-scattering process at the surface of nanoparticles is assumed to be unity. Volume fraction is varied from 0% to 30% while NPs radius is maintained constant (25 nm). Multiple oscillatory peaks are seen in the lower wavelength region upon the addition of Au nanoparticles. A shift in the original peak of SiC at ≈13 μm is seen when volume fraction is large (30%). While the presence of a peak at ≈10.33 μm and a peak at ≈13 μm can be related to n and κ plots shown in **Figure 3(a)** and **(b)**, multiple peaks between 0.35 and 8 μm cannot be explained using the refractive index characteristics. While the change in refractive index is seen around 500 nm which corresponds to surface plasmon resonance of Au, one may expect a peak around

500 nm. Multiple peaks are observed instead. **Figure 2(c)** and **(d)** shows emission spectra of polystyrene thin film doped with BN nanoparticles and Au nanoparticles, respectively. The dielectric function of PS is in the form of [36]

$$\varepsilon(\omega) = 1 + \sum_{i=1}^{i=4} \left(\frac{f_i}{w_i^2 - \omega^2 - jg_i\omega} \right) \tag{11}$$

where the parameters f_i, w_i, and g_i are, in the units of eV, given by f_i = [14.6, 96.9, 44.4, 136.9], w_i = [6.35, 14.0, 11.0, 20.1], and g_i = [0.65, 5.0, 3.5, 11.5], respectively. In case of BN, the appearance of new peaks is quite similar to that in **Figure 2(a)** and its relation with the refractive indices shown in **Figure 4(a)** and **(b)** is obvious.

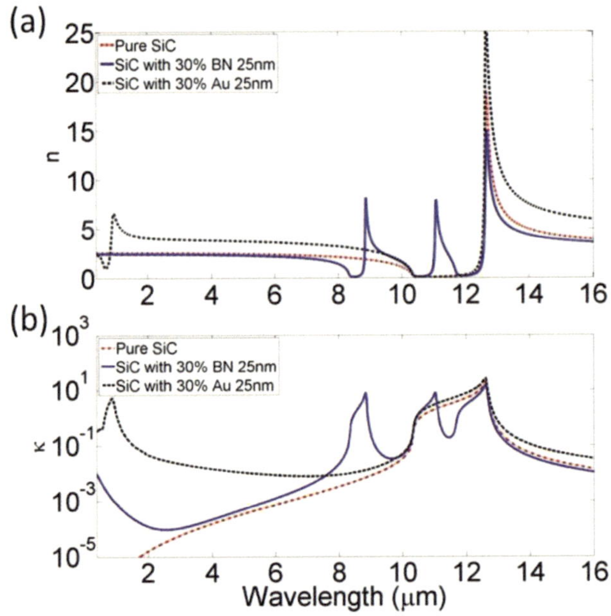

Figure 3. Refractive index of SiC and SiC film doped with 30% BN or Au nanoparticles. (a) Real part of refractive index and (b) imaginary part of refractive index.

The appearance of emission peaks at the locations of $\lambda_k^{\text{mixture}}$ and $\lambda_n^{\text{mixture}}$ is evident. However, when polystyrene film is doped with Au nanoparticles, we once again see multiple peaks produced in the region 0.35–6 μm that are not related to the refractive index characteristics. Since polystyrene does not support either of SPhPs or of SPPs, the interaction between SPPs of Au and surface polariton of the host is not responsible for the multiple peaks. We hypothesize that the origin of multiple peaks is due to the interaction of SPPs of Au and the boundaries of the thin film. In case of SiC film doped with Au nanoparticles, the shift in the peak of ≈13 μm is due to the interaction between SPPs of Au and SPhPs of SiC; in either cases (SiC and PS) the inclusion material does not produce new polaritons as seen in refractive index characteristics.

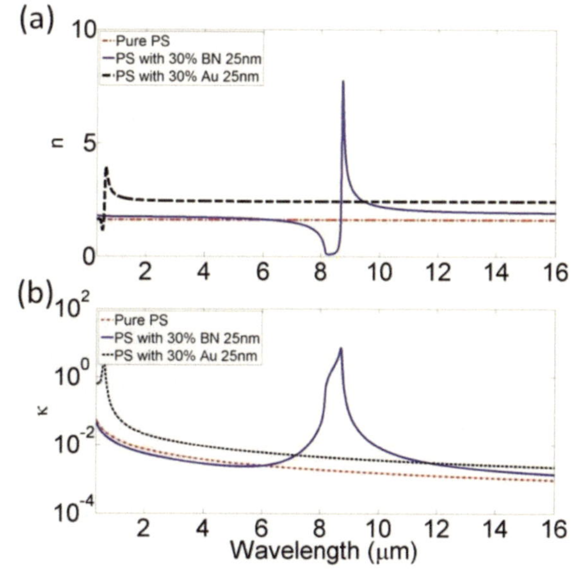

Figure 4. Refractive index of PS and PS film doped with 30% BN or Au nanoparticles. (a) Real part of refractive index and (b) imaginary part of refractive index.

Figure 5 shows the effect of NPs size on the emission spectra. In **Figure 5(a)**, SiC film of 0.4 μm is doped with BN nanoparticles and volume fraction of BN NPs is maintained constant at 10% and the radius is varied from 1 to 50 nm. The majority of emission spectrum shows no effect of BN particle size. However, the effect of size is noticeable at wavelengths less than

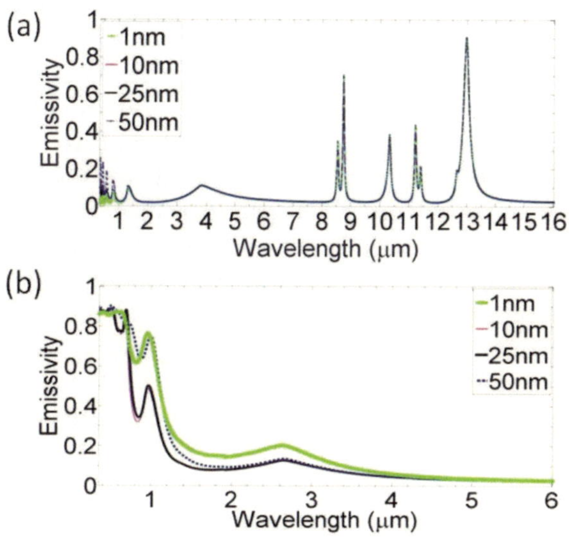

Figure 5. Emissivity spectra for SiC or PS thin film of thickness 0.4 μm embedded with BN or Au nanoparticles of volume fraction 10% and various radii. (a) BN nanoparticle-embedded SiC thin film doped and (b) Au nanoparticle-embedded PS thin film.

1 μm. This is due to the fact that Mie scattering becomes important at shorter wavelengths giving rise to higher peaks for larger particles. **Figure 5(b)** presents the calculation of emissivity for 0.4-μm thick polystyrene (PS) film doped with Au NPs. The volume fraction of NPs is fixed at 10% and the particle size is changed from 1 to 50 nm. Unlike **Figure 5(a)**, **Figure 5(b)** shows a strong influence of particle size on the emissivity. While the spectrum in the visible region shows a negligible response to particle size, gold NPs greatly influence the near-infrared region between 1 and 4 μm. As the NP size is increased from 1 nm, the emissivity peaks reduce in magnitude, showing smaller peaks for 10 and 25 nm. Emissivity for larger particles of 50 nm, however, is increased again and is comparable to that of NPs of 1 nm. This is due to the presence of two counteracting phenomena here. First is the change in dielectric function of Au NPs leading to decreased emissivity of larger particles and the second being Mie scattering of electromagnetic (EM) waves in the host causing an increased emissivity of larger particles.

Next, we present the effect of the doped nanoparticles on radiative heat transfer. We analyze radiative heat transfer between two identical multilayered structures at 300 and 301 K as shown in **Figure 1(b)**. Each structure has a top layer of 0.4 μm deposited on 1 μm of Au. The top layer is doped with nanoparticles of 25 nm and different volume fractions. **Figure 6** shows radiative heat flux versus distance between the structures and the normalized spectral density (defined as the ratio of $dq/d\omega$ to the maximum value over the range of wavelengths considered) at a distance of 100 nm is shown in the inset of figures. Consider a structure with SiC layer doped with BN nanoparticles. **Figure 6(a)** shows very little change in overall heat transfer. While thin film of pure SiC shows nearly monochromatic heat transfer, selectivity is seen at additional bands of wavelength. These locations are wavelengths where the effective refractive index of the mixture becomes zero (**Figure 3(a)** and **(b)**). While the locations of the new peaks depend on the volume fraction, the peak corresponding to the host material is relatively unchanged. In **Figure 6(b)**, SiC film is doped with Au nanoparticles of radius 25 nm with different volume fractions. The change in total heat transfer characteristics is not significant with the addition of nanoparticles. Selectivity is observed near $\lambda = \lambda_1 \approx 10.33$ μm as in Mulet et al. [10], which is one of characteristic wavelengths of SiC. Moreover, the inclusion of Au nanoparticles has only a small impact on the selectivity in the near-field limit and this can be related to the refractive indices of the mixture (**Figure 3(a)** and **(b)**). When the top layer is polystyrene doped with BN nanoparticles (**Figure 6(c)**), the near-field heat flux is clearly dependent on the volume fraction of the inclusion in both the near-field and the far-field regime. Since polystyrene does not support SPPs/SPhPs, the inclusion of BN clearly makes significant enhancement in heat transfer. The surface becomes selective at the wavelengths at which the real part of the effective refractive index becomes zero. When the PS film is doped with Au nanoparticles instead, the radiative heat transfer in **Figure 6(d)** shows an increment with an increase in NPs volume fraction, in both the near-field and the far-field limit. However, the normalized spectral density does not show any selectivity in the near field. In summary, the wavelength selectivity of thin films in the near field can be related to its effective refractive index in all four cases. This is logical as the selectivity is due to the

http://dx.doi.org/10.5772/67395

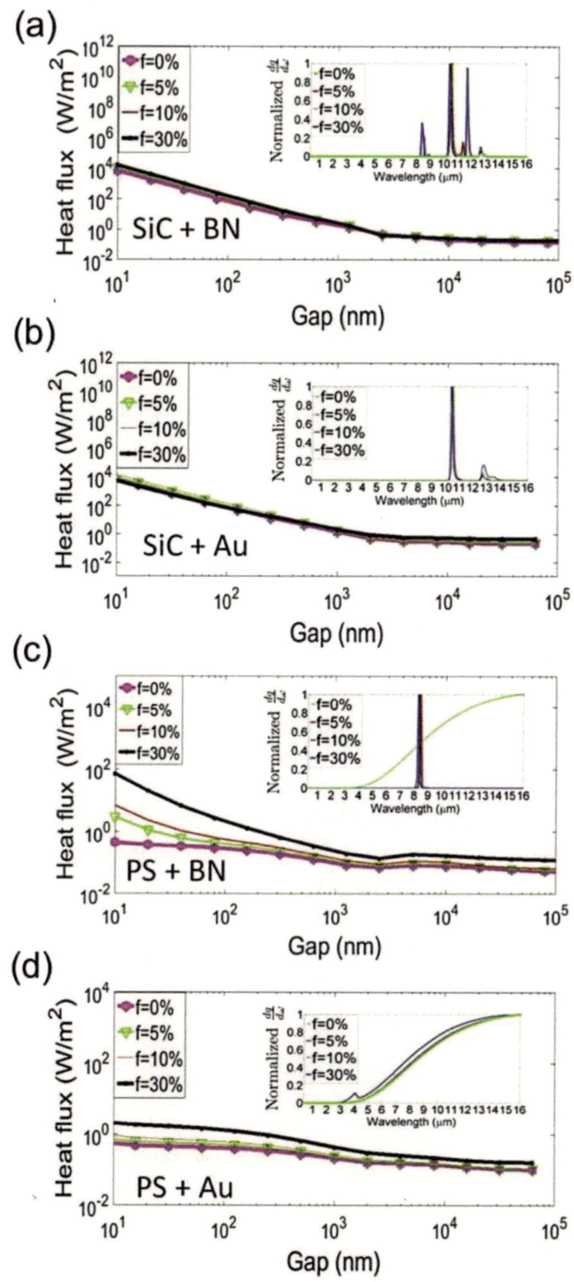

Figure 6. Heat flux of microscopic-layered media doped with BN or Au nanoparticles of radius 25 nm and various volume fractions due to near-field radiative effect. Inset: normalized spectral heat flux at a 100-nm gap between each half space with 0.4-μm thick nanoparticle-embedded layer on 1-μm Au layer. (a) BN nanoparticle-embedded SiC thin film, (b) Au nanoparticle-embedded SiC thin film, (c) BN nanoparticle-embedded PS thin film, and (d) Au nanoparticle-embedded PS thin film.

presence of SPPs/SPhPs across the interfaces. It is interesting to note that, unlike in the far-field regime, the selectivity is affected only when BN particles are used as inclusions. The addition of Au particles shows little or no impact on the selectivity in the near field. This supports the idea that metallic nanoparticles do not induce new SPPs/SPhPs in the surfaces while dielectric nanoparticles such as BN produce new SPhP in the material.

4. Conclusion

We have demonstrated that nanoparticles influence the emission spectra of the multilayered structures. Wavelength selectivity can be altered and controlled by the size and/or volume fraction of the NPs. The presence of NPs in a host material gives rise to an appearance of new emission peaks and a shift in the existing peaks in the emission spectra. When the metallic NPs are used, the effect of size is stronger as the dielectric function of metallic NPs has a strong dependence of particle size due to electron scattering. We have also shown that the volume fraction of the nanoparticles plays an important role in the near-field radiative heat transfer. If the NPs support SPhP, wavelength selectivity of thin films in the far field is at the locations where the real part of effective refractive index of the mixture becomes zero or the imaginary part of refractive index is small while the real part of the index is large. If the material of inclusion supports SPPs, as in metallic nanoparticles multiple emission peaks are seen which cannot be related to n- and κ-values of the mixtures. (Our observation is limited to the case where the host material is thin film.)

In the near field, for NPs supporting SPhPs or SPPs the heat transfer is nearly monochromatic around the wavelength at which n for the mixture becomes zero. It is observed that only SPhP supporting inclusions can influence the location of λ_n of the mixture; hence wavelength selectivity of thin films in the near field has little or no effect due to the presence of metallic nanoparticles. This can be understood as the presence of NPs in the thin film does not induce new kind of SPPs/SPhPs resonance across the interfaces. This work broadens the range of designs and methods for wavelength-selective emitters in both the far-field and the near-field regime.

Acknowledgements

This work is partially funded by the Rhode Island STAC Research Grant under grant number AWD05085 and NIH RI-INBRE Pilot Research Development Award supported in part by an Institutional Development Award (IDeA) Network for Biomedical Research Excellence from the National Institute of General Medical Sciences of the National Institutes of Health under grant number P20GM103430.

http://dx.doi.org/10.5772/67395

Author details

Yi Zheng

Address all correspondence to: zheng@uri.edu

Department of Mechanical, Industrial and Systems Engineering, University of Rhode Island, Kingston, RI, USA

References

[1] S. Basu, Y.-B. Chen, and Z. Zhang, "Microscale radiation in thermophotovoltaic devices—a review," International Journal of Energy Research 31, 689–716 (2007).

[2] C. McDonagh, C. S. Burke, and B. D. MacCraith, "Optical chemical sensors," Chemical Reviews 108, 400–422 (2008).

[3] M. Pralle, N. Moelders, M. McNeal, I. Puscasu, A. Greenwald, J. Daly, E. Johnson, T. George, D. Choi, I. El- Kady, and R. Biswas, "Photonic crystal enhanced narrow-band infrared emitters," Applied Physics Letters 81, 4685–4687 (2002).

[4] B. Guha, C. Otey, C. B. Poitras, S. Fan, and M. Lipson, "Near-field radiative cooling of nanostructures," Nano Letters 12, 4546–4550 (2012).

[5] A. Narayanaswamy and G. Chen, "Thermal emission control with one-dimensional metallodielectric photonic crystals," Physical Review B 70, 125101 (2004).

[6] A. Narayanaswamy, J. Mayo, and C. Canetta, "Infrared selective emitters with thin films of polar materials," Applied Physics Letters 104, 183107 (2014).

[7] Y. Zheng and A. Narayanaswamy, "Patch contribution to near-field radiative energy transfer and van der Waals pressure between two half-spaces," Physical Review A 89, 022512 (2014).

[8] H. Sai, H. Yugami, Y. Akiyama, Y. Kanamori, and K. Hane, "Spectral control of thermal emission by periodic microstructured surfaces in the near-infrared region," JOSA A 18, 1471–1476 (2001).

[9] P. Bermel, M. Ghebrebrhan, M. Harradon, Y. X. Yeng, I. Celanovic, J. D. Joannopoulos, and M. Soljacic, "Tailoring photonic metamaterial resonances for thermal radiation," Nanoscale Research Letters 6, 1–5 (2011).

[10] J.-P. Mulet, K. Joulain, R. Carminati, and J.-J. Greffet, "Enhanced radiative heat transfer at nanometric distances," Microscale Thermophysical Engineering 6, 209–222 (2002).

[11] Y. Zheng and A. Ghanekar, "Radiative energy and momentum transfer for various spherical shapes: a single sphere, a bubble, a spherical shell and a coated sphere," Journal of Applied Physics 117, 064314 (2015).

[12] M. Francoeur, S. Basu, and S. J. Petersen, "Electric and magnetic surface polariton mediated near-field radiative heat transfer between metamaterials made of silicon carbide particles," Optics Express 19, 18774–18788 (2011).

[13] S. J. Petersen, S. Basu, B. Raeymaekers, and M. Francoeur, "Tuning near-field thermal radiative properties by quantifying sensitivity of Mie resonance-based metamaterial design parameters," Journal of Quantitative Spectroscopy and Radiative Transfer 129, 277–286 (2013).

[14] H. Gonome, M. Baneshi, J. Okajima, A. Komiya, N. Yamada, and S. Maruyama, "Control of thermal barrier performance by optimized nanoparticle size and experimental evaluation using a solar simulator," Journal of Quantitative Spectroscopy and Radiative Transfer 149, 81–89 (2014).

[15] H. Gonome, M. Baneshi, J. Okajima, A. Komiya, and S. Maruyama, "Controlling the radiative properties of cool black-color coatings pigmented with CuO submicron particles," Journal of Quantitative Spectroscopy and Radiative Transfer 132, 90–98 (2014).

[16] J. M. Garnett, "Colours in metal glasses, in metallic films, and in metallic solutions. II," Philosophical Transactions of the Royal Society of London. Series A, Containing Papers of a Mathematical or Physical Character, pp. 237–288 (1906).

[17] W. T. Doyle, "Optical properties of a suspension of metal spheres," Physical Review B 39, 9852 (1989).

[18] C.-H. Wang, W.-Y. Shen, P.-S. Sheng, C.-Y. Lee, and H.-T. Chiu, "Deposition of mesoporous silicon carbide thin films from (me3si) 4sn: tin nanoparticles as in situ generated templates," Chemistry of Materials 19, 5250–5255 (2007).

[19] H. Wei and H. Eilers, "Electrical conductivity of thin-film composites containing silver nanoparticles embedded in a dielectric fluoropolymer matrix," Thin Solid Films 517, 575–581 (2008).

[20] W. C. Chew, "Waves and fields in inhomogeneous media," vol. 522. New York, USA: IEEE Press (1995).

[21] M. Moharam, D. A. Pommet, E. B. Grann, and T. Gaylord, "Stable implementation of the rigorous coupled-wave analysis for surface-relief gratings: enhanced transmittance matrix approach," JOSA A 12, 1077–1086 (1995).

[22] M. Planck, The theory of heat radiation (Dover Publications, 2011).

[23] A. Narayanaswamy and Y. Zheng, "A Green's function formalism of energy and momentum transfer in fluctuational electrodynamics," Journal of Quantitative Spectroscopy and Radiative Transfer 132, 12–21 (2014).

[24] V. Myroshnychenko, J. Rodr'ıguez-Ferna'ndez, I. Pastoriza-Santos, A. M. Funston, C. Novo, P. Mulvaney, L. M. Liz-Marza'n, and F. J. G. de Abajo, "Modelling the optical response of gold nanoparticles," Chemical Society Reviews 37, 1792–1805 (2008).

[25] U. Kreibig and M. Vollmer, Optical properties of metal clusters, vol. 25 (Springer, Berlin, 1995).

[26] M. S. Wheeler, "A scattering-based approach to the design, analysis, and experimental verification of magnetic metamaterials made from dielectrics," PhD thesis (2010).

[27] X. Liu, T. Bright, and Z. Zhang, "Application conditions of effective medium theory in near-field radiative heat transfer between multilayered metamaterials," Journal of Heat Transfer 136, 092703 (2014).

[28] B. Liu and S. Shen, "Broadband near-field radiative thermal emitter/absorber based on hyperbolic metamaterials: Direct numerical simulation by the wiener chaos expansion method," Physical Review B 87, 115403 (2013).

[29] E. D. Palik, Handbook of optical constants of solids, vol. 3. London, UK: Academic Press (1998).

[30] B. Neuner III, D. Korobkin, C. Fietz, D. Carole, G. Ferro, and G. Shvets, "Midinfrared index sensing of pL-scale analytes based on surface phonon polaritons in silicon carbide," The Journal of Physical Chemistry C 114, 7489–7491 (2010).

[31] P. B. Johnson and R.-W. Christy, "Optical constants of the noble metals," Physical Review B 6, 4370 (1972).

[32] V. C. Nguyen, L. Chen, and K. Halterman, "Total transmission and total reflection by zero index metamaterials with defects," Physical Review Letters 105, 233908 (2010).

[33] Y. He and T. Zeng, "First-principles study and model of dielectric functions of silver nanoparticles," The Journal of Physical Chemistry C 114, 18023–18030 (2010).

[34] K. L. Kelly, E. Coronado, L. L. Zhao, and G. C. Schatz, "The optical properties of metal nanoparticles: the influence of size, shape, and dielectric environment," The Journal of Physical Chemistry B 107, 668–677 (2003).

[35] H. Hö vel, S. Fritz, A. Hilger, U. Kreibig, and M. Vollmer, "Width of cluster plasmon resonances: bulk dielectric functions and chemical interface damping," Physical Review B 48, 18178 (1993).

[36] V. A. Parsegian, Van der Waals forces: a handbook for biologists, chemists, engineers, and physicists Cambridge, UK: Cambridge University Press (2006).